Wildlife and Natural Resource Management

THIRD EDITION

Delmar Cengage Learning is
proud to support FFA activities

Join us on the web at

agriculture.delmar.cengage.com

Wildlife and Natural Resource Management

THIRD EDITION

Kevin H. Deal

DELMAR
CENGAGE Learning

Australia • Brazil • Japan • Korea • Mexico • Singapore • Spain • United Kingdom • United States

Wildlife and Natural Resource Management, Third Edition
Kevin H. Deal

Vice President, Career and Professional
Editorial: Dave Garza

Director of Learning Solutions: Matthew Kane

Acquisitions Editor: Benjamin Penner

Managing Editor: Marah Bellegarde

Product Manager: Christina Gifford

Editorial Assistant: Scott Royael

Vice President, Career and Professional
Marketing: Jennifer McAvey

Marketing Director: Debbie Yarnell

Marketing Manager: Erin Brennan

Marketing Coordinator: Jonathan Sheehan

Production Director: Carolyn Miller

Production Manager: Andrew Crouth

Senior Content Project Manager:
Elizabeth C. Hough

Senior Art Director: David Arsenault

Technology Project Manager: Patricia Allen

Production Technology Analyst: Thomas Stover

For product information and technology assistance, contact us at
Cengage Learning Customer & Sales Support, 1-800-354-9706

For permission to use material from this text or product,
submit all requests online at **www.cengage.com/permissions**.
Further permissions questions can be e-mailed to
permissionrequest@cengage.com

Library of Congress Control Number: 2009928004

ISBN-13: 978-1-4354-5397-5

ISBN-10: 1-4354-5397-2

Delmar
5 Maxwell Drive
Clifton Park, NY 12065-2919
USA

Cengage Learning is a leading provider of customized learning solutions with office locations around the globe, including Singapore, the United Kingdom, Australia, Mexico, Brazil, and Japan. Locate your local office at: **international.cengage.com/region**

Cengage Learning products are represented in Canada by Nelson Education, Ltd.

To learn more about Delmar, visit **www.cengage.com/delmar**

Purchase any of our products at your local college store or at our preferred online store **www.ichapters.com**.

Notice to the Reader

The Publisher does not warrant or guarantee any of the products described herein or perform any independent analysis in connection with any of the product information contained herein. The Publisher does not assume, and expressly disclaims, any obligation to obtain and include information other than that provided to it by the manufacturer. The reader is expressly warned to consider and adopt all safety precautions that might be indicated by the activities described herein and to avoid all potential hazards. By following the instructions contained herein, the reader willingly assumes all risks in connection with such instructions. The publisher makes no representations or warranties of any kind, including but not limited to, the warranties of fitness for particular purpose or merchantability, nor are any such representations implied with respect to the material set forth herein, and the publisher takes no responsibility with respect to such material. The publisher shall not be liable for any special, consequential, or exemplary damages resulting, in whole or part, from the readers' use of, or reliance upon, this material.

Printed in the United States of America
1 2 3 4 5 6 7 14 13 12 11 10 09

Contents

SECTION
TWO
Wildlife and Fish Identification / 137

SECTION THREE

Careers / 397

Preface

It is my opinion that our fish, wildlife, land, forests, and water resources are our most valuable natural assets. The beauty of our great country arises from this wonderful natural abundance. Most of the wealth and prosperity we Americans have enjoyed over the past 200 years and more is due to our abundant natural reserves.

A great deal has been learned about our wildlife and other natural resources and their management in the past 60 to 70 years. Although the natural world is not as sensitive in some areas as was once thought, it is still very vulnerable to abuse. Water and soil are easily polluted and wildlife habitat and forests destroyed in the name of "progress." In the past 30 years we have come a long way, particularly in terms of managing fish and wildlife, conserving soil, and protecting our water supply. However, we must be extremely careful to not take a single step backward.

This textbook presents the very broad and complex subject of wildlife, fisheries, and natural resource management in a manner that is interesting and understandable to the student. As our human population continues to grow, the pressure on natural resources will inevitably increase. It is therefore imperative that our young people develop an understanding and appreciation of our fish, wildlife, and other natural resources. It will be their responsibility to conserve and preserve them for future generations.

Each chapter includes a series of objectives, key terms, study questions, and student activities. Each of these features is designed to help the student understand the chapter and retain the information in it. There is a complete alphabetized glossary at the end of the book as well as several useful appendices. We have also included references to some of the more useful Web sites.

A serious attempt has been made to present an unbiased view of wildlife, fisheries, and natural resource management. We have developed a disturbing tendency in this country in recent years to let emotionalism and "political correctness" dictate wildlife management policy. This is foolish and dangerous. Natural resource managers, from wildlife biologists to foresters, are professionals. Most have university degrees and years, if not decades, of experience in their fields. We must remember that these professionals are privy to all the scientific data and experience to make careful, scientific, and not emotional, decisions.

Acknowledgments

There are many people involved in the completion of a project this size. Dozens of individuals contributed ideas and materials and provided assistance. I would like to express my gratitude to everyone who has had a part in the completion of this book.

Ms. Karen Rike

Ms. Jasmine Wooten

Ms. Jennifer Pilcher

Texas Parks & Wildlife Department, Austin, Texas
Mr. Steve Hall, Mr. Paul Montgomery, Ms. Grace A. Perez, Mr. Glenn Mills, and most especially Chase Fountain for all his hard work in tracking down photographs for this edition.

Delta Waterfowl Foundation, Deerfield, Illinois

Quail Unlimited, Augusta, Georgia

Pheasants Forever, Inc., St. Paul, Minnesota

The United States Department of the Interior, Washington, D.C. The United States Department of Agriculture, Washington, D.C.

National Wildlife Federation, Washington, D.C.

Kansas Department of Wildlife Parks, Pratt, Kansas

Oklahoma Department of Wildlife & Conservation, Oklahoma City, Oklahoma

Iowa Department of Natural Resources, Des Moines, Iowa

Ducks Unlimited, Long Grove, Illinois

John D. Willson, University of Georgia, Savannah River Ecology Laboratory

Leonard Rue Enterprises

Brian E. Small Photography

I also wish to express my appreciation to my wife, Deanna, and children, Emily, Amanda, and Spencer, for their patience and understanding during the completion of this manuscript.

The author and Delmar would like to express their appreciation to the reviewers of this revision for their valuable input and content expertise.

Amanda Briggs
Madison County High School
Madison, IN

Tonny Hamby
Canadian High School
Canadian, TX

Jim Satterfield
Jefferson County High School
Dandridge, TN

Barry Fry
Milford Senior High School
Milford, DE

Also available for the Instructor:

Instructor Guide
ISBN: 1 4354 5398 0
ISBN 978 1 4354 5398 2
The Instructor's Guide is made up of chapter objectives, key term list, and answer key to the end-of-chapter questions.

Student Workbook
ISBN: 1 4354 5399 9
ISBN 978 1 4354 5399 9
This comprehensive workbook tests students' knowledge and reinforces learning of text content.

Instructor's Guide to Accompany Student Workbook
ISBN: 1 4354 5400 6
ISBN 978 1 4354 5400 2
The Instructor's Guide provides answers to workbook exercises.

Class Master to Accompany Wildlife and Natural Resource Management, 3E
ISBN: 1 4354 5401 4
ISBN 978 1 4354 5401 9
This powerful electronic resource is an essential tool. In it an instructor will find everything needed to prepare for, teach, present, and test the concepts of wildlife and natural resource management.

This resource includes:
A PDF of the full instructor's guide with objectives, key terms, and an answer key to the end-of-chapter questions;
Computer Testbank containing more than 900 questions with answers and Internet testing capability;
PowerPoint Presentations containing over 400 slides that focus on each chapter's key points to facilitate classroom presentations; and
Image Library including all figures appearing within the text.

Section
ONE

Introduction

CHAPTER 1

The History of Wildlife Management in America

OBJECTIVES

After completing this chapter, you should be able to

- Describe the development of wildlife management in America.
- List specific actions that led to modern wildlife management.
- Understand the role that wildlife has played in the development of America.
- Describe the era of exploitation of America's wildlife.
- Understand the role outdoor enthusiasts have played in the conservation movement in America.

INTRODUCTION

Wildlife management, as we know it today, has **evolved** during the past 75 to 100 years. Before this time **wildlife** was exploited, ignored, and pushed aside in the rush to settle America. This resulted in a great loss of wildlife, wildlife **habitat**, and the complete loss of some **species**, such as the carrier or passenger pigeon. However, some attempts at wildlife management in America began in the early 1800s. These efforts were not very successful until the beginning of the twentieth century, but they were an important beginning. People began to realize that the vast amount of wildlife that had been present since colonial times was not **inexhaustible**. In fact, Massachusetts declared

(Source: Courtesy of Photodisc.)

FIGURE 1-1 When the first Europeans arrived in North America, the land was raw and unspoiled.

a **closed season** on white-tailed deer as early as 1694. It was largely ignored, but it was an early effort at wildlife management.

When the first Europeans arrived in America with the intention of staying, the land was raw, unspoiled, and generally teeming with wildlife (Figure 1-1). These early arrivals had no idea how abundant wild **fauna** was in this "new" land. They also had no idea how to utilize it, as evidenced by the deaths of most of the original 600 Jamestown colonists. These deaths were largely due to **starvation** and diseases brought on by their weakened and **malnourished** condition. These early colonists were, for the most part, from English cities. They had little hunting, fishing, or farming experience. It is no wonder that the Native Americans considered them a "weak, unworthy people." However, it took less than two generations for the colonists to learn to hunt and fish with great skill. They went from starvation to **prosperity** by learning how to **exploit** the abundant native wildlife. As the settlers multiplied and more people came over from Europe, the pressure on America's wild fauna slowly increased.

It took nearly 200 years for agriculture to get a firm grip on American soil. Even after European farmers began to arrive in large numbers, they had to deal with a different **climate** and different soils. Many agricultural skills had to be learned by trial and error. There were no large research facilities or universities to test and discover new crops, so it was a slow process. During this period, and afterward to some extent, American wildlife fed, clothed, and put money in the pockets of most early Americans.

The period from 1700 to 1900 saw the greatest abuse of wild animals in North America. Vast numbers of animals were shot to provide for a growing population. Millions of acres of habitat were altered or destroyed to make room for towns and farms. In 1748 alone, South Carolina traders shipped 160,000 deer skins to England. No laws restricted the harvest of wildlife at this time, and bucks, does, and fawns were shot and trapped 365 days a year. The resulting drop in numbers was staggering. By 1776 every southeastern state except Georgia had established a closed season on deer. These seasons were largely ignored. Deer were major pests in most farmers' fields, and settlers believed they had the right to shoot what they needed to feed their families.

Rural families constituted more than 90 percent of the population. They lived off what they shot or trapped. The belief that the land "owed" them a living and that wildlife could be used as they saw fit persists in some extent to this day. They shot not only deer but turkeys, raccoons, waterfowl, bobcats, and panthers. Anything that would feed or clothe them or compete with them for a **livelihood** was fair game.

Many of the settlers' land practices were actually beneficial to some species of wildlife. As tracts of forest were cleared for cropland, white-tailed deer benefited from the resulting edges and cleared areas. At the same time many of the deer's **predators** were being systematically eliminated. As a result, deer numbers increased in some areas.

The elk and woods bison were not so lucky. The woods bison, larger and darker than its plains cousin, never existed in the enormous numbers seen on the plains. However, both the elk and woods bison were quite common at one time. By the mid-1800s, market and pioneer hunters had exterminated them east of the Mississippi. The same land practices that were beneficial to the whitetail could be extremely **detrimental** to other species, such as turkeys and passenger pigeons. As increasingly larger tracts of forest were cut and burned, these two birds had less and less habitat in which to reproduce. At the same time, they were being shot at a tremendous rate to feed a growing population.

EXPLOITATION

From the very beginning in Jamestown, the English settlers had no intention of adapting to the **wilderness**. Unlike the French to the north, who largely adapted to their new **environment**, the predominantly English colonists to the south set about clearing the forests and building towns, cities, and roads. This process would forever alter North America and its wildlife. The French became hunters, trappers, and gatherers of the wilderness bounty, at peace with the wildness of their new home. Not so the English. They changed their new environment to suit their own ideals. This process is still going on today, with shopping malls, parking garages, highways, and subdivisions being constructed at a great rate (Figure 1-2). At least today some thought is given to the damage being done to wildlife. Once a piece of land is paved, its use as a source of wildlife habitat is over.

FIGURE 1-2 Modern development has severely affected wildlife habitat and many species of wildlife.

It may be impossible for a modern American to imagine the tremendous numbers of wild creatures this country once held. In 1804, when Meriwether Lewis and William Clark began their epic journey into America's heartland, 40 to 60 million plains bison roamed the Great Plains. Passenger pigeons numbered in the billions, and an estimated 40 million pronghorn antelope ran with the vast herds of buffalo. One colony of pigeons, roosting near present-day Marietta, Ohio, reportedly covered 1,000 acres of beech and oak forests.

An early observer of American wildlife was Alexander Wilson. A native of Scotland who lived in Philadelphia, Wilson spent many years studying America's wildlife, particularly birds. In 1810 Wilson observed a flock of passenger pigeons in Kentucky flying overhead, "at a height beyond gunshot . . . from right to left as far as the eye could reach" for five hours. He estimated the total number of pigeons in that flight to be 2,230,272,000! The following quote is taken from Wilson's description of the enormous flocks of passenger pigeons.

The most remarkable characteristic of these birds is their associating together, both in their **migrations** and also during the period of **incubation**, in such **prodigious** numbers as almost to surpass belief. . . . These migrations appear to be undertaken rather in quest of food, than merely to avoid the cold of the climate, since we find them lingering in the northern regions around Hudson's Bay so late as December.

. . . I have witnessed these migrations in the Genessee country—often in Pennsylvania, and also various parts of Virginia, with amazement; but all that I had then seen of them were mere straggling parties, when compared with the congregated millions which I have since beheld in our western forests, in the states of Ohio, Kentucky, and the Indiana territory. These fertile and **extensive** regions abound with the nutritious beech nut, which constitute the chief food of the Wild Pigeon. . . . It sometimes happens that having consumed the whole produce of the beech trees in an extensive district, they discover another at the distance perhaps of sixty or eighty miles, to which they regularly repair every morning and return as regularly . . . in the evening, to their . . . roosting place. . . . The ground is covered to the depth of several inches with their dung; all the tender grass and under wood destroyed; the surface strewed with large limbs of trees broken down by the weight of the birds . . . and the trees themselves, for thousands of acres, killed as completely as if **girdled** with an axe.

These passenger pigeons may have been the largest concentration of birds of one species ever. Though they were shot, clubbed, and netted by the millions, it was only the gradual destruction of the hardwood forests that doomed the pigeon. The last big pigeon nesting occurred in the mid-1870s, in central Michigan. An estimated 136 million birds were **devastated** by **market hunters** from all over the country (Figure 1-3). However, it was the cutting of the hardwood forests that concentrated the birds into smaller and

FIGURE 1-3 Market hunters were driven to kill waterfowl, passenger pigeons, bison, and numerous other species for profit.

(Source: Courtesy of U.S. Fish and Wildlife Service.)

smaller remnants of their habitat. The Chicago Fire of 1871 may have sped up their demise when the forests of the Great Lakes states were turned into lumber to rebuild the city.

The last passenger pigeon died in the Cincinnati Zoo in 1914. In the meantime, the slaughter of the great herds of bison began with the end of the beaver trade in the mid-1830s. In 1840 the successor to the Rocky Mountain Fur Company shipped more than 65,000 buffalo skins down the Missouri River. The buffalo, antelope, and other plains game got a brief respite during the American Civil War years of 1861 to 1865. However, with the war over and the South devastated, the tide of immigrants swelled westward. When the Transcontinental Railroad was completed in 1869, the huge main herd of buffalo was cut in two. Although Native Americans had hunted buffalo for many years, sometimes quite wastefully, the damage to the vast herds was not serious. The railroads brought hunters with modern rifles and ammunition westward in increasingly large numbers. Buffalo hides, tongues, and, after a few years, buffalo bones, to be ground up and used as fertilizer, were shipped eastward. In the early 1870s, buffalo hides and tongues sold for $1.25 and $.25, respectively. In 1872 and 1873 the railroads shipped an estimated 1,250,000 buffalo hides out of Kansas and nearby territories.

Even at this great rate of slaughter, there were still large numbers of buffalo. In 1869 a Kansas Pacific train was kept in one spot for nine hours while a herd crossed the tracks. The U.S. government actually encouraged the slaughter, although not openly, as a means of defeating the Native Americans of the Great Plains. The Native Americans depended on the buffalo for their very existence, and the government felt, quite accurately, that they could not survive without them. By 1878 the vast herds of buffalo were gone forever. With the invention of the first chilled-steel plow in 1868 by James Oliver of South Bend, Indiana, the hard sod of the Great Plains could be plowed. It took all too few years for the

(Source: © David Sucsy, 2009. Used under license from iStockphoto.com)

FIGURE 1-4 Vast expanses of wheat feed America and the world; however, wheat fields are no substitute for the wildlife habitat once provided by millions of acres of native prairies.

Great Plains grasses to be replaced by a sea of wheat (Figure 1-4). The pronghorn antelope was a victim of fences, settlers' rifles, and the modern steel plow. Today, the Great Plains of the United States is the world's breadbasket, with much of the wheat and other crops produced there being exported to feed a growing world population.

Although one of these species, the passenger pigeon, is lost forever, both the plains bison and the pronghorn are with us today. Though they will never number in the millions, as they did in the 1800s, it is a tribute to early **conservationists** that the bison and pronghorn are here at all. Through the efforts of sport hunters and conservationists, many species of wildlife that were all but extinct in the early part of the twentieth century are abundant today.

SERIOUS CONSERVATION EFFORTS BEGIN

Beginning in the early 1870s, sport hunters began a serious effort to protect and **preserve** the remnants of America's wild fauna. They began these efforts by working to change American views on the **commercialization** of wild game. For almost 400 years, Americans had exploited wildlife for profit. Market hunters shot waterfowl during the spring and fall migrations, netted thousands of carrier pigeons and shore birds on their nests, and killed hundreds of elk, deer, bison, and antelope. Market hunters played a major role in eliminating much of America's wildlife. **Sport hunting**, however, has never endangered a single species. Most sport hunters realized that as long as it was profitable to deal in wildlife, some people would seek those profits. They also realized that even with adequate habitat, species such as elk, bighorn sheep, and most waterfowl could be over-harvested to the point of **extinction**. Sport hunters reasoned that allowing wildlife to be killed for market was the same as putting a bounty on their heads.

All wildlife produces a **surplus**, which allows for some harvest without damaging the overall population. Market hunters, with the motivation to kill for money, quickly slaughter the surplus and start cutting into the breeding stock of many species. While sport hunters use only sporting methods and regulate their harvest, market hunters had no such scruples. Where a market hunter might shoot several hundred ducks, a sport hunter might shoot a dozen.

All species of wildlife, no matter how **secretive** they may appear to be, are extremely **vulnerable** when unsporting tactics are used. For example, ducks, geese, and wild turkeys are relatively easy to bait with grain. Deer were commonly driven into the water with dogs and clubbed to death by men in canoes. Commercial pigeon hunters would congregate at the breeding grounds and fill freight cars with the bodies of their prey. This heavy commercialization of wildlife led to a dramatic decline in the numbers of many species. The reports of these declines in early hunting magazines such as *Forest and Stream* and *American Field* (which later became *Field and Stream*) helped to awaken the public to the plight of America's wildlife. Sport hunters were not only interested in the protection of **game animals**. In the late 1870s it became stylish for women to wear hats heavily decorated with feathers or even whole birds. This led to the exploitation of many nongame species, such as egrets. Because many types of birds are concentrated during the breeding season and because their feathers are most beautiful at this time, many nesting colonies were invaded. The adult birds were killed and the young left to starve, bake in the sun, or be eaten by predators. This led to the elimination of colony after colony.

Because the commercialization of any species is against the code of sport hunters, they quickly led the fight to save many of these species. Through such organizations as the Boone and Crockett Club and on the editorial pages of *Forest and Stream* and *American Field*, America's sport hunters pushed for change. These were the first efforts at wildlife **conservation** and management as we know them today. These early calls for wildlife management were largely for wildlife protection. Sport hunters wanted an end to the commercialization of wildlife, better enforcement of existing laws and the passage of new ones, and the outlawing of certain methods used to harvest game. Despite fierce opposition from market hunters, sport hunters and the sporting journals were united in their efforts to save America's wildlife. The tide of public opinion was finally turning.

PROGRESS IS FINALLY MADE

The passage of the Lacey Act in 1900 ended market hunting and the interstate shipment of wildlife and wildlife products, such as feathers, taken in violation of state law. This act was the work of John F. Lacey, who was an avid hunter and fisherman. Another hunter who had a tremendous impact on the development of wildlife conservation was Theodore Roosevelt. Roosevelt had a deep love of big-game hunting and felt a strong responsibility to preserve what was left of America's wildlife. Building on the base that sport hunters had created in the 1870s to 1890s, Roosevelt's administration created more than 50 wildlife **refuges**, 5 national parks, and 17 national monuments. Even at this stage Roosevelt could not openly establish national parks, monuments, or refuges to protect wildlife. To do so would have been political suicide. The people were not yet ready to accept the idea that wildlife needed to be saved. However, they supported the idea that the forests needed to be preserved to ensure a supply of lumber for building. Roosevelt reasoned that if he could set aside large areas of habitat under the guise of protecting America's forests, wildlife might also survive in these areas. He was correct. To this day a large portion of the elk, bison, mule deer, and bighorn sheep and virtually all the grizzly bears in the lower 48 states are found in national forests and national parks. In fact, Yellowstone National Park was recently the site for the extremely successful reintroduction of gray wolves. Because of his work to preserve our wildlife and their habitat, Theodore Roosevelt has been called the father of American conservation.

The man given credit for the development of modern wildlife management, Aldo Leopold, was also an outdoorsman and hunting enthusiast. Leopold has been called the first professional wildlife manager. His first book, *Game Management*, published in 1933, is still considered the definitive work on modern wildlife management. In *Game Management*, Leopold advanced many concepts that were new at the time and quite different from previously accepted theories. Practices such as taking a **census** of the game animals in an area and the theory that animals have a "home range" from which they seldom stray were two of the ideas advanced by Leopold. He developed management plans around factors such as the potential for reproduction of a population and explained how to manage a population by controlling predators (both human and animal), food, water, cover, and other factors. Leopold was a dedicated and innovative wildlife scientist; consequently, he is often called the father of modern wildlife management.

Yellowstone National Park, the nation's first and for many years only national park, was founded in 1872 (Figure 1-5). It was established through the efforts of sport hunters to preserve a unique piece of the American West. It took

FIGURE 1-5 Yellowstone National Park, established in 1872, serves as home to many forms of wildlife, such as this elk.

(Source: © Photo courtesy of U.S. Fish and Wildlife Service. Photo by James Leupold.)

more than 20 years of agitation by sport hunting groups and environmentalists before the park's wildlife was afforded the protection it enjoys today. On paper, Yellowstone's wildlife had been protected since President Grant signed the bill establishing the park. In fact, until the late 1890s, there was no one in the park to protect the wild creatures from **poachers**. Pressure from sport hunters and sporting publications led to the national park system. It is taken for granted today that all our national parks are wildlife and wilderness **sanctuaries**. This was not always the case. When serious **restocking** efforts began at the turn of the twentieth century, much of the wildlife for these efforts came from our national parks and other sanctuaries. Without the breeding stock found within such areas, restocking of some species might not have been possible. America now has a system of national parks and national forests that virtually ensures the preservation of at least some wildlife as long as these refuges remain undeveloped. What must be of major concern to all Americans today is the destruction of habitat outside these refuges.

Wildlife management as we know it today has been evolving since the mid-1800s. It was not until the early 1900s that the work of early conservationists really began to pay off. The passage of the Lacey Act in 1900 was a major step toward wildlife management as we have come to know it today. It spelled the beginning of the end for the commercialization of wildlife in America. Sport hunters pushed for and got spring waterfowl seasons closed, as well as more restrictive seasons and bag limits for many other game species. Almost all conservation efforts during the early 1900s were prohibitive ones. They were designed to protect wildlife and to increase its numbers.

The passage of the Migratory Bird Conservation Act in 1929 was important to the recovery of waterfowl populations. This act was a cooperative agreement between the United States and Canada (Mexico joined the effort a short time later) to manage waterfowl populations. In 1934, the Migratory Bird Hunting Stamp Act was passed. Commonly called the Duck Stamp Act, this law has provided many millions of dollars for waterfowl management through the sale of stamps to hunters. In 1937, sport hunters, conservationists, and the firearm and ammunition industries succeeded in getting the Federal Aid in Wildlife Restoration Act passed. This legislation, which was sponsored by Senator Key Pittman and Congressman A. Willis Robertson, is more commonly known as the Pittman-Robertson Act. The act called for a 10 percent **excise tax** on firearms and ammunition. This tax is collected directly from the manufacturers, and it provides most of the funding for the wildlife departments in all 50 states. These funds are distributed to the various states by the secretary of the interior. The tax was raised to 11 percent in 1941 and remains at that rate today. In 1970 a 10 percent tax was added to handguns and in 1971 an 11 percent tax was added to archery equipment. Today these excise taxes contribute more than $175 million per year to state fish and wildlife programs. A similar excise tax on fishing equipment contributes an additional $200 million per year. To date, the Pittman-Robertson Act has provided over $4 billion for wildlife management, research, protection, and the **acquisition** of millions of acres of habitat (Figure 1-6). With funding, wildlife managers went to work. From an estimated low of fewer than 500,000 animals in 1920, white-tailed deer have rebounded to more than 14 million today. Pronghorn antelope once numbered around

(Source: Photo courtesy of USDA/NRCS.)

FIGURE 1-6 Wooded wetlands are just one type of habitat purchased and protected with Pittman-Robertson funds.

25,000; today the population exceeds 750,000. These are just two of the many success stories that wildlife managers and sport hunters can be proud of. Most of the suitable wildlife habitat in the United States is currently stocked at capacity levels.

SUMMARY

The future of wildlife management is the management of populations within the available habitat. White-tailed deer populations in many areas of the Northeast, Southeast, and portions of Texas have already reached the saturation point. These deer have become serious pests of residential landscaping. It has been proven time and again that, given a **modicum** of protection, many species of wildlife are very adaptable. The challenge for today's wildlife managers is to balance wildlife populations with a shrinking amount of habitat. America was once graced with a wealth of wildlife. This abundance was not seriously reduced until the mid-1800s. Between 1850 and 1900, America's wildlife was exploited, and many species were pushed to the verge of extinction. Some, such as the passenger pigeon, were pushed beyond the brink and will never be seen again.

By the turn of the century the pendulum was swinging in favor of wild creatures. With the protection and funding that became available between 1900 and 1940, America's wildlife began a comeback. Though species such as the whooping crane are still precariously close to extinction, many more species are found in greater numbers than they were at the turn of the century. America's sport hunters have played a vital role in this comeback.

REVIEW QUESTIONS

Fill in the Blank

Fill in the blanks to complete these statements.

1. Modern _____ has evolved only since the early 1900s.

2. _____ had a closed season on white-tailed deer as early as 1694.

3. America was blessed with a tremendous abundance of _____ .

4. Early settlers in America went from starvation to prosperity by learning how to _____ wildlife.

5. During the first 200 years of our nation, before _____ became well established, wildlife provided food and clothing for a growing nation.

6. Early English colonists, unlike the French to the north, had no intention of adapting to the _____ .

7. Many species of wildlife were found in great numbers; it is estimated that there were _____ million plains bison and billions of _____ .

8. An early observer of American wildlife was _____ .

9. The last _____ died in the Cincinnati Zoo in 1914.

10. The U.S. government unofficially encouraged the slaughter of the _____ as a means of defeating Native Americans on the Great Plains.

11. Through the efforts of _____ and conservationists, many species of wildlife are plentiful today.

12. Early sport hunters worked to change American attitudes toward the _____ of wild game.

13. _____ hunters played a major role in eliminating much of America's wildlife from its original range.

14. All wildlife produces a _____ , which allows for some harvest without damaging the overall population.

15. All wildlife are extremely vulnerable when _____ tactics are used.

16. Market hunters were motivated to kill for _____ .

17. In the late 1870s it became stylish for women to wear hats heavily decorated with _____ .

18. The passage of the _____ in 1900 effectively ended the market hunting era.

19. _____ , the nation's first national park, was established in 1872.

20. An avid big game hunter, as well as president of the United States, _____ has been called the father of American conservation.

Short Answer

1. Why did wildlife management evolve? How?

2. What were some of the obstacles to early wildlife management? How were they overcome?

3. What factors contributed to the decline of many wildlife species in America?

4. What group(s) led the fight for conservation? Why?

Discussion

1. In your opinion, what factors led to the exploitation of wildlife in America during the 1700s and 1800s?

2. Who was instrumental in leading the fight to conserve America's natural resources, particularly wildlife? Why did these people speak out in an era when their views were neither popular nor commonly held?

Learning Activities

1. Using additional references as needed make a chronological list of the dates and events that were important to the development of wildlife conservation and management in America.

2. Split into groups. Have each group select a person who was active in early conservation and wildlife management efforts, such as Theodore Roosevelt, Aldo Leopold, or John F. Lacey. Research the person's accomplishments and conduct a class discussion of each one's contributions to conservation and management.

CHAPTER 2

The Importance of Natural Resources

TERMS TO KNOW

abundant
aesthetic
arable land
beaver
Cape Horn
dependent
derived
fertilizers
finite
forest resources
fossil fuels
fur-bearers
Great Plains
hospitable
hunt
immigrants
insatiable
irrigation
mineral resources
natural resources
nongame species
nonrenewable
processing
pulpwood
renewable
sod
windmills

OBJECTIVES

After completing this chapter, you should be able to

- Identify the common natural resources people use.
- Understand the concepts of renewable and nonrenewable natural resources.
- Identify renewable and nonrenewable natural resources.
- Understand the importance of our natural resources.
- List the uses of natural resources.

INTRODUCTION

What is a **natural resource**? Most define natural resources as items found in or on the earth that are of use to humans as fuel, food, shelter, or a source of wealth. Some common natural resources are oil, coal, water, **mineral resources**, **arable land**, wildlife, and **forest resources**. Our existence depends on these resources.

Can you imagine life without automobiles or airplanes, without electricity or running water, without supermarkets full of food? Without steel created from mineral resources there would be no planes, trains, or automobiles. Without oil there would be no fuel to run them, nor would we have the many plastic products we have come to rely on. Without oil, coal, and water, we could not produce the **abundant**, cheap electricity we all take for granted. Without forest products (Figure 2-1A) you would most likely not be reading this text, which is, like most printed material, produced on paper **derived** from **pulpwood**. Nor would you be able to live in your current home, which almost certainly contains a large amount of wood (Figure 2-1B). Without arable land we would not enjoy an abundant food supply. In fact, arable land alone is not enough. To feed the billions of people on Earth requires oil for fuel and mineral resources for the steel in the equipment to produce and process the harvest. Modern agriculture also requires billions of gallons of water for

FIGURE 2-1(A) Loggers harvest and transport logs from forests.

(Source: © Kelly Boreson, 2010. Used under license from Shutterstock.com)

FIGURE 2-1(B) Finished lumber used to construct homes and businesses is the product of lumber mills.

(Source: © Stephen Coburn, 2010. Used under license from Shutterstock.com)

the **irrigation** and **processing** of crops. Mineral resources, oil, and forest resources are used to make cans, plastic bags, and boxes in which food is stored after it is processed.

Humans and most animals and plants could not survive without fresh water. Water is vital, not only to our quality of life, but for our very survival. Life as we have come to enjoy it in America is not possible without our use of natural resources. Actually, no human life could exist without natural resources; even early humans, huddled in cave shelters, required them. They used wildlife and wild plants for food and drank from rivers and streams. As humans developed from early hunter-gatherers into farmers, their food supply became more stable and they had more free time on their hands. With time to think and experiment, humans learned, over hundreds of years, to use more and more of what nature provided. Today humans are more **dependent** on a wider variety of natural resources than were our early ancestors. We also live longer, more comfortable lives.

NATURAL RESOURCES— RENEWABLE OR NONRENEWABLE?

Natural resources can be divided into two categories: **renewable** and **nonrenewable**. Resources in the nonrenewable category are mineral resources, such as iron, copper, lead, uranium, and aluminum; and **fossil fuels**, such as oil, coal, and natural gas (Figure 2-2A). Although we may have an ample supply of these nonrenewable resources for many years to come, they are **finite**. A nonrenewable resource is one which, once the available supply has been used, cannot be replenished. Conversely, renewable resources are those which, with proper management and care, can be utilized indefinitely. Forests, wildlife, water, and

FIGURE 2-2(A) One nonrenewable natural resource is oil. Oil is taken from the earth by oil-drilling rigs.

(Source: Courtesy of U.S. Fish and Wildlife Service. Photo by Scott Swanson.)

FIGURE 2-2(B) America has millions of acres of forested land, which, if managed properly, will remain a vital natural resource.

soil are renewable natural resources. However, all renewable natural resources are vulnerable to abuse. For example, once a wildlife species' habitat becomes fragmented (i.e., broken up into small, isolated pockets) and its population drops below a certain number, it is almost certainly doomed to extinction. An example of this is the passenger pigeon, mentioned in Chapter 1. Unfortunately, hundreds of other species have already gone the way of the passenger pigeon. However, wildlife in general is a renewable natural resource. With proper management of wildlife and wildlife habitat, we can enjoy an abundance of wildlife for generations to come. Forest resources, or timber, are also renewable resources. If we manage our forests so as to plant as many or more acres than we harvest, we will continue to have a bountiful timber harvest (Figure 2-2B).

THE ROLE OF NATURAL RESOURCES IN AMERICA'S DEVELOPMENT

People are drawn to a country or area by its abundant natural resources. Plentiful natural resources attracted people to North America. Early settlers came here hoping to find freedom as well as land. Land, that most precious natural resource, was in short supply in Europe. Land had for centuries been the property of the nobility, and it was virtually impossible for a commoner to own even a small piece of it. In America there was land to farm, timber to build houses, and wildlife to eat. To those early adventurers in the New World it must have seemed too good to be true. People coming from disease-ridden, crowded European cities quickly adapted to their new surroundings. They used the available resources, timber, land, and wildlife for food, shelter, and wealth. Natural resources played a vital role in the development of America.

Let's take a more in-depth look at the role of natural resources in the development of America. About 90 percent of all the early settlers in America were farmers. They depended on land, a natural resource. They were also dependent on the forest to provide logs with which to build shelters and the native wildlife to provide food between harvests. These early Americans could not go to the local supermarket or lumberyard. They would clear a piece of ground, sometimes using the trees to build a cabin, and then plow and farm it. However, there were no commercial **fertilizers** available at the time. This meant that in one to three years their crops would have taken the available nutrients from the soil, requiring them to clear another field. The problem was made worse by the fact that the coastal northeastern areas of the United States, the first areas settled, did not have particularly deep or productive soils. Thus, the farmers practiced what has become known as slash and burn agriculture. This method of farming requires a great deal of land. It was practiced until after the Civil War, especially in the South, where cotton was the main crop. More and more land was required, which was one of the first forces causing people to move west. Slash and burn agriculture is still practiced in parts of the world today, most notably in the South American rainforests (Figure 2-3).

Natural resources of all types were vital to the settling and development of America. The first Spanish explorers came to America in the hope of finding great wealth in the form of minerals, such as gold, silver, and precious jewels. Although they did not find the mineral wealth they had hoped for, they did discover a land rich in wildlife, forests, and an enormous amount of sparsely inhabited land. As we will see, the mineral resources were here, but early settlers did not find or develop them. The tremendous abundance of wildlife in North America was another force in the development of the United States. Early settlers soon discovered that Europeans had a seemingly **insatiable** appetite for furs and hides, using them for hats, gloves, coats, and leather

FIGURE 2-3 Slash and burn agriculture is still practiced today in some parts of the world. Notably, it is used to clear land in South American rainforests.

(Source: Photo courtesy of Photodisc.)

goods. As early settlers attempted to meet these demands and their own needs for wild game, they quickly exhausted coastal game supplies, so they went farther and farther inland. These early settlers may have searched for gold, diamonds, silver, and jewels; however, these resources were not plentiful in the eastern half of what was to become the United States. What these early explorers did discover, as they moved into western Pennsylvania and the Ohio and Tennessee River valleys, was rich, deep soil, and a wealth of fur- and meat-bearing wildlife. The search for more and better land to farm and the need to find additional sources of furs for the European market were primary forces in the spread of European settlers in America. It was to remain so for more than 200 years.

Settlers leapfrogged across eastern America in search of land and wildlife resources. However, the pattern changed dramatically in 1849, when the first major gold deposits in North America were found. With the discovery of gold at Sutter's Mill, California, the character of settlement in America was changed forever. Until this time only a few hardy souls had crossed the **Great Plains**, then commonly called the "Great American Desert." They went in search of **beaver** and other **fur-bearers** located in the Rocky Mountains region, and they often had to deal with hostile conditions for an uncertain reward. Gold, however, was a more certain reward. The prospects of becoming rich and famous almost overnight changed the character of the game. Men who had been content to farm a piece of ground for four or five years and then move on suddenly sold everything except what was needed for the trip, threw the family in the wagon, and headed for California (Figure 2-4). It became an enormous stampede of humanity. Thousands perished on the Great Plains, the victims of thirst, hunger, cold, increasingly hostile Indians, disease, or just poor planning. Hundreds more

FIGURE 2-4 Many people sold their land and rushed to California to make their fortunes in gold.

FIGURE 2-5 Windmills allowed settlers on the Great Plains to pump water from deeper in the earth than hand-dug wells.

perished at sea, while attempting to sail around **Cape Horn**. There was an influx of **immigrants** greatly increasing the population of the United States.

Throughout this westward movement, few stayed on in the "Great American Desert" that was soon to become America's "bread basket." People were bound for the gold fields of California and later, the rich farming country of Oregon. Most did not realize that the sea of grass would soon be replaced by an equally imposing sea of wheat. All too soon the Rocky Mountains and all areas westward were being searched for gold, silver, copper, tin, and other minerals that could be exploited for wealth.

After the invention of the steel plow, the importation of Russian seed wheat, and the widespread availability of **windmills**, the Great Plains became more **hospitable** (Figure 2-5). The steel plow offered a means by which settlers could break up the thick prairie **sod**. The Russian wheat provided a crop that could be grown in an area that received limited rainfall, and the windmills presented a reasonably secure source of drinking water for the settlers and their livestock.

The invention of the internal combustion engine in the late 1800s led to the "oil rush" of the early 1900s. The increasing needs of a growing population fueled the exploration of the oil fields of Texas, Oklahoma, and Louisiana during this period. The search for oil and natural gas continues to this day. We continue to use them for fuel, and for the production of plastics, chemicals, tires, and other products.

We are more dependent on a wider variety of natural resources than early Americans were. Their needs were simple, but ours have grown complex with the contributions of technology and science. However, we are not as dependent on wildlife as our early ancestors were. We have developed a very efficient agricultural system, and we shop for clothing in department stores. Why is it, then, that wildlife is such a valuable natural resource? It is very difficult to place a dollar value on wildlife as you can on oil or timber or land. Wild creatures mean different things to different people. Wildlife has a certain **aesthetic** value, that is, value related to beauty and appreciation. Observing wildlife is the number one outdoor form of recreation. Wildlife is considered a natural wonder by many people. Who can put a value on the sight of a white-tailed deer gently licking her fawn? How much is the sight of 100,000 snow geese in flight over the South Texas rice country worth? Is the cardinal eating at your bird feeder worth anything? These are questions that each person must answer.

Although it is difficult to place a monetary value on wild animals, it is considerably easier to establish the value of a species as a game animal. The majority of Americans do not need to **hunt** in order to feed themselves and their families; however, a large number do choose to hunt. The economic impact of more than 15 million hunters is considerable. Sportsmen and women spend in excess of $950 million per year on license fees alone. Excise taxes on sporting arms,

(Source: Courtesy of U.S. Fish and Wildlife Service. Photo by Ryan Hagerty.)

FIGURE 2-6 Hunting and fishing are popular recreational uses of wildlife.

ammunition, and archery equipment contribute an additional $200+ million. The economic contribution of sport hunting to the U.S. economy each year is estimated at more than $6 billion. In many rural areas, hunting season may be the major income producer of the year. The majority of funding for wildlife conservation of both game and **nongame species** comes from hunter-generated dollars (Figure 2-6).

SUMMARY

Human life cannot survive without natural resources. Our use of natural resources has grown throughout the country's history. All the natural resources mentioned in this chapter are very important, but some might be considered more so than others. Water, soil, wildlife, and forest resources are musts for survival, providing water, food, and fuel. These basic resources sustain life. Natural resources support billions of humans. We must conserve and protect all our natural resources, both renewable and nonrenewable, for future generations of Americans.

REVIEW QUESTIONS

Fill in the Blank

Fill in the blanks to complete these statements.

1. A _____ is an item found on or in the earth that is of use to humans.
2. _____ is a land that is suitable for crops.
3. Oil and coal are _____ fuels.
4. Mineral resources such as _____ and iron are used extensively by humans.
5. Modern agriculture uses billions of gallons of _____ .
6. _____ is used to provide crops with water if there is inadequate rainfall.
7. List three examples of renewable natural resources.
    (a) _____
    (b) _____
    (c) _____
8. List three examples of nonrenewable natural resources.
    (a) _____
    (b) _____
    (c) _____
9. Throughout history humans have been drawn to areas with abundant _____ .
10. Early settlers were drawn to America primarily by its abundant natural resources
    of _____ , _____ , and timber.
11. Early American farmers practiced _____ agriculture.
12. The need for more and more _____ was a primary driving force in the spread of settlers in America.
13. Another driving force in the spread of settlers was their search for fur-bearing wildlife, such as _____ , to supply the European market.
14. In 1849 _____ , a mineral resource, was discovered at Sutter's Mill, California.
15. Windmills helped people settle on the Great Plains because they provided a more reliable source of _____ .
16. Observing _____ is the number one form of outdoor recreation.

Short Answer

1. Why do you feel our natural resources are important?
2. How are natural resources important?
3. Why did early American farmers practice slash and burn agriculture?

Discussion

1. Discuss the many ways that natural resources affect your life. How do you use natural resources each day?
2. What would be the impact on your life if just one natural resource, such as oil, were not available?

Learning Activities

1. Select a nonrenewable natural resource and list all the major uses we have for it. Present your findings to your class.

2. Select a renewable natural resource and list all the major uses we have for it. Present your findings to your class.

3. For activities 1 and 2, ask the class how the loss of these resources would affect them.

CHAPTER 3

Conservation: Wise Use of Natural Resources

TERMS TO KNOW

accelerate

alternative

conserve

desalination

drought

Dust Bowl

encroachment

energy

erosion

forests

fragile

Great Depression

polluted

productivity

recycling

reforestation

replenished

significant

soil conservation

Soil Conservation Service

topsoil

USDA

vocation

OBJECTIVES

After completing this chapter, you should be able to

- Understand the importance of conservation in the use of natural resources.
- Identify common conservation techniques.
- Understand the positive effects conservation has on the environment.
- Understand the negative effects a lack of conservation can cause.

INTRODUCTION

In Chapter 2 we discussed the importance of natural resources. Their importance dictates that we work to **conserve** them. Some natural resources can be **replenished** through human efforts, but most cannot. It is vital that we conserve the resources we have (Figure 3-1). Future generations may develop **alternatives** to many of the resources we currently use, but for the foreseeable future we have no viable substitutes for most of the natural resources we use today. Oil, natural gas, coal, and water (used to turn turbines to produce electricity) will remain our main sources of **energy**. Forests will continue to produce a major portion of our building materials, and soil will grow our crops for many years to come. Therefore, it is in our own interest to conserve existing natural resources, work to develop

(Source: Courtesy of USDA.)

FIGURE 3-1 Some natural resources, such as wetlands, are difficult and expensive to replenish. Plants and wildlife need a long time to reestablish themselves.

effective substitutes, and continue to look for ways to renew natural resources. The outlook for natural resources is not as bad as many in the media would have you believe. For example, we have effectively managed our soils, forests, and wildlife during the past 50 to 60 years. However, our efforts in other areas of natural resource conservation have not been nearly as effective.

SOIL CONSERVATION

Work in the field of **soil conservation** became a serious **vocation** during and immediately after the **Great Depression** and **Dust Bowl** era of the 1930s. When dust from the western states began to darken the skies over New York City during the Dust Bowl, people realized that soil is a **fragile** resource.

What led to soil or "dust" in the air over New York? **Erosion** did. Erosion is the most serious threat to our soil (Figure 3-2). Erosion is a natural process that occurs continually to most soils, but abuse of the soil, by not leaving any grass or crop cover, for example, can greatly accelerate soil erosion. Erosion is not usually a serious problem when the soil is properly protected by a covering of plant growth. Plant leaves slow the impact of raindrops, and their roots help hold and stabilize the soil. Acute problems can occur when the protective plant cover is removed in order to build a house, a highway, a shopping mall, or to plant a crop.

Erosion comes in two main forms: water erosion and wind erosion. Each of these can be broken down into several different types. Water erosion occurs in two ways—from water runoff and from raindrops striking the earth. A raindrop may not seem very destructive to the soil, but there are many billions in just one summer shower. Each of these strikes the earth at 20 to 25 mph. If they strike bare earth, they tear up the soil on impact. Look at your garden plants, your yard fence, or the side of your house after the next rain. If these areas border bare ground, they will be splattered with soil. This is caused by raindrop impact. However, this form is not the most destructive type of water erosion. Water

FIGURE 3-2 When the wind is particularly strong at soil level, it can cause erosion.

flows along the path of least resistance. This is, of course, generally downhill. As rainfall runs off, it gathers speed, picking up soil particles as it does. If there are no plants and plant roots to slow this process, the damage done by the run-off can be **significant**. Water is unbelievably powerful. The next time there is a heavy rain in your area look, from a safe distance, at a stream or river. You will likely see many small plants and maybe a tree or two go by. You will also notice that after a heavy rain, even in areas with adequate plant cover, the water in the ditches, streams, and rivers will turn a dirty brown color. This "muddy" color to the water is because of dissolved and suspended soil particles in the water.

Wind erosion is caused when the wind picks up soil particles and moves them from their original location. In areas where the wind blows steadily, there can be significant loss of soil if plant cover is inadequate. Sand and other soil particles picked up and carried by the wind become tiny missiles. They strike and erode other soil particles and are capable of peeling the paint off objects in their path. In fact, a young crop can be lost or severely damaged by blowing soil. The soil particles are quite abrasive and will cut a young, tender plant off right at ground level. The effect of severe wind erosion is what the citizens of New York, and much of the rest of America, witnessed in the 1930s. Vast areas of the West and the Great Plains were being farmed. With no plant residue or cover crops to protect it, the soil was very vulnerable. After a prolonged **drought**, high winds severely eroded the soil.

The tragedy of the Dust Bowl called attention to the need for increased soil conservation (Figure 3-3). The **Soil Conservation Service** was formed to educate America's farmers about good soil management practices. Erosion is very damaging to the **productivity** of the soil. It removes valuable **topsoil**, lowering the productivity of the soil, thus making it more difficult for the farmer to produce a bountiful harvest. Through education, the Soil Conservation Service works to see that the Dust Bowl is not repeated. Such things as shelter belts (rows

(Source: Courtesy of NRCS.)

FIGURE 3-3 The tragedy of the Dust Bowl highlighted the need for soil conservation.

of trees planted close together) help break the force of the wind and thereby slow down wind erosion. Instead of plowing wheat or grain stubble under, today farmers leave it to catch blown soil and runoff water. Contour plowing, which is plowing around a slope instead of up and down it, is now normal practice. Plowing the furrows around a slope notably slows runoff water, allowing it to soak into the soil. These and many other practices have combined to greatly reduce soil erosion. These practices also make a soil more productive by keeping the rich topsoil intact and improving moisture levels in the soil, making more water available to the crop.

Government programs such as the Soil Bank in the 1950s and the Conservation Reserve Program (CRP), which came into being with the 1985 farm bill, have been very successful. These programs encouraged farmers to set aside, or not plow, marginal farmland. Land that received too little rainfall or had shallow soils was planted in native grasses, shrubs, or trees. Land enrolled in the CRP is set aside for 10 years. This program has taken millions of acres of marginal farmland out of production and prevented the erosion of many tons of soil. At the same time, it has created millions of acres of wildlife habitat. Unfortunately, today we are seeing more acreage being taken out of the CRP and returned to crop production. Although much of this land is marginal, the price of grain has steadily increased, making it much more attractive to the farmer to attempt to increase production by cultivating all his or her acreage. Increased demand for corn, to be used in ethanol production, and increased world demand for food grains, such as wheat, are also fueling the return of millions of acres to grain production. It is extremely important to remember that minimizing erosion also greatly increases the quality of our water supply. While we have made great strides toward conserving our valuable soils, we must be extremely careful that we do not lose what we have gained in the rush to produce more grain.

FOREST CONSERVATION

As with farmland, we have been learning how to conserve our **forests**. America's forests are one of our greatest resources. When Europeans first reached the shores of North America, most of the eastern half of America was forest. Today there are approximately 738 million acres of forests left in the United States. This is roughly 70 percent of the amount that was here 500 years ago. Considering the wasteful methods used to harvest timber during the early stages of America's development, this is an impressive percentage. However, our forests are vital to our continued prosperity and to the well-being of hundreds of species of wildlife. We must manage them very carefully. Great progress has been made in forest conservation in the past few decades.

Forest products, of which wood is the most obvious, are used to produce a large variety of items that we use every day. Wood is used for furniture, boxes, framing, cabinets, and flooring. It is also used for fencing material, railroad ties, telephone or electric poles, plywood, and wafer-board. However, wood products may not be so obvious. Wood that has been mechanically or chemically altered is referred to as converted wood. Converted wood products include paper, wood fiber, charcoal, plastics, and explosives. There are dozens of others, such as shatterproof glass, sponges, rayon, and photographic film.

Reforestation is the future. The major timber companies must replant at least as many acres as they harvest if we are to continue to have wood products to use in the future. Although we are much more efficient with our forests today, we have fewer acres of forest to work with but greatly increasing demands. Each year several thousand acres of forest are cleared to make way for housing developments, shopping malls, parking lots, and office parks (Figure 3-4). Proper forest management will become increasingly important in the twenty-first century. We

FIGURE 3-4 Continued human expansion will make proper forest management increasingly crucial in the twenty-first century.

(Source: Courtesy of NRCS. Photo by Lynn Betts.)

have made great progress in forestry management since World War II; however, we must **accelerate** our management efforts if we are to have the forest products we will need for the future.

WILDLIFE CONSERVATION

Wildlife is another natural resource we have learned to manage successfully. Although several species are now extinct, many others are found in greater numbers than in colonial America. Wildlife is constantly losing the fight for habitat, which is needed for their survival. Wildlife management techniques of the future will likely be based on minimizing the effects of human **encroachment** on wildlife habitat. Wildlife managers spent the majority of the twentieth century reestablishing wildlife populations in suitable habitats. Now they must try to protect the forests, grasslands, and meadows that are home to these animals.

FOSSIL FUEL CONSERVATION

Conservation efforts directed toward fossil fuels, mainly oil, natural gas, and coal, have been essentially pathetic. For many years America had oil and gas to burn, literally. Natural gas from oil wells was routinely burned off. No serious conservation efforts were aimed at reducing our use of these fuels. The world and America in particular, enjoyed abundant, relatively cheap fuel.

In 1960 four Persian Gulf states and Venezuela formed the Organization of Petroleum Exporting Countries (OPEC). OPEC was created to halt the decrease in crude oil prices that the seven largest oil companies had attempted to impose on these oil-producing countries. Although OPEC succeeded, it was unable to significantly raise the prices these countries received for their crude oil. This changed in 1973. After the Arab–Israeli war, the Arab oil-producing countries cut back production and stopped oil shipments to the United States, which sharply limited supplies and had the desired effect on prices. The price for a gallon of gas more than doubled. Supplies were limited, resulting in long lines at gas stations (Figure 3-5). The oil embargo caused several things to happen. Americans had been accustomed to spending 30¢ to 35¢ for a gallon of gas. When prices shot up and supplies became limited, people were shocked. The big, heavy American car that used 8 to 10 miles per gallon suddenly became much more expensive to drive. Fuel economy became more important. Until this time, few small cars were sold in the United States. American automobile companies were unprepared for the crisis and had few small, efficient models to offer the public. This allowed the foreign makers of automobiles, primarily the Japanese, to acquire a share of the American market. Their market share has continued to increase to this day.

The oil embargo also demonstrated to our leaders in Washington just how dependent we are on foreign oil imports, and some efforts at conservation were initiated. For example, the federally mandated 55 mph speed limit was

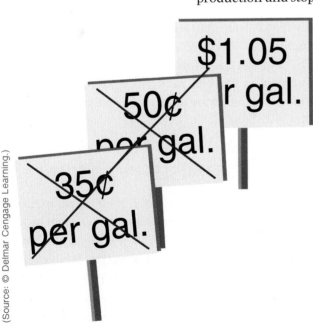

(Source: © Delmar Cengage Learning.)

FIGURE 3-5 Gas prices escalated during the oil embargo of the 1970s.

placed on most highways as this speed was determined to offer the best fuel economy. Unfortunately, a few years later, after everyone had forgotten the gas lines and shortages of 1973–1974, the 55 mph speed limit was largely dropped. This was especially true in western states, as many times vast distances must be covered between towns. Another result of these efforts is the vastly improved fuel economy of most American automobiles. Although transportation represents only about 25 percent of the total energy used in the United States it accounts for 60 percent of U.S. oil consumption. Congress passed a law in 1975 requiring new car fuel efficiency to be doubled by 1985. Gasoline shortages in 1974 and 1979, coupled with much higher prices, forced the increase in fuel efficiency of the average U.S. car by 40 percent by 1990. Improvements have continued since then, but these have been largely offset by the growing numbers of vehicles on the road. We have also seen an increase in car pooling and use of public transportation. Although both methods are fairly efficient, the sprawling design of most American cities makes their use difficult.

We are beginning to see alternatives to gasoline in the form of electric, propane, and natural gas. City governments in particular are using vehicles powered by propane and natural gas because there is no significant decrease in power, they generate less pollution, and they are cheaper than gasoline. Until these fuels are available to the average driver, however, their use will be limited.

We still rely on large amounts of foreign oil, but the situation has improved somewhat in the past 25 years. Although the amount of oil we use continues to increase, it is increasing at a slower rate. This is due to such factors as better fuel economy in automobiles, improved insulation in our homes and offices, more efficient air conditioning and heating units, and the use of alternative energy sources. Unfortunately, the world demand for energy has greatly increased, most notably in developing countries like China and India. With over one billion people, China is a potentially huge consumer market. As the Chinese government has moved from a communist economy to a more open, free market economy, the country's demand for consumer goods, all of which require petroleum products in their production, has soared. This worldwide increase in demand, coupled with a lack of refinery capacity in the United States and a lack of exploration and production in North America, is largely responsible for the tremendous increase in fuel prices seen in 2006–2008.

MINERAL CONSERVATION

Conservation of mineral resources is not the pressing matter that conserving fossil fuels is. There are still vast reserves of most mineral resources around the world. Still, it is important that we have made some progress toward conservation. **Recycling** of some metals has been practiced for years. As early as the late 1930s, Japan, a country with few mineral resources, was buying large quantities of American scrap metal to produce war materials. More recently, recycling of metals, principally aluminum and steel, has been increasing (Figure 3-6). With the advent of

(Source: © Katrina Brown, 2010. Used under license from Shutterstock.com)

FIGURE 3-6 Recycling is one way we can help to conserve nonrenewable resources.

the aluminum beverage can, recycling became more practical. Manufacturers of aluminum cans discovered that it is more economical to recycle aluminum cans than to make new ones. Crushed cans are worth enough that people are recycling in record numbers. Paper and plastics recycling has also increased greatly. Some form of recycling is now available in most areas of the United States.

WATER CONSERVATION

Water is a natural resource without which humans cannot survive. Water is essential for all living organisms. With approximately 70 percent of the Earth covered by water, it might seem that a lack of water would not be a problem. Unfortunately, 97 percent of the water on Earth is salt water, located in oceans and seas. Approximately 3 percent is fresh water, more than half of which is frozen in glaciers and icecaps. Without expensive **desalination** being done, the water located in the oceans and seas cannot be used as drinking water as it contains too much salt. Salt water is also unsuitable for most industries and for irrigation. Of the 3 percent freshwater on Earth, less than 2 percent is available for our use.

Water was for many years taken for granted, wasted, and abused. We **polluted** rivers, streams, and lakes by pumping untreated waste into them. The human body requires fresh, clean water to function properly. It has been estimated that Americans drink about 100 million gallons of water per day. Although this is a lot of water, it is a tiny amount compared to the total amount of water used in this country each day. Industry, agriculture, domestic users, and fish and wildlife are all major users of water. The United States Department of Agriculture (**USDA**) estimates that industry uses 40 billion gallons of water per day, but only about 2.5 billion gallons are actually lost or consumed. The remainder is returned to rivers and lakes where it can be used again, if it is not too polluted. Strong pollution controls have helped to improve our water supplies by ensuring water quality. Industry uses water in the process of manufacturing goods. For example, paper is made in water, and water is used to cool, clean, and/ or carry the product in many factories.

The agricultural industry uses water primarily for crop irrigation. Irrigation is not a new technique; it was used thousands of years ago by the Egyptians and by the Hopi tribe of Arizona. Irrigation is used in parts of the United States where annual rainfall is inadequate. Huge amounts of water are used to irrigate crops, and the many wells that have been drilled to supply this water have severely lowered the water table in many western states. Millions of people rely on ground water for their supply of water. When water tables fall, it is a very serious matter. At least one major metropolitan area, San Antonio, Texas, relies on ground water to supply the needs of its citizens. San Antonio draws its water from the Edwards Aquifer, a large underground lake. Increased use of water by a growing population and for agriculture has caused a dramatic drop in the level of the aquifer.

We are all major users of fresh water. Few people realize just how much water they use in a day. Each person in the United States uses approximately 150 gallons of water each day. This is more water, on average, than is used by any other people in the world. We use much more water than we need to survive. We use water for bathing, cooking, cleaning, and carrying away our wastes. An average bath uses 20 to 30 gallons of water. A shower typically uses

FIGURE 3-7 Water is a precious resource that is essential to the survival of all species.

(Source: Courtesy of U.S. Fish and Wildlife Service.)

about 5 gallons per minute. When we flush a toilet we use 3 gallons of water. Because of conservation legislation, newer toilets require only 1 to 1.5 gallons of water per flush. However, getting the desired results often requires multiple flushes, thus using the same or more water than older models. In addition to these uses of water, we wash cars, water lawns, and bathe pets. We tend to take water for granted (Figure 3-7). We cannot continue to waste this most precious commodity and hope to have enough for the future.

Water is very important to fish and wildlife as well. Fish require relatively clean water with few pollutants. We damage habitat when we dam or otherwise alter a river or stream. All species of wildlife need water. River-bottom and stream-side habitat supports a great variety of wildlife, from white-tailed deer to wild turkeys.

SUMMARY

We need to be aware of growing demands on natural resources brought about by increasing population. There are known reserves of oil in the world to last about 100 years. It is estimated that unknown reserves will increase this period substantially. Our country has great reserves of natural gas, and coal should not be in short supply in the foreseeable future, but we must conserve fossil fuels and find alternatives to them. Air pollution and the potential for global warming are two additional reasons we need to find alternatives to fossil fuels. Due to our reliance on individual automobiles for personal transportation in the United States, we must accelerate the research and development of alternative fuel sources, such

SUMMARY (*continued*)

as hydrogen and electricity. While there are a number of hybrid gas or electric vehicles on the market today, there are woefully few choices available in a nation with hundreds of millions of vehicles on the roads. Although, there is a rapid expansion in the use of wind and solar power to produce electricity, in the immediate future, coal, oil, and water will continue to produce the bulk of our electricity. Our forest and mineral resources will continue to be viable if we manage and conserve them with care. Forests are vulnerable to overharvest and must

be replanted at least at the same rate they are harvested. Our wildlife resources will survive if we can slow the human encroachment on their habitat. Many species are on the verge of extinction; we must decide how far we are willing to go to save them. We have adequate supplies of fresh water in this country, but they are not always where they are needed. Some areas of the country already have or will have water shortages. We must develop water conservation measures and strive to protect our existing water supplies from pollution.

REVIEW QUESTIONS

Fill in the Blank

Fill in the blanks to complete these statements.

1. During the 1920s and 1930s America's western states experienced terrible wind erosion. This period became known as the _____ Bowl era.

2. _____ is a natural process that occurs continually in most soils.

3. Erosion comes in two main forms: _____ and _____ erosion.

4. The most destructive form of water erosion is caused by _____ water.

5. _____ erosion is a serious problem in areas with inadequate plant cover and consistently high winds.

6. Wind erosion is particularly dangerous during periods of _____.

7. Erosion removes valuable _____ and decreases soil's productivity.

8. List three common methods used to control or slow erosion.

    (a) _____

    (b) _____

    (c) _____

9. A government program that encouraged farmers to set aside marginal farmland in the 1950s was called the _____.

10. Its current equivalent is the _____, more commonly called CRP.

11. These types of programs also create millions of acres of _____ habitat.

12. America's _____ were and are one of our greatest resources.

13. _____ products are the ones most commonly associated with forests.

14. List three products commonly made from wood.

(a) _____

(b) _____

(c) _____

15. List three products that are not commonly thought of as wood products.

(a) _____

(b) _____

(c) _____

16. _____ is the future of our forests.

17. The most serious threat to wildlife is the effect of human _____ on wildlife habitat.

18. Our conservation efforts directed at _____ have been weak at best.

19. The Arab oil _____ caused people to demand more efficient automobiles.

20. Recycling of some _____ has been practiced for many years.

21. The most commonly recycled metals today are steel and _____.

22. _____ is essential for all living organisms.

23. Approximately _____ percent of the Earth is covered by water.

24. Ninety-seven percent of the Earth's water is _____, found in the oceans and seas of the world.

25. Only about 3 percent of the Earth's total water is _____ water.

26. Major users of water include industry, _____, domestic users, and fish and wildlife.

27. Americans drink about _____ million gallons of water per day.

28. The USDA estimates that _____ uses 40 billion gallons of water each day.

29. Agriculture uses water primarily for crop _____.

30. The average person uses _____ gallons of water per day.

31. List five ways you use water each day.

(a) _____

(b) _____

(c) _____

(d) _____

(e) _____

32. Although we have adequate supplies of fresh water in this country, they are not always where they are _____.

Short Answer

1. What are natural resources?

2. Why must we conserve natural resources?

3. What is the Soil Conservation Service? Why was it formed?

Discussion

1. Why is water conservation important? In your daily use of water, how can you be more conservative in its use?
2. What can be done to help conserve all natural resources? How would you implement conservation measures?

Learning Activities

1. Using the same renewable natural resource you chose in Chapter 2 Learning Activities, research the efforts being taken to conserve and renew the resource. Share your research with the rest of your class.
2. Using the same nonrenewable natural resource, research conservation efforts. Are we likely to exhaust our supplies of the resource in the near future? Again, share your findings with the rest of your class.

CHAPTER 4

The Administration of Wildlife Management

OBJECTIVES

After completing this chapter, you should be able to

- Identify the major federal agencies directly involved in wildlife management.
- Describe the methods these agencies use to protect wildlife resources.
- Understand the need for these agencies.
- Discuss these agencies and their roles in protecting wildlife resources.

INTRODUCTION

Today many government agencies affect wildlife, but only a few have a direct impact on wildlife in America. One agency at the federal level is directly responsible for managing fish and wildlife resources, the U.S. Fish and Wildlife Service. Several other federal agencies play important roles in wildlife management. The U.S. Department of Agriculture, the Bureau of Land Management, and the U.S. Forest Service all have an impact on wildlife resources. Each state also has its own wildlife and fisheries agency responsible for managing the wildlife within that state. Many more agencies and departments directly or indirectly affect wildlife than can be covered in one chapter. (Appendix B of this text contains a complete list of state agencies.) It is strongly recommended that students contact the agency responsible for wildlife and fisheries management in their state. The agency will be happy to supply information about the agency and its efforts to conserve wildlife resources.

THE U.S. FISH AND WILDLIFE SERVICE

In 1871 the U.S. Congress established the U.S. Fish Commission. This body was established to study the decrease in the nation's **food fishes** and to recommend ways to reverse the decline. Americans were beginning to take notice of the critical decline of **inland fisheries** and wanted the trend to stop. The Fish Commission was the first formal effort by the federal government to manage fisheries. It was placed under the Department of Commerce in 1903 and was renamed the Bureau of Fisheries.

In 1885 Congress created the Office of Economic Ornithology in the Department of Agriculture. The responsibilities of this office included the study of food habitats and **migratory** patterns of birds, especially those that had a direct impact on agriculture. The responsibilities of this office gradually increased, and it went through several name changes. In 1905 it was renamed the Bureau of Biological Survey and was responsible for managing the nation's first wildlife refuges and studying the distribution and habits of birds and mammals. It was also responsible for managing dwindling populations of **herons**, **egrets**, **waterfowl**, and other **migratory birds**; enforcing wildlife laws; and controlling predators.

The bureaus of Fisheries and Biological Survey were transferred to the Department of the Interior in 1939. In 1940 they were combined and named the Fish and Wildlife Service. In 1956 the Fish and Wildlife Act created the U.S. Fish and Wildlife Service and established within the agency two separate bureaus, Commercial Fisheries and Sport Fisheries and Wildlife. The Bureau of Commercial Fisheries was transferred to the Department of Commerce in 1970 and is now known as the National Marine Fisheries Service. The Bureau of Sport Fisheries and Wildlife remained a part of the Department of the Interior. Since 1974, when "Bureau" was dropped, the agency has been called the U.S. Fish and Wildlife Service.

The mission of the U.S. Fish and Wildlife Service is to conserve, protect, and enhance fish and wildlife and their habitats for the continuing benefit of the American people. The primary responsibilities of the agency are protecting and enhancing migratory birds, **endangered species,** freshwater and **anadromous** fisheries, and certain marine mammals. The agency employs more than 6,000 people at facilities across the United States (Figure 4-1). It maintains a nationwide network of law enforcement agents, research laboratories, wildlife refuges, fish hatcheries, seven regional offices, and a headquarters in Washington, D.C.

(Source: Courtesy of NRCS. Photo by Lynn Betts.)

FIGURE 4-1 NRCS District Conservationist Mark Schutt observes wildlife habitat improvements at a WRP site in Worth County, Iowa.

ENDANGERED SPECIES

The Fish and Wildlife Service is the agency charged with the protection and preservation of America's endangered species. It also leads the effort to restore and protect plants and animals that are considered to be in danger of extinction in the United States and around the world. The Fish and Wildlife Service uses the best scientific evidence available to identify species that appear threatened or endangered. The evidence is reviewed by scientists, and public comment is solicited. Species that meet the criteria of the Endangered Species Act are placed on the official "List of Endangered and Threatened Wildlife and Plants" maintained by the Interior Department. Fish and Wildlife Service biologists work with universities, other federal and state agencies, and private organizations to develop a plan to save these species (Figure 4-2). Recovery plans may include habitat preservation and management, **captive breeding**, research, and law enforcement. It might also include the **reintroduction** of a species into suitable habitat that is devoid of the species or in which the species are present in very small numbers. Service biologists recommend ways that development projects can lessen their impact on endangered species. The Service also advises other federal agencies on the impact of federal projects on endangered species.

These efforts have been successful in bringing some species back from the edge of extinction. Recent success stories include such diverse animals as the American alligator and the bald eagle. Exploitation for its valuable hide and habitat destruction brought the alligator dangerously close to extinction. However, habitat conservation and strict law enforcement have allowed the alligator to increase to the point of being pests in some southern states. The bald eagle population has been increasing at a steady pace since the use of the pesticide DDT was banned in this country. Strict protection, captive breeding, reintroduction, and habitat conservation have also played important roles in the bald eagle's comeback.

(Source: Courtesy of NRCS. Photo by Paul Fusco.)

FIGURE 4-2 Fish and wildlife biologists work with universities, other federal and state agencies, and private organizations to educate the public about the needs of many species.

Unfortunately, other species are not doing as well. The manatee, California condor, red wolf, and Kemp's Ridley sea turtle are among more than 500 species of mammals, birds, reptiles, amphibians, and fishes listed as threatened or endangered in the United States.

The U.S. Fish and Wildlife Service also works with other countries to help preserve their native wildlife. More than 90 countries now participate in the Convention on International Trade on Endangered Species of Wild Fauna and Flora (CITES). This is an international treaty organization designed to prevent the exploitation of rare wildlife from commercial trade. The United States is a member of CITES, and the U.S. Fish and Wildlife Service is responsible for its implementation in the United States.

NATIONAL WILDLIFE REFUGES

The U.S. Fish and Wildlife Service also supervises the nation's wildlife refuge system. President Theodore Roosevelt established the nation's first national wildlife refuge in 1903 by designating tiny Pelican Island, Florida, as a bird sanctuary. The system has since grown to more than 90 million acres of prime wildlife habitat located in more than 450 parcels from Maine to the Caribbean and from the Arctic Ocean to the South Pacific. These refuges vary in size from half an acre to several thousand square miles. Alaska contains the bulk of this acreage, but refuges are found in all 50 states and several territories. Many early refuges were created for egrets, herons, and other water birds (Figure 4-3). Some were established to protect large mammals, such as elk and bison. The majority have been created to protect migratory waterfowl. Because migratory birds may cross several international borders in the course of migration, they must be protected jointly. To this end the United States has treaties with countries that migratory birds, particularly waterfowl, inhabit during the course of

FIGURE 4-3 The U.S. Fish and Wildlife Service maintains and improves key refuge areas along waterfowl migration routes.

(Source: Courtesy of U.S. Fish and Wildlife Service. Photo by Bob Ballou.)

a year. To meet these treaty responsibilities, the United States has created many refuges specifically for waterfowl.

National refuges are also critical to the survival of many endangered species. In some cases they provide the only habitat in which a given species is found. Many refuges protect endangered species, and they all provide stable and secure habitat for many species of resident mammals, fish, reptiles, amphibians, insects, and native plants. Our national wildlife refuges also provide an opportunity for millions of people to experience and enjoy wildlife. Nearly 25 million Americans fish, hunt, observe, and photograph wildlife, or participate in interpretive activities on our refuges each year.

MIGRATORY BIRDS

The U.S. Fish and Wildlife Service has international responsibilities for migratory bird conservation. The United States has treaties with Canada, Mexico, Japan, and Russia on the protection of migratory birds. These agreements are vital to the welfare of migratory birds, especially waterfowl. Some species of waterfowl spend the spring nesting in Arctic Canada. They then spend the fall and early winter migrating across the United States and Mexico before reaching their final wintering grounds in South America. It is therefore impossible for any one nation to protect a species effectively. This places the responsibility for more than 800 species of migratory birds squarely on the shoulders of the U.S. Fish and Wildlife Service.

With the use of satellite imagery, air and ground surveys, and **birdbanding** (Figure 4-4), the Service assesses habitat conditions and estimates population levels and trends. Hunters play a vital role in the information-gathering process. Selected hunters are sent questionnaires that ask for information on the number of waterfowl taken. Hunters who are successful are also asked to send the Service the

(Source: Courtesy of U.S. Fish and Wildlife. Photo by Neesha Wendling.)

FIGURE 4-4 Banding waterfowl allows researchers to determine migration routes and reproduction habits for many species of ducks, geese, and shorebirds.

wings of ducks taken and the tail feathers of geese. This allows the Service to identify the age, species, and sex of the birds taken, which gives them information about the previous year's waterfowl production and harvest.

The Service has coordinated bird banding since 1920. Approximately 4,000 certified banders band more than 1 million birds each year. The Service has a bird banding laboratory in Laurel, Maryland, that maintains records on more than 45 million birds. Some 50,000 bands are recovered each year. The data gathered presents an invaluable source of information for researchers, helping them determine migratory routes, distribution, breeding age, and other life history information of various species. The Service also conducts an annual Breeding Bird Survey to obtain information about songbirds and many non-game species. Recent declines in some songbird numbers have increased efforts to determine effective management techniques for these species.

FISHERIES

The U.S. Fish and Wildlife Service is also actively involved in managing the nation's fisheries. The Service works to restore fisheries that have been depleted and are considered of national importance. Fish species can become depleted in several ways, but the most common culprits are pollution, other habitat damage, and overfishing. Restoring these fisheries is a major effort of the Fish and Wildlife Service. The Service currently focuses on four important species: the major **salmonid** species of the Pacific Northwest (coho, chinook, and steelhead), the striped bass of the Chesapeake Bay region and the Gulf Coast, the Atlantic salmon of New England, and the lake trout in the upper Great Lakes.

To help in these efforts the Service maintains research laboratories to study fish genetics, health, nutrition, and other aspects of fish ecology. These laboratories provide vital information needed to restore wild fish populations. The information is also very important to the over 70 national fish **hatcheries** the Service utilizes (Figure 4-5). These hatcheries produce more than 160 million

FIGURE 4-5 Modern fish hatcheries allow wildlife professionals to replenish and supplement freshwater and saltwater fish populations.

fish each year of 50 different species, which the Service stocks in our nation's waters. Of course, the ultimate goal of Service biologists is for these fish populations to once again become self-sustaining.

HABITAT PROTECTION

Another important aspect of the U.S. Fish and Wildlife Service's job is conserving wildlife habitat. The Service works with other federal agencies, industry, the states, and members of the public to provide vital biological advice concerning the effect of development activities on wildlife habitat. Service employees work out of field offices located throughout the country to assess the harmful effects of development. Projects such as dams and reservoirs, dredging, oil leasing and energy projects, and interstate highways, all of which may require federal permits or funding, are studied by Service personnel to assess their potential effect on wildlife species. Service biologists make recommendations on ways to minimize, avoid, or compensate for projects that will have a harmful impact on wildlife or fish resources. Service personnel also work directly with the Department of Agriculture and farmers to restore damaged or drained wetlands and to conserve the remaining acres of wetlands.

RESEARCH

Conducting research may be the most important job the U.S. Fish and Wildlife Service performs. Research is vital to conserving fish and wildlife resources. Biologists must know the impact chemicals, pesticides, heavy metals, and other pollutants have on fish and wildlife. Researchers also study endangered species and their habitats; fish health, genetics, and nutrition; and migratory bird habitat, diseases, and populations. To meet these research needs the Service maintains large research laboratories and field operations and works with universities across the country. The Service provides funding to many state universities to train graduate students in fish and wildlife biology. These students become involved in research projects on issues important to fish and wildlife managers.

LAW ENFORCEMENT

The Service is also charged with enforcing federal wildlife laws. These laws protect endangered species, **marine mammals**, migratory birds, and fisheries. The Service is also responsible for enforcing wildlife laws that arise from international agreements. Wildlife and plants covered by these treaties and laws require federal permits for their **exportation** or **importation**. This permit system is an important deterrent to those who would illegally traffic in wildlife. Upward of 200 special agents and inspectors are employed by the Service to help enforce permit requirements and wildlife laws. Wildlife inspectors stationed at major points of entry, such as airports, ports, and international borders, check permits and inspect shipments of live animals and wildlife products. These inspections are necessary to ensure that protected species are not imported or exported illegally. Agents investigate large-scale poaching and the commercial trade in protected wildlife as well as individual violations of migratory bird–hunting regulations. These efforts are critical to the future of wildlife in America and the world.

ADMINISTERING FEDERAL AID

Federal aid for fish and wildlife comes from two laws administered by the U.S. Fish and Wildlife Service. The Federal Aid in Wildlife Restoration Act, commonly known as the Pittman—Robertson Act after its congressional sponsors, provides money to support a wide variety of wildlife projects. The states and territories use these funds to acquire land for wildlife habitat, conduct research, manage and maintain wildlife habitats, provide access to hunting areas, and carry out hunter education and hunter safety training. The money comes from federal excise taxes on sporting arms, ammunition, and archery equipment. The states first received these funds in 1939. The other major funding law that the Service administers is the Federal Aid in Sport Fisheries Restoration Act. It too is named after its congressional sponsors and is commonly called the Dingell—Johnson Act. The money for the Dingell—Johnson Act comes from an excise tax on sport fishing tackle and was first allotted to the states in 1952. In 1984 this funding was supplemented by the Wallop—Breaux amendments, which increased revenue for sport fish restoration by extending the excise tax to include sporting equipment that had not been taxed previously. The amendment also designated a portion of the existing federal tax on motorboat fuels and import duties on pleasure boats and fishing tackle to fish restoration. These monies are used for fisheries research, access to fishing and boating areas, managing and maintaining fish habitat (Figure 4-6), restoring depleted fisheries, and carrying out aquatic education. These funds are distributed using a formula that considers the number of fishing and hunting license holders within a state or territory and the state's or territory's area. American sportsmen and sportswomen have provided in excess of $10 billion for fish and wildlife conservation programs since these laws were enacted. Many species that were in danger of extinction at the turn of the twentieth century, such as the white-tailed

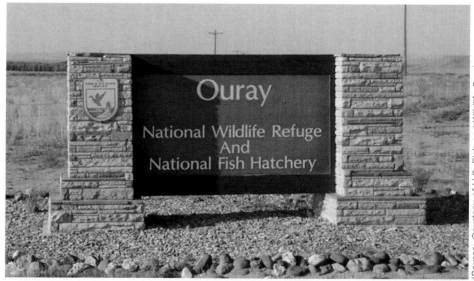

FIGURE 4-6 Federal aid pays for the restoration and maintenance of wildlife and fish habitat.

(Source: Courtesy of U.S. Fish and Wildlife Service. Photo by Robert H. Pos.)

deer, black bear, and wild turkey, have been successfully restored thanks to the money provided by the Pittman—Robertson Act. The Dingell—Johnson Act has enhanced populations of such fish species as the largemouth bass, channel catfish, and rainbow trout, thereby giving millions of Americans fishing opportunities.

(Source: Courtesy of U.S. Fish and Wildlife Service. Photo by John and Karen Hollingsworth.)

FIGURE 4-7 Nesting boxes are an innovative way to restore shelter in critical management situations.

THE U.S. FOREST SERVICE

No other federal agency has as direct or far-reaching an impact on wildlife and fish resources as the U.S. Fish and Wildlife Service, but several have less direct impacts on wildlife. The U.S. Forest Service manages some 190 million acres of public forest land in the United States. Most of this land is vital wildlife habitat. The rivers, streams, and lakes within the national forest system are important fisheries. The Forest Service and the public have become more aware of the importance of our forests to many wildlife species. Forests are vital to the survival of species such as the northern spotted owl. Recognizing this, the Forest Service has expanded programs to protect and improve fish and wildlife habitat. Each year the Forest Service completes wildlife and fish improvement projects on national forest land. Thousands of acres are reforested and hundreds of structures, such as nest boxes (Figure 4-7) and fish passages are constructed for the benefit of wildlife and fish each year. The National Wild and Scenic Rivers System continues to increase, currently exceeding 4,300 miles in the national forest system. The National Wilderness Preservation System contains more than 34 million acres. All this acreage is vital to the future of fish and wildlife resources.

THE BUREAU OF LAND MANAGEMENT

The Bureau of Land Management (BLM) is another government agency responsible for managing public lands. The 270 million acres of public lands the agency oversees contain a tremendous variety of invaluable wildlife habitats. Some 3,000 species of plants, birds, reptiles, fish, and mammals make their homes on land the BLM manages. In recent years increased concerns for the wildlife on the BLM's lands have increased efforts to manage the lands with wildlife in mind. Responding to recent declines in the **neotropical** and nongame migratory bird populations, the BLM has developed its own Nongame Migratory Bird Habitat Conservation Strategy Plan. The BLM continues to make progress managing and restoring riparian and wetland areas. Riparian areas are lands located next to streams, rivers, creeks, and lakes (Figure 4-8). Wetlands include such areas as marshes, estuaries, bogs, and wet meadows, which are frequently saturated or inundated by surface or ground water. In the arid West, where the vast majority of BLM-managed lands are, these areas provide crucial wildlife habitat. The BLM develops dozens of riparian and wetland improvement projects each year and maintains hundreds of existing projects. Through the management of millions of acres of public land, the BLM will continue to have an impact on many species of plants and animals for years to come.

THE U.S. DEPARTMENT OF AGRICULTURE

The U.S. Department of Agriculture does not have direct responsibilities for wildlife conservation and management, but it does administer a program that has been very beneficial to wildlife. From a wildlife management standpoint, the Conservation Reserve Program (CRP) was the most important part of the 1985 farm bill. The CRP was (and is) a soil conservation measure. Its purpose was to remove highly **erodible** farmland from production. Lands enrolled in the program were planted in permanent vegetative cover for a period of time. There have been additional enrollments and reenrollments of CRP acreage, as well as acreage removed from the program, since its inception. It is estimated that the CRP has successfully reduced soil erosion by many millions of tons of soil per year. By reducing the amount of **sediment** in waterways, water quality has been improved in many areas. The big benefit for wildlife has come from the additional habitat available. Some 30 million acres are enrolled in the program. The grassy cover provided by these acres is utilized by a tremendous variety of ground-dwelling wildlife. Rabbits, badgers, pheasants, quail, meadowlarks, and roadrunners are just a few of the species that have benefited from the CRP. With these additional acres to nest, forage, and hide in, grounddwelling bird populations have increased greatly in many areas. The CRP should continue to be extremely **beneficial** to wildlife all across America.

(Source: Courtesy of Cathy Espert.)

FIGURE 4-8 The Bureau of Land Management maintains and improves hundreds of riparian and wetland areas on the 270 million acres the agency oversees.

SUMMARY

America has an effective and efficient wildlife and fisheries management system. We have an abundance of fish and wildlife resources, and if we continue to support these agencies, our children will also have a wonderful wildlife **heritage.** However, we must be alert to the destruction of wildlife habitat. Loss of habitat is our greatest challenge in conserving fish and wildlife resources.

REVIEW QUESTIONS

Fill in the Blank

Fill in the blanks to complete the statements.

1. The primary federal agency responsible for wildlife management in America is _____ .
2. The Bureau of _____ and the U.S. _____ Service have wildlife management responsibilities.
3. Each _____ has an agency that manages wildlife.
4. In 1956 the _____ Act created the U.S. Fish and Wildlife Service.
5. The mission of the U.S. Fish and Wildlife Service is to conserve, _____ , and enhance fish, wildlife, and their_____ .
6. The primary responsibilities of the Service are protecting and enhancing migratory birds, _____ , freshwater and fisheries, and certain marine mammals.
7. Recovery plans established to help save endangered species may include habitat preservation and management, _____ , research, and _____ .
8. Two recent success stories in the endangered species battle include the _____ and the bald eagle.
9. The Fish and Wildlife Service supervises the nation's _____ system.
10. The nation's first wildlife refuge was established in 1903 by President _____ .
11. Although the first refuge consisted of only 3 acres, the National Wildlife Refuge System has since grown to more than _____ million acres.
12. Refuges are located in all 50 states and several _____.
13. The majority of National Wildlife Refuges have been created to protect _____ .
14. Refuges are enjoyed by more than _____ million Americans each year.
15. The United States has treaties with Canada, _____ ,_____ , and Russia on the protection of migratory birds.
16. The U.S. Fish and Wildlife Service has responsibility for more than _____ species of migratory birds.
17. The Service supervises the banding of more than _____ birds each year.
18. Banding helps the Service determine migratory routes, _____ , breeding age, and other life history information of various species.
19. Restoring fisheries is a major effort of the _____ Service.
20. The Service uses more than 70 national fish _____ to produce more than 160 million fish each year.
21. The U.S. Fish and Wildlife Service also works to conserve wildlife _____ .
22. Conducting _____ may be the most important task of the Fish and Wildlife Service.
23. Biologists must know the impact _____ , pesticides, heavy metals, and other _____ have on fish and wildlife.
24. Researchers study endangered species and their _____ ; fish health, genetics, and nutrition; and migratory bird habitat, _____ , and populations.
25. The Service is also responsible for enforcing federal _____ .

26. The Federal Aid in Wildlife Restoration Act, commonly known as the _____ Act, provides money to support a wide variety of wildlife projects.

27. The Federal Aid in Sport Fisheries Restoration Act, commonly known as the _____ Act, provides money for fishery research and restoring depleted fisheries.

28. Money for conservation programs comes from federal _____ taxes on sporting arms, ammunition, and sport fishing tackle.

29. Since these laws were enacted, American sportsmen and sportswomen have paid more than $ _____ billion for fish and wildlife conservation.

30. The U.S. _____ Service manages more than 190 million acres of public forest land.

31. The _____ (BLM) oversees approximately 270 million acres of public lands.

32. More than 3,000 species of plants, _____ , reptiles, _____ , and mammals make their home on BLM's land.

33. The U.S. Department of Agriculture administers the _____ (CRP), which has been very beneficial to wildlife.

34. Ground-dwelling species have benefited the most from the permanent vegetative cover established on the _____ million acres enrolled in the CRP.

Short Answer

1. What are migratory birds? Why is it important to have international treaties on migratory birds?

2. Discuss three of the major responsibilities of the U.S. Fish and Wildlife Service.

3. In what ways can the Bureau of Land Management and the U.S. Forest Service benefit wildlife?

4. What is the Conservation Reserve Program? How has it benefited wildlife?

Discussion

1. What do you think needs to be done to protect fish and wildlife habitat?

2. In what ways might we improve the management of fish and wildlife?

Learning Activities

1. Construct a flow chart that follows Pittman–Robertson funds from their source to their final destination. Include a graph that shows what percentage of the funds goes to each program (research, law enforcement, hunter education, and other activities). Information may be obtained by calling or writing the nearest U.S. Fish and Wildlife Service office (see Appendix A).

2. Write and present to your class a paper discussing in detail the steps, methods, and techniques used to manage an endangered species. You may use any of the 500-plus species on the endangered or threatened list. Procedures for each species will vary. Additional information may be obtained from the U.S. Fish and Wildlife Service.

3. Draw a map of the North and South American continents. Illustrate each of the major migratory waterfowl flyways. List the major species of waterfowl using each flyway and the approximate numbers of each species migrating on each flyway.

Useful Web Sites

UNITED STATES FISH AND WILDLIFE SERVICE
http://www.fws.gov

UNITED STATES FOREST SERVICE
http://www.fs.fed.us

BUREAU OF LAND MANAGEMENT
http://www.blm.gov

UNITED STATES DEPARTMENT OF AGRICULTURE
http://www.usda.gov

CHAPTER 5

Habitat Requirements of Wildlife

OBJECTIVES

After completing this chapter, you should be able to

- Describe the basic habitat requirements of all wildlife.
- Describe wildlife ecosystems.
- List the elements necessary for ideal wildlife habitat.
- Explain why each element in ideal wildlife habitat is of equal importance.
- Discuss the many things that affect wildlife populations.

INTRODUCTION

All wildlife, regardless of species, have the same basic habitat needs. An animal's habitat is its home, the area where it eats, rests, and reproduces. Wildlife require food, water, cover, and space, all suitably arranged, for their survival. Naturally, different species require different types of food, different types of cover, and varying amounts of water (Figure 5-1). However, each of these basic elements must be available to a species for it to survive. In later chapters we will discuss individual species in detail. In this chapter we cover each of the components of good wildlife habitat and why each is important in its own right.

It is difficult to **prioritize** these basic habitat needs. Food, water, cover, and space, in a suitable arrangement, are essentially of equal importance. Imagine these basic habitat requirements as a chain. If you remove just one link from a chain, the chain is broken

(Source: Photo courtesy of NRCS. Photo by Gary Kramer.)

FIGURE 5-1 Habitat with some variety is what most species prefer. Mixed brush land with grass and forb filled clearings makes excellent habitat for a wide range of species, from white-tailed deer to songbirds.

and it becomes worthless. If you remove any of the **essential** elements in an animal's habitat, that habitat becomes much less valuable to the specie. In a similar way, you may have all the elements necessary for excellent wildlife habitat within the normal **range** of a species, for example, white-tailed deer. However, if that habitat is split by a four-lane highway, it becomes more difficult, if not **hazardous,** for the deer to use it. Hence, the *arrangement* of the four essential elements—food, cover, water, and space—is very important.

FOOD

All living beings require food. Food is anything an animal consumes and digests to provide energy. The animal uses energy provided by food to meet its daily needs. Animals require varying amounts of food during different times of the year. For example, an animal's food requirements generally increase during extremely cold weather, reproduction, migration, or any other situation that causes stress in the animal. Animals use the energy they get from food just as humans do: to stay warm, walk and run, and reproduce (Figure 5-2). If for some reason the **quality** or **quantity** of food is **inadequate**, animals will likely suffer from weakness and malnutrition. An animal that is not getting enough to eat, either in quality or quantity, is more likely to **succumb** to sickness, disease, or predators. If the nutritional deficiency is severe or occurs during a period of increased stress, such as a severe winter, the animal may starve to death. Lack of adequate nutrition almost always results in decreased reproduction; weak, small offspring that are much more

(Source: Courtesy of U.S. Fish and Wildlife Service. Photo by R.H. Barrett.)

FIGURE 5-2 A coyote listens for the sound of rodents scurrying through a field. To survive, the coyote must get an adequate supply of food and water.

vulnerable to predators, disease, and sickness; smaller litter or clutch size; or in extreme cases, no offspring at all.

The actual amount of food required by an animal depends on a variety of factors: its age, sex, size, location, and the season of the year, for example. An animal may not necessarily have its greatest nutritional need during the winter months of December, January, and February. In the Southwest, for example, August and September can be very hard on wildlife. Often the green spring and early summer **forage** dies and dries up before the fall **mast** crop and **forbs** are available. If a drought limits the mast crop and there is an early hard winter, wildlife can suffer terribly. Ideally food is present in a habitat in adequate quantity, quality, and variety to sustain a population of wildlife without their expending valuable energy traveling to and from cover to get it. The less distance an animal has to travel to secure its meal, the less vulnerable the animal is to predators. It can also use the energy not spent searching for food for other needs, such as keeping warm or cool. As a general rule, the less time an animal spends searching for food and the less distance traveled, the better.

COVER

Cover or shelter serves several important functions for wildlife. It provides a relatively safe haven from predators and shelter from **inclement** weather, such as rain, snow, or excessive heat. It must also give them a place to reproduce and raise their young. Wildlife must have places to rest, feed, and sleep, as well as places to retreat to if threatened by a predator. Ideally, cover is so arranged that an animal need not overexpose itself to predators or the elements as it goes about its daily business. A wildlife population can suffer unusually high **mortality** rates if it is forced to feed, water, or travel too great a distance from its escape cover.

Of course, shelter takes many forms for wildlife. Although it is typically some form of vegetation, it also can be a pile of rocks, a cliff overhang, a hole in the ground, or a simple depression in the earth. **Vegetative** cover comes in countless variations (Figure 5-3). What works or is preferred by one species may not work for another. Some species, such as white-tailed deer, are very adaptable and can use a wide variety of habitats. One can find white-tailed deer in mature hardwood forests, marshes, farmlands, grasslands, suburban yards, and other habitats. Other species, such as the northern spotted owl and the red-cockaded woodpecker, for example, are found only in very specific habitats. The northern spotted owl is found in the Pacific Northwest and the red-cockaded woodpecker in the South and Southeast, both in old-growth forests.

Many species have never been very numerous and probably never will be because they cannot **adapt** to a wide variety of cover or habitat situations. Each species occupies a specific niche or spot in a given habitat for which it is adapted. The more specific or restricted the animal, the more vulnerable it becomes to changes in habitat. These changes may occur naturally or they may be human made. An animal species can often adapt to natural changes to its environment because such changes usually occur very slowly, over many years. However, human-made changes usually happen much faster, giving the animal little time to change and adapt. Species such as the Attwater's prairie chicken, which

(Source: Photo courtesy of USDA/NRCS.)

FIGURE 5-3 Dense cover, like the tall grass this cock pheasant is using, makes excellent resting and escape cover for most ground-dwelling species, from mice to rabbits to songbirds.

inhabits the native coastal prairie region of south Texas, face a very unsure future. Only a few thousand acres of such native pasture remain, with the rest paved, plowed, grazed, or otherwise occupied by humans. The Attwater's prairie chicken will have to adapt to changing habitat, survive in a tiny **remnant** of its former habitat, or perish. Wildlife managers are faced with the **unenviable** task of saving many species in much the same situation as the Attwater's prairie chicken. Habitat loss, or human-made dramatic, sudden change, has historically been the number-one reason for species becoming endangered or extinct.

WATER

Water is a resource that most wildlife cannot live without. There is tremendous **variation** in the amount of water used by different animals. Some animals require bodies of standing water, such as lakes and ponds, but many do not. Most large land mammals, such as bear, deer, elk, bison, and antelope, require standing water. They typically drink two to three times per day, usually in the early morning and late evening (Figure 5-4). Other species, such as the white-tailed deer, may drink during the middle of the day. This is particularly common during the summer and early fall in southern and western states, where midday temperatures may reach 100°F.

Many smaller mammals and birds obtain water from the food they consume. Most songbirds and upland game birds do not require large bodies of standing water to survive. They use it if it is available, but it is not essential. They can drink the **dew** off vegetation in the mornings or they may drink from puddles during the day. The wild turkey is a notable exception among upland game birds; it requires standing water and often prefers river and stream **bottoms** as habitat. These areas provide water, cover, and food. In the Southwest, where standing water is scarce,

FIGURE 5-4 Most large land mammals require fresh water in order to survive.

many species get moisture from native vegetation or from the prey they consume. Although water may not be a serious limiting factor in an area that receives 30 or 40 inches of rainfall per year, it can be a factor in an area that receives only 10 inches. In areas with limited surface water, wildlife managers often construct artificial watering devices, commonly called guzzlers. These structures catch and store rainwater that would otherwise run off and be unavailable to wildlife. In many areas of the arid Southwest, these artificial waterers have been helpful in restoring and maintaining wildlife numbers.

Of course, many species live in or close to water and cannot survive elsewhere. America's rivers, streams, marshes, swamps, lakes, and potholes once teemed with a wide variety of wildlife. In most cases, **wetlands** still do. However, for centuries we have dammed, drained, channeled, diked, polluted, and otherwise altered wetland areas. Only recently have we begun to realize the damage we are causing. If we do not stop the loss of wetlands, many species of wildlife may be lost as well.

SPACE

All animals have a home range—an area within which they live that provides their daily needs. The size of this area varies greatly for different animals. Some animals, such as wolves or mountain lions, require many square miles of habitat. Others, such as cottontail rabbits, may spend their entire lives within a 1- to 2-acre area (Figure 5-5). As a general rule, small animals have small territories, and larger animals have larger home ranges. Species such as migratory waterfowl utilize a combination system: They have a small home range during the

FIGURE 5-5 With the human encroachment on habitat evident in the background, we have to wonder if these waterfowl and shorebirds will have a marsh to return to next fall.

breeding season and a much larger one during the fall and winter. While waterfowl are migrating to the wintering grounds and to some extent all-winter, they have several temporary home ranges. As the available food supply in an area is consumed, the birds move to other areas, usually fairly close by, that still provide food. Mule deer are good examples of a species that may have two or more home ranges, depending on the time of year. Mule deer typically spend the summer at higher elevations, where forage tends to stay green and lush longer. When winter arrives and the snow begins to fall, they tend to migrate to lower elevations, where forage is usually easier to find. They establish a home range or territory within each of these areas. Thus home ranges may be temporary during the course of the year or during a migration.

Some animals tolerate other members of their species to a much greater extent than others do. Geese, for example, are quite **gregarious** and usually travel to their wintering grounds in large flocks (Figure 5-6). Once there, they feed and rest together. In the spring, they even nest in close proximity to one another, but they defend the area immediately around their nests ferociously. In contrast, some animals, most notably large predators such as wolves and cougars, may not tolerate another member of their species within an area of several square miles! In any habitat it is important that animals not be overcrowded. If a species is too heavily represented in an area, the resulting competition for food can be a serious problem. Wildlife populations can severely damage their habitat if they are overcrowded. The results can be poor reproductive rate, increased death rates due to malnutrition and stress, and generally poorer health for the entire species. Another serious problem when wildlife become too numerous is disease. For example, in areas of the South Texas coast where hundreds of thousands of geese spend the winter, an outbreak of **avian cholera** can devastate the population. In any situation in which large numbers of a species are in close proximity, disease tends to spread rapidly. For these reasons adequate space is an important component of an animal's habitat.

FIGURE 5-6 Snow geese are a gregarious species. They often migrate in flocks of several thousand birds.

(Source: Courtesy of Texas Parks & Wildlife Department © 2002.)

ARRANGEMENT

Last, but certainly not of least importance, is the arrangement of habitat components. As stated earlier, all wildlife require food, water, cover, and space, suitably arranged. If these components are present in a habitat, regardless of their arrangement, that habitat will almost certainly be used by wildlife. However, **optimum** populations will not be achieved unless the food, water, cover, and space are appropriately arranged. Let's take a covey of bobwhite quail as an example. They have escape cover, water, and space all within a reasonable distance of one another. However, their primary feeding areas are located across an area of native pasture that is not sufficiently dense or tall to protect them from overhead predators. Such a covey will likely suffer an **inordinately** high death rate due to the number of birds that end up as hawk dinners! Another example, which is much more likely to occur these days, is habitat fragmented or split up by malls, road construction, subdivisions, and parking garages. On the outskirts of an imaginary midsize American town we'll call "Wildlifeville" is a lake of several acres. Two hundred yards away is an old-growth forest of several acres. Numerous species of wildlife use this habitat. The mud around the pond is thick with the tracks of deer, rabbit, and great blue heron, just to name a few, but the majority of the tracks belong to the large population of raccoons that live in the wooded area and come to the pond to feed every night. The raccoons thrive on the small fish, frogs, and shellfish they catch in the shallows of the pond. Wildlifeville is growing, however, and city officials become alarmed at the volume of traffic traveling through their city. A four-lane loop around the town is the solution. When it is finished, it splits off the lake from the woods. Because their main food source is now located across a four-lane highway, there is an immediate rise in the number of "road-killed" raccoons (Figure 5-7). This simple example shows how arrangement of food, water, cover, and space is a crucial part of wildlife habitat.

FIGURE 5-7 Cities and the transportation arteries that link them often fragment what was once good wildlife habitat.

(Source: Photo courtesy of USDA/NRCS.)

SUMMARY

Each habitat element is of equal importance. A lack of food, water, cover, or space prevents wildlife from reaching their optimum population within the habitat. Similarly, a less than ideal arrangement of the habitat can also restrict wildlife populations (Figure 5-8). A serious shortage of food during critical periods such as mating season or winter may lead to population declines. A lack of good cover can also lead to population decline through increased mortality from weather and predators. Water can be a limiting factor in arid areas; where potholes, marshes, and swamps have been drained; or if water is polluted to the point of being harmful or unusable.

Space or a lack of it can affect wildlife in several ways. Most animals, if their population is dense, become stressed from the overcrowding. Disease can be a real enemy of wildlife that are too densely concentrated for the available habitat. America has millions of acres of excellent wildlife habitat. We have also destroyed millions of acres of excellent wildlife habitat and continue to do so at an alarming rate. Human development in the United States alone consumes about 2 million acres each year. If wildlife is to have a place in our future, we must protect their habitat. Humans and wildlife can coexist, but we must leave wildlife somewhere to live.

(Source: Photo courtesy of Texas Parks & Wildlife Department © 2002.)

FIGURE 5-8 This wildlife habitat has been split up by plowed fields. Wildlife left in the small island of cover are isolated from other unplowed areas.

REVIEW QUESTIONS

Fill in the Blank

Fill in the blanks to complete these statements.

1. An animal's habitat is its_____ .

2. _____ require food, water, cover, and space for their survival.

3. Food, water, cover, space, and the arrangement of each of these elements are of equal _____ .

4. Food is anything an animal consumes and digests to provide _____ .

5. Animals that do not obtain _____ in adequate quantity and quality may suffer from malnutrition.

6. List five things that affect the actual amount of food required by an animal.

    (a) _____

    (b) _____

    (c) _____

    (d) _____

    (e) _____

7. As a general rule, the less time an animal spends searching for food and the less _____ traveled, the better.

8. Cover provides wildlife with a relatively safe haven from _____ .

9. Cover provides wildlife with a place to_____ and raise their young.

10. Cover is typically some form of _____ , but it may take many forms.

11. Some species are very _____ and are found in many types of cover.

12. Some species are found only in specific _____ .

13. Species that have specific habitat needs are vulnerable to_____ in their habitat, especially human-made ones.

14. _____ is a resource that most wildlife cannot live without.

15. Most large mammals require_____ water and will typically drink _____ times per day.

16. Many smaller mammals and birds receive water from the _____ they consume.

17. All animals have a _____ , an area within which they live, that provides them with their daily needs.

18. Some animals, such as wolves, require many square _____ of habitat.

19. Some species have _____ or more home ranges.

20. _____ can be a serious problem for wildlife that are too numerous for their habitat.

21. Optimum wildlife populations cannot be achieved unless food, water, cover, and space are appropriately _____ .

22. Each habitat element is of _____ importance.

Short Answer

1. Why are each of the habitat elements of equal importance?
2. Why is it important that an animal's habitat be suitably arranged?
3. Examine each of the requirements for good wildlife habitat (food, water, cover, space, and their proper arrangement). Where in your area might all these factors be found? List some of the animals that might use this area.

Discussion

1. Discuss areas near your home that have fragmented wildlife habitat. How has this affected wildlife in your area?
2. In your opinion, what can be done to lessen the impact of habitat fragmentation on wildlife? How would you accomplish this?

Learning Activities

1. Select a wildlife species and describe its habitat requirements. Analyze the type of habitat in which this species might be found and compile a list of additional species one might expect to find in this habitat or area.
2. Take a trip to a park, national wildlife refuge, or farmer's field. List the types of habitat you observe and the elements that make up the habitat, such as trees, grassland, brush, and streams. Now observe and list as many species as possible that are using the area. Look for signs such as tracks and droppings of species that use the area. Report your results to class. (Hint: The soft ground around ponds or along stream sides is an excellent place to find wildlife tracks.)

CHAPTER 6

The Human Impact on Wildlife Habitat

OBJECTIVES

After completing this chapter, you should be able to

- List the ways humans have affected wildlife habitat.
- Describe human activities that destroy or harm wildlife habitat.
- Describe some things humans do that benefit wildlife.
- Explain how specific human activities harm wildlife and wildlife habitat.

INTRODUCTION

Humans affect wildlife habitat, and therefore wildlife, in many ways—more than we can cover in one chapter. However, we discuss here the most obvious human impact on our world.

Habitat destruction is the single greatest threat facing wildlife today and for the foreseeable future. Habitat destruction results from a variety of human activities, including construction, farming, mining, and timber harvesting. All are related to human encroachment on wildlife habitat. Pollution of Earth's air, water, and land is certainly a serious threat. We humans generate a tremendous amount of **chemical pollution**, as well as garbage, every day (Figure 6-1).

Increasing human populations require more and more living space each year. As human communities spread, they **encroach** on wildlife communities. We are converting farmland into subdivisions, supermarkets, office complexes, and shopping centers at an alarming

(Source: Photo courtesy of Texas Parks & Wildlife Department © 2002.)

FIGURE 6-1 Garbage dumps pollute the environment and use land that wildlife once used.

rate. It is estimated that 2 million acres of farmland are lost in the United States each year due to human expansion. Much of this land is also excellent wildlife habitat. As a result of human encroachment, wildlife often takes a double hit. Faced with the loss of 2 million productive acres each year and an increasing demand for food, farmers are forced to find additional acreage to bring into production. In an effort to recover the lost acreage and meet our needs for food and fiber, farmers often utilize land once used **exclusively** by wildlife.

WETLANDS LOSS

America's wetlands are some of our most productive wildlife "factories." They support an astounding variety of wildlife and are often the only habitat in which many species are found. When we use the term *wetlands*, we are referring to a variety of wet environments. Wetlands include wet meadows, ponds, fens, bogs, coastal and inland marshes, wooded swamps, mudflats, and bottomland **hardwood** forests. These areas have traditionally been looked on as having little **economic value**. In fact, throughout the past 200 years we Americans have encouraged the draining and development of these areas. Laws and programs were actually designed to benefit these developments. The results of this development have been devastating to America's wetlands (Figure 6-2). Over 100 million acres of wetlands have been lost, which is more than half of all of America's wetlands.

In the past few years, however, people have realized what great value these wetland areas hold. They are precious **ecological resources**. They are extremely important to wildlife; they slow the destructive power of floods and storms, purify polluted waters, and provide a variety of recreational opportunities. Over the past 20 years, our change of attitude toward wetlands has resulted in federal and state laws and programs designed to protect and preserve wetlands.

(Source: © Zacarias Pereira da Mata, 2010. Used under license from Shutterstock.com)

FIGURE 6-2 Millions of acres of wetlands have been lost to refineries and other forms of coastal development.

One such law, written into the 1990 farm bill, is commonly known as the Swampbuster Act. Swamp buster imposes fines ranging from $750 to $10,000 on farmers who drain wetlands. Farmers may also lose a portion of their federal crop subsidy benefits. Should they violate the law twice in a 10-year period, they could lose all federal crop benefits. Although Swampbuster was a step in the right direction, its effectiveness depends on strict enforcement.

AMERICA'S WETLANDS

In 1979 the U.S. Fish and Wildlife Service began a massive study called "the National Wetlands Trends Analysis." It was the first study designed to determine the **status** of America's wetlands and related trends in terms of gains and losses. This study concentrated on the 20-year period from the mid-1950s to the mid-1970s. Aerial photographs taken in the 1950s were compared to aerial photographs taken 20 years or so later. Analysts answered many important questions by measuring the changes that had taken place.

It was discovered that during the 20-year period, the annual loss of wetlands averaged 575,000 acres. Our most valuable wetlands, inland marshes, and swamps made up the majority of the areas that were drained. The figures are shocking—6 million acres of forested wetlands, 4.7 million acres of inland marshes, 400,000 acres of shrub swamps, and 400,000 acres of coastal marshes and mangrove swamps destroyed in just 20 years (Figure 6-3). This adds up to an area twice the size of New Jersey, or 11.5 million acres.

The majority of this loss can be attributed to the demands of agriculture. Over 11 million acres of wetlands were drained for crop production. During this period, agricultural development and construction projects created some 3 million acres of lakes and ponds. These areas, however, are generally not as valuable as wildlife habitat as are **vegetated wetlands**. In addition, many wetlands are damaged by pollution, highway and railroad construction, and urban sprawl.

(Source: Photo courtesy of Texas Parks & Wildlife Department © 2002.)

FIGURE 6-3 Bottomland hardwood forest, or forested wetland, is some of the most valuable wildlife habitat in America. So much of it has been drained, however, that many states have lost 80 percent of their forested wetlands.

PRAIRIE POTHOLES

Perhaps our most valuable wetlands are the prairie potholes of southern Canada, Iowa, Minnesota, Nebraska, North and South Dakota, and northern Montana. The most recent ice age created thousands of shallow depressions as glaciers withdrew from these areas. These potholes filled with water and have become a major breeding ground for North American waterfowl. Prairie potholes are a very important nesting area for many waterfowl, especially ducks. These inland marshes provide food, water, and cover for a great variety of wildlife (Figure 6-4A). Their importance to hundreds of species cannot be overemphasized.

FIGURE 6-4(A) This is a healthy prairie pothole wetland area.

(Source: Courtesy of U.S. Fish and Wildlife.)

FIGURE 6-4(B) This area was once similar to the one shown in Figure 6-4(A), but this wetland has been ditched and drained for farming purposes.

Unfortunately, these prairie potholes are spread across much of our most productive farmland. Consequently, many hundreds, perhaps thousands, have been drained and turned into cropland (Figure 6-4B). For example, Nebraska's Rainwater Basin has seen 90 percent of its original wetlands drained. This basin is a major staging area for millions of geese, ducks, and cranes during their annual migration. Draining the basin forced waterfowl onto just 10 percent of their original habitat. Should disease break out within such a greatly reduced area, the effects on a population could be devastating.

The inland marshes of Florida provide another good example of our effect on this vital type of wildlife habitat. The Florida marshes have provided feeding grounds for wading birds and winter homes for migratory waterfowl for many years. They are also the breeding grounds for such species as the endangered everglade kite, the mottled duck, and a variety of wading birds, such as rails. In addition, these marshes provide a home to raccoons, muskrats, alligators, and dozens of other species. However, human encroachment in the form of agriculture and development is seriously affecting waterfowl and other wildlife populations.

FORESTED WETLANDS

The U.S. Fish and Wildlife Service reports that nearly 80 percent of the bottomland hardwood forests of the lower Mississippi Valley have been destroyed. These **periodically** flooded hardwood forests are prime, **irreplaceable** wildlife habitat. Where there were once 25 million acres of bottomland hardwood forests, less than 6 million acres exist today. The remaining forests are still the winter home for the majority of the continent's mallards and virtually all the wood ducks in the central United States. They also provide **spawning** and nursery areas for fish as well as abundant habitat for many other species. These areas continue to be drained, leveled, and converted to farmland at an alarming rate.

The Pocosin wetlands of coastal North Carolina are another example of wetlands under great pressure from humans. The two main reasons are **peat** mining and agriculture. These wetlands are covered with **evergreen** shrubs and trees and serve a vital function by regulating the flow of freshwater to nearby coastal **estuaries.** Pamlico Sound's productive fisheries rely on this flow of freshwater.

COASTAL WETLANDS

America's coastal wetlands are important breeding grounds for wading birds, as well as wintering areas for migratory waterfowl. The mangrove swamps and coastal salt marshes that make up these wetlands are also important spawning and nursery grounds for the majority of our **commercial** and **sport fish**, including **shellfish**. These valuable coastal marshes were disappearing at an alarming rate before 1960 (Figure 6-5). The losses were particularly troubling in Florida, Louisiana, and Texas. Some of this loss can be attributed to natural causes, such as sinking terrain, which allows the marsh to become flooded. This can occur for a variety of reasons, such as the **subsidence** of the coastal plain or a rise in sea level. Much of Louisiana's losses are due to these factors. Some of the human activities that contribute to coastal marsh flooding are oil and gas **extraction**, **levee** construction, and **channelization**. Most of the coastal marsh losses that could be directly attributed to human activity were caused by urbanization, primarily dredging and filling for developments.

CAN WETLANDS BE SAVED?

Over the past 50 years, more than 3 million acres have been preserved using "Duck Stamp" dollars. Federal waterfowl stamps, more commonly called Duck Stamps, must be purchased by all waterfowl hunters. The money raised from these stamps is used to buy or lease prime wetland habitats. These stamps can be purchased at your local post office, and each person who purchases a Duck Stamp directly contributes to preserving our wetlands (Figure 6-6).

The pressures on wetlands will continue to increase. As urban areas expand and the need for farmland increases, wetlands will continue to disappear. It is therefore important to monitor wetlands and any proposed changes to them. This vital job falls to the U.S. Fish and Wildlife Service. With the use of such techniques as wetland habitat acquisition, wetlands preservation easements, federal permit reviews, and environmental education, the Service will continue to protect wetlands.

FIGURE 6-5 The variety of bird life using this coastal marsh indicates that this type of habitat is extremely valuable to wildlife.

(Source: Courtesy of USDA/NRCS. Photo by Lynn Betts.)

FIGURE 6-6 Private conservation organizations like Ducks Unlimited have restored thousands of acres of drained potholes and marshes. Dozens of species use these wetland habitats.

Regulations and laws that were passed in the1960s and early 1970s have already benefited coastal wetlands. Federal protective laws and those of several states enacted during this period have helped to slow the destruction of these vital wetlands. We must continue to educate ourselves about the critical role all wetlands play in our environment.

NATIVE PRAIRIES

America's native prairie once stretched from the Mississippi River to the Rocky Mountains and from southern Canada to Texas. The prairie once covered an area larger than France. Today only about 1 percent of the native prairie still exists. America's native prairies are much more endangered than the more pub-licized South American rain forest. Millions of bison and pronghorn antelope once roamed the prairies. Prairies were also home to millions of small animals, including prairie dogs, coyotes, rabbits, snakes, and many species of birds. Some species, such as the greater and lesser prairie chickens, Attwater's prairie chicken, and the black-footed ferret are found nowhere else. Consequently, both the Attwater's prairie chicken and the black-footed ferret are on the Endangered Species List, as are numerous other prairie-dwelling species. Native prairies are probably our most endangered ecosystem.

Prairies have been destroyed for many of the same reasons as wetlands, using many of the same methods. Early on in the development of America, settlers realized that any soil that could grow grass as tall as a horse should be able to produce bountiful crops of grain. As quickly as the buffalo and Native Americans were removed from the prairie, the prairie grasses were plowed under. Development, in the form of subdivisions, factories, and metropolitan areas, has also contributed to the loss of our native prairies (Figure 6-7). For example, it is estimated that metropolitan Houston covers over 1 million acres of Gulf Coast

FIGURE 6-7 Millions of acres of native prairies have been converted to wheat fields and urban areas. The few remaining acres are often bisected by highway and utility lines.

(Source: Photo courtesy of USDA/NRCS.)

prairie habitat that was once available to the endangered Attwater's prairie chicken. If such species as black-footed ferrets and Attwater's prairie chickens are to continue to share this planet with us, we must preserve what is left of their habitat. Through such programs as habitat acquisition, incentives to private landowners to preserve their native pastures, and curbing or stopping development in these areas, we can save what is left of the prairies. One law that was passed as a part of the 1990 farm bill, commonly known as the Sodbuster Act, should help to protect America's remaining native prairies. Designed to discourage the cultivation of fragile soils, Sodbuster carries much the same penalties as Swampbuster. Fines range from $500 to $5,000, and those who violate the provisions of Sodbuster twice in a 5-year period could lose all federal crop subsidy benefits.

AMERICA'S FORESTS

When the colonists arrived, eastern North America was largely forested. These forests were of tremendous value to early Americans. Within 200 years, however, vast areas of forest had been cut to provide lumber for ships, logs for homes, and firewood, or they were simply cut, cleared, and burned to make way for the plow. As a result, many species of wild animals that relied on the forests for a home were adversely affected. Species such as the wild turkey and white-tailed deer became scarcer and more widely scattered. With continued clearing of the forests and pressure from market hunting, some species, such as the passenger pigeon, became extinct.

Most trees, particularly hardwoods, grow relatively slowly. Consequently, many of the trees in old-growth forests are hundreds of years old. It will take a similar period for these forests to regenerate. Another factor that adversely affected wildlife was the way forests were harvested. Until recently, the most common method of harvest was clear-cutting. This technique calls for all the marketable trees in a large area to be cut at the same time. Although it is the most economical method for the logger, it is not necessarily the best method for the environment or wildlife.

When large areas are clear-cut, no cover is left for wildlife and no protective covering left for the soil. The results can be very detrimental to soil, wildlife, and to nearby **watersheds**. Wide-scale erosion can result from the unprotected soil. Streams and rivers in the area fill up with silt and soil eroded from the surrounding hillsides. Wildlife populations can be seriously affected by the disruption and destruction of their habitat. Clear-cutting remains a common logging technique, but reforestation and fast-growing cover crops have helped to reduce erosion. Fast-growing seedlings and the grasses used as cover crops provide good wildlife habitat and help to keep rivers and streams clear.

Much of the original forest in the Northeast has recovered from past logging. Vast tracts of upper New York and western Pennsylvania are covered with beautiful forests. There are large forested areas in the Upper Midwest and in many western states. The southern pine forests are also reestablished in many states. Modern forestry techniques include replanting harvested acres with faster-growing varieties of trees. Areas that were clear-cut 20, 30, 40, or even 50 years ago are now covered with new-growth forests. These areas will be harvested as the trees mature. Our forests are a wonderfully renewable natural resource, and we have become considerably better at managing them in recent years.

HABITAT ALTERATIONS THAT BENEFIT WILDLIFE

Agriculture, the very thing that has been so destructive to wildlife habitat, can also be beneficial. The most beneficial aspect of agriculture for wildlife is providing them with food. With millions of acres of grain to harvest each fall, it is not surprising that some of that grain is eaten by wildlife. Many species, from upland game birds, such as bobwhite quail and pheasants, to songbirds, such as cardinals and meadowlarks, benefit from this plentiful food supply (Figure 6-8). Large and small mammals, such as white-tailed deer and cottontail rabbits,

FIGURE 6-8 Many wildlife species, such as these sandhill cranes, use farm fields at various times of the year.

(Source: Photo courtesy of Texas Parks & Wildlife Department © 2002.)

also use farm fields and their produce. Of course, anywhere prey species can be found, predators like coyotes, bobcats, and birds of prey will also be found. It is primarily because of this rich food supply that many species, such as white-tailed deer, are more numerous today than in the past.

The corners of fields, fence rows, irrigation ditches, and anywhere else that plows cannot reach also provide cover for many species of wildlife. Standing crops such as corn, milo, and alfalfa also provide cover for wildlife. Hay fields and cropland left fallow are used by many wild animals. Species such as barn owls, mice, snakes, and skunks actually move into farm buildings and make themselves at home. These are just a few of the ways that wildlife species benefit from agriculture.

Clear-cutting tracts of forest can also benefit wildlife. Species such as mule deer and elk prefer a mixture of forest and clearings. Clearing old-growth forests allows a variety of grasses and forbs that would not be able to grow in the forest to take hold and grow. These plants are often the preferred food of browsing and grazing animals such as deer and elk. If the forest is harvested so as to leave wildlife corridors between the clear-cut areas, these areas can actually become more attractive to wildlife. On the other hand, species such as the northern spotted owl apparently require large tracts of unbroken, old-growth forests. It is therefore important that an area of forest be **evaluated** before it is disturbed.

SUMMARY

Humans can be, and often are, destructive to the environment. Destruction of wildlife habitat follows from many human actions, with the most common being agriculture, construction, and other development. Something as basic as a dam constructed to provide water and power for human consumption can be detrimental to wildlife, as thousands of acres of wildlife habitat are buried under millions of gallons of water. The fish, amphibians, and invertebrates that inhabit the river or stream that was dammed lose their homes. The effect on wildlife must be considered before construction of these types of projects can be allowed to begin.

Inland wetlands are perhaps our most valuable type of wildlife habitat, providing food, water, and cover to an enormous variety of wildlife (Figure 6-9). Wetlands provide the

(Source: Photo courtesy of USDA.)

FIGURE 6-9 Many species, including these red-eared turtles, make their homes in wetlands.

SUMMARY (*continued*)

breeding habitat for 12 million ducks nesting in prairie potholes in the United States each year. They also are home to millions of wading birds and countless invertebrates and amphibians. Wetlands provide temporary homes to millions of migratory waterfowl, such as ducks, geese, and cranes. They also serve as giant sponges that absorb and help control flood waters. Wetland areas such as swamps and marshes also serve as natural filters, purifying water that flows into and through them. We must work to lessen the pressures on our wetlands and continue to work to preserve those that remain.

REVIEW QUESTIONS

Fill in the Blanks

Fill in the blanks to complete these statements.

1. _____ is the single greatest threat facing wildlife today.

2. _____ of Earth's air, water, and land is also a serious threat.

3. As human communities spread, they encroach on _____ communities.

4. _____ million acres of farmland are lost each year, primarily to urban expansion.

5. America's _____ are some of our most productive wildlife "factories."

6. List the nine common types of wetlands.

   (a) _____

   (b) _____

   (c) _____

   (d) _____

   (e) _____

   (f) _____

   (g) _____

   (h) _____

   (i) _____

7. It is estimated that more than _____ of America's wetlands have been lost.

8. Wetlands are precious _____ resources.

9. List three reasons wetlands are important.

   (a) _____

   (b) _____

   (c) _____

10. During the 20-year period from the mid-1950s until the mid-1970s, America's loss of wetlands averaged _____ acres per year.

11. _____ million acres of forested wetlands were drained during the same 20-year period.

12. From the mid-1950s until the mid-1970s, more than _____ million acres of America's wetlands were drained.

13. _____ are a very important nesting area for waterfowl, especially ducks.

14. There were once about 25 million acres of bottomland hardwood forests in the lower Mississippi Valley. Today less than _____ million acres remain.

15. America's coastal wetlands are vital breeding grounds for _____ .

16. Coastal wetlands also serve as _____ for migratory waterfowl.

17. _____ must be purchased by all waterfowl hunters.

18. Money raised from the sale of duck stamps has been used to preserve over _____ million acres of wetlands during the past 50 years.

19. The American prairies once covered an area larger than _____ .

20. Today only about _____ percent of native prairies remain.

21. America's eastern forests were originally cleared to produce _____ for ships, logs for _____ , and fields to be farmed.

22. _____ calls for all marketable timber in a large area to be cut at the same time.

23. Clear-cutting can result in wide-scale _____ due to the unprotected soil.

24. America's forests are _____ natural resources.

25. American agriculture not only feeds millions of people, it also feeds a tremendous variety of _____ .

26. Standing crops such as _____ , milo, and alfalfa provide cover for wildlife.

27. Species such as barn owls, mice, snakes, and skunks may move into _____ buildings.

28. _____ tracts of forest can also benefit wildlife.

29. Humans can be, and often are, destructive to the _____ .

30. Our _____ are perhaps our most valuable type of wildlife habitat.

Short Answer

1. In what ways do humans affect wildlife habitat?
2. Why do humans alter the environment to such an extent?
3. How can we expand wetlands acquisition and protection?
4. Why are wetlands vital to wildlife? How are they beneficial to humans?

Discussion

1. In your opinion, in what ways might humans lessen our impact on wildlife habitat?
2. If you had to choose one thing that would most help protect wildlife habitat, what would it be? How would you implement it?

Learning Activities

1. Research methods for restoring drained wetlands and present your findings to your class.
2. Research road construction designs and techniques. Write a report or build a scale model of a four-lane highway to be built across a wetland. Construct the roadway in such a manner as to do

minimal damage to the wetland. Keep in mind such factors as cost and construction materials. Your state department of transportation and the U.S. Department of Transportation are good sources of information.

3. Design a timber harvest plan for 10,000 acres of western forest. Your plan should have the least impact on wildlife, yet be economical, limit erosion, and take into account any sensitive species present.

CHAPTER 7

Wildlife and American Sport Hunting

OBJECTIVES

After completing this chapter, you should be able to

- Explain the primary sources of funding for wildlife and habitat enhancement in the United States.
- List the acts and legislation that have been passed to aid wildlife.
- Describe the relationship between sport hunters and wildlife populations.
- List some of the major private wildlife conservation organizations.
- Explain how funding generated by sport hunters also benefits nongame species.

INTRODUCTION

Since the fight to conserve America's wildlife began in the late 1800s, it has been led by American sport hunters. In contrast to the "market hunters" of the late 1800s and early 1900s, sport hunters were concerned about America's wildlife resources. Market hunters showed little regard for the wildlife that provided for their livelihoods. They harvested wildlife by the millions, basically all year long (Figure 7-1). They took migratory waterfowl and shorebirds during both fall and spring migrations and from their nesting colonies as well. This was an **appalling** situation to sport hunters, who tried for many years to pass on their own **ethics** to the market hunters. Sport hunters and conservationists soon realized that as long as wildlife was commercialized and could be shipped and sold for profit, it was in

(Source: Courtesy of U.S. Fish and Wildlife Service.)

FIGURE 7-1 This market hunter shot more than 100 ducks and geese in a single morning's hunt. It was this type of abuse that sport hunters worked to stop.

grave danger of being overexploited. In fact, such species as the **heath hen** and passenger pigeon were exploited to extinction.

Sport hunting publications such as *Forest and Stream* and *American Field* and conservation groups such as the Audubon Society, the League of American Sportsmen, the American Ornithologist's Union, and the Boone and Crockett Club were **instrumental** in the passage of the Lacey Act of 1900. This act banned the interstate transportation and sale of most wildlife and wildlife by-products, such as feathers. It was the beginning of the end for the market hunting era. Once it was no longer profitable to slaughter wildlife, responsible management could begin. Even though some progress was being made in regard to wildlife, the early 1900s was not a particularly good period for America's wildlife. Habitat destruction in the form of cleared forests, plowed grasslands, and dammed rivers was widespread. There were few wildlife laws and fewer game wardens to enforce them. There was no system for funding wildlife management or habitat preservation. The 1930s probably represented the low point for America's wild creatures. Several species were already extinct, and many more were on the same road. People began to realize that serious action had to be taken. The adage "Leave the wildlife alone and it will be all right" was not working. In many areas the sight of once-common animals such as white-tailed deer was something that only the old-timers could remember.

FUNDING AT LAST

Fortunately for America's wildlife, some dedicated sport hunters and conservationists organized the firearms and ammunition industries and proposed a remarkable plan. They encouraged Congress to continue collecting a 10 percent excise tax on firearms and ammunition used for sporting purposes. However, they

(Source: Photo courtesy of Texas Parks & Wildlife Department.)

FIGURE 7-2 Pittman-Robertson funds have been used to restock many species of wildlife, such as wild turkeys, to their former range.

wanted those funds earmarked for wildlife **restoration**. These funds were to be used for wildlife research and habitat management to help stabilize and increase animal populations. This effort resulted in the Federal Aid in Wildlife Restoration Act, more commonly known as the Pittman-Robertson Act. President Franklin D. Roosevelt signed it into law in 1937. The excise tax was raised to 11 percent in 1941, and it remains at that level today.

Progress was slow at first. Just as the program was getting started, World War II began, and millions of sport hunters joined the armed forces. This sharply reduced the amount of money raised by the excise tax. However, by the early 1950s millions of dollars began to flow into the management of America's wildlife. Many species of wildlife, both game and nongame species, have benefited from the Pittman-Robertson Act. In 1930, for example, wild turkeys were rare in all but a few southern states (Figure 7-2). Today turkeys number in the millions and they can be found in nearly every state. Wood ducks were thought to be past saving because of extensive habitat destruction. Today they are one of our more numerous ducks, perhaps the most common breeding duck in the East. The pronghorn antelope population has increased from fewer than 30,000 to more than 1 million. These are just a few of the wildlife management success stories made possible by Pittman-Robertson funding.

HOW THE ACT WORKS

The equipment excise taxes collected directly from **manufacturers** and importers go to the U.S. Treasury's Trust Fund Branch. The money is then distributed to the states using a formula that takes into account both the number of hunting licenses sold in the state and the state's area. Additional funding was provided in 1970, when a 10 percent tax on handguns was added to the original act, and in 1972, when an 11 percent tax was placed on **archery** equipment. Both of these amendments provided significant additional funding. It is estimated that these taxes, paid by sportsmen and sportswomen, have produced over $10 billion for wildlife conservation since the program began. Today taxes collected from sportsmen and sportswomen contribute over $400 million per year for wildlife conservation programs. This steady, earmarked supply of money has allowed wildlife managers to undertake the long-term projects essential to successfully manage wildlife. The states and territories use these funds where they are most needed, typically to purchase land for wildlife habitat, manage and maintain existing habitat, and conduct research. Perhaps the best thing about the Pittman-Robertson Act is that many people do not realize they are paying a tax because it does not occur at the cash register when they check out, like a sales tax does.

THE ECONOMIC IMPACT OF HUNTING

In addition to the money generated by the Pittman-Robertson Act each year, sportsmen and sportswomen spend millions more. Hunters must buy **licenses** and **tags** for the wildlife they attempt to harvest. This provides millions of dollars to the wildlife departments in most states. In 2007 a hunter in Kansas, for example, would have spent $20.50 for a resident hunting license. This entitled the hunter to hunt small game such as quail, rabbits, and pheasants. To hunt waterfowl required a federal waterfowl stamp costing $15.50 and a state waterfowl stamp for $6.75. This totals $42.75 in license fees to hunt birds and small game. To hunt white-tailed deer or other large game, additional license fees would apply. These fees are typical of the license fees paid by hunters throughout the country. License fees currently supply state wildlife agencies with more than $900 million each year for wildlife management, research, habitat preservation, restoration of species, and other tasks.

Many sportsmen and sportswomen travel out of state to hunt. Some species of game animals may not be available in their state or may not be as abundant as in a neighboring state (Figure 7-3). When hunters are not residents of the state or states in which they intend to pursue game, they must pay nonresident fees. These fees are usually considerably higher than resident fees. For example, a nonresident small-game license in Kansas is $72.50, compared to a cost for residents of $20.50. A general nonresident hunting license in Texas would cost $300. It is important to note that license fees provide a significant portion of each state's wildlife conservation budget. In western states such as Colorado, Wyoming, Montana,

(Source: Photo courtesy of Texas Parks & Wildlife Department.)

FIGURE 7-3 Elk are popular game animals whose range extends through Canada and the western region of the United States.

and Idaho, a permit to hunt elk or mule deer may cost $400 or more. As states tighten their budgets, these license fees have become even more important to the wildlife conservation efforts of each state.

License fees are only part of the economics of hunting. Sport hunters spend many billions of dollars each year on hotel rooms, meals, gas, clothing, and other equipment. In some states, such as Texas, where leasing of land for hunting is commonplace, the **lease** fees received by the landowner are very important. Leases are becoming more and more common in many Midwestern states, and leasing also helps landowners realize the economic value of their wildlife resources. In some cases lease fees might be the difference between profit and loss for the landowner. Many farmers and ranchers have been helped through tough economic times by lease fees. When cattle or crop prices are low, fees paid by hunters to landowners can make the difference between survival and bankruptcy. Many rural communities, already hard hit by tough economic times, are dependent on money generated by hunters for their survival. In addition, lease fees encourage farmers and ranchers to leave some areas for wildlife. Once landowners realize the economic value of wildlife, they are more likely to manage and protect that resource. It is estimated that Texas hunters spend over a billion dollars per year to pursue game animals. Of course, many species other than game species benefit from hunter-generated funds. The total economic impact of hunting in the United States easily exceeds $12 billion per year. It is easy to see how important hunting is to wildlife and to the economy in general.

PRIVATE CONSERVATION ORGANIZATIONS

There are dozens of privately funded and managed wildlife conservation organizations. The majority of those that take an active role in wildlife management and conservation were begun by sport hunters. These organizations spend millions of dollars each year on habitat acquisition and improvement, research, and management of a variety of wildlife species. It is impossible to discuss each of the several hundred private conservation organizations in the United States. However, a brief description of a couple of these conservation organizations is in order.

Although some conservation organizations are designed to help a specific species, projects they undertake benefit a variety of nontargeted wildlife. For example, a restored freshwater marsh helps not only the target species, such as ducks, but also dozens of other animals, such as muskrats, raccoons, shellfish, predators, and fish. Similarly, it is unlikely that bobwhite quail would be the only species to use a food plot planted for their benefit. Many species of seed-eating birds, small mammals, and their predators would use such a food plot. In this way hundreds of nongame species benefit from the management of a few game species. The millions of dollars that sportsmen and sportswomen spend to help game animals actually benefits many more nongame animals.

DUCKS UNLIMITED

Waterfowl are some of the most intensely managed species in the entire world. Many species fly thousands of miles each year between nesting and wintering grounds. Thus, waterfowl require not only national but international management. Ducks Unlimited (DU), one of the oldest conservation organizations in the United States, deals exclusively with waterfowl. Since its incorporation in 1937, DU has been a pioneer in waterfowl research, habitat conservation, and habitat improvement (Figure 7-4). An avid duck hunter, Joseph Palmer Knapp of New York is considered the father of DU. Upset by a serious decline in waterfowl numbers during the early Dust Bowl years; Knapp formed the More Game Birds in America Foundation in 1930. The board of directors of More Game Birds soon realized that to improve duck numbers significantly in the fall, nesting habitat and thereby nesting success would have to be improved. To achieve this goal DU was formed in 1937. The majority of the ducks in North America originate in the prairie pothole region of southern Canada, so all of DU's early efforts were aimed at this region.

By 1940 DU was raising about $140,000 per year for waterfowl management, and More Game Birds phased itself out, giving DU all its assets. In 1943 DU had 103 projects on 1 million acres. In 1966, DU had its first $1 million fundraising year. Ducks Unlimited established Ducks Unlimited de Mexico (DUMAC) in 1974, with the goal of protecting critical winter waterfowl habitat south of the border. Today DU has some 500,000 members and has raised at least $750 million for wetland conservation. This is a far cry from the $90,000 and the 6,000 supporters DU had in its first year of existence. DU has a mission statement that reads: "The mission of Ducks Unlimited is to fulfill the annual life cycle needs of North American waterfowl by protecting, enhancing, restoring, and managing important wetlands and associated **uplands**." The importance of DU's work on behalf of wetland and waterfowl conservation cannot be

FIGURE 7-4 Major efforts have been made to preserve and maintain wetlands.

(Source: Courtesy of U.S. Fish and Wildlife Service.)

overestimated. With the continued efforts of thousands of conservation-minded sportsmen and sportswomen, our **dwindling** waterfowl habitat and the many species that live there have a fighting chance.

DELTA WATERFOWL

Dozens of other conservation organizations work to protect and enhance wildlife resources and habitats. Delta Waterfowl is dedicated to waterfowl and wetlands research and education. Delta Waterfowl is a division of the North American Wildlife Foundation, established in 1911. The Delta Waterfowl and Wetlands Research Station had its beginnings in the 1930s and today is considered one of the premier waterfowl and wetlands research centers in the world. For half a century the Station, which is located on the 50,000-acre Delta Marsh in central Manitoba, Canada, has conducted scientific research and trained biologists.

Recognizing that the majority of ducks grow up on private prairie potholes, Delta Waterfowl launched a program known as Adopt a Pothole. Within the framework of its Prairie Farm Program, Delta contracts with farmers to protect and enhance private potholes throughout the prairie pothole region. Farmers are paid incentives to maintain their wetland areas, and nesting structures known as hen houses are erected. Research has indicated that up to 90 percent of all duck nests may be destroyed by predators. The hen house is designed to reduce these losses to predators. We have covered only two conservation organizations in any detail, but there are dozens more. Many, such as Pheasants Forever, Quail Unlimited, the Rocky Mountain Elk Foundation, the National Wild Turkey Federation, and the Desert Bighorn Sheep Council, for example, are concerned with a specific species of wildlife. Others, such as the Boone and Crockett Club and the Foundation for North American Sheep, are involved with multiple species. The Boone and Crockett Club, for example, was founded by Theodore Roosevelt and other concerned sport hunters to promote hunting ethics and establish wildlife conservation practices, which led to the recovery of many big-game species in North America. Additional information on these and other conservation organizations can be found in Appendix C.

SUMMARY

America's sport hunters are largely responsible for the recovery and current abundance of many species of wildlife (Figure 7-5). The funding and leadership they provide are crucial to the continued well-being of our wildlife resources. Regulated sport hunting has never threatened or endangered a species. Habitat destruction and competition from introduced species are responsible for the bulk of our endangered and threatened species. The protection of wildlife habitat should concern everyone who cares about America's wild animals.

SUMMARY (*continued*)

Money generated from the Pittman-Robertson Act is also used to fund hunter education and firearms safety classes in every state. All states have some requirements for hunter education, with most being mandatory. Contact your state's wildlife conservation agency (see Appendix B) for more information on hunter education and firearms safety.

(Source: Photo courtesy of Texas Parks & Wildlife Department.)

FIGURE 7-5 Sporting organizations have installed thousands of nesting boxes. These boxes, complete with predator guards, have brought the wood duck back from the edge of extinction. Today the wood duck is one of the most common ducks in the eastern United States.

REVIEW QUESTIONS

Fill in the Blank

Fill in the blank to complete the statements.

1. American _____ have led the fight to conserve wildlife.

2. Species of wildlife such as the _____ and the passenger pigeon were exploited to extinction.

3. The _____ Act of 1900 banned the interstate transportation and sale of most wildlife and wildlife products.

4. The _____ Act, better known as the Pittman-Robertson Act, provided funding for wildlife management.

5. By the early _____ , millions of dollars began to flow into the management of America's wildlife.

6. Many species of wildlife, both _____ and _____ , have benefited from the Pittman-Robertson Act.

7. Excise taxes on firearms and ammunition are collected directly from _____ and importers.

8. Pittman-Robertson funds are distributed to states based on their area and the number of _____ sold in them.

9. Additional funding was provided in _____ , when a 10 percent tax on handguns was added to the original Pittman-Robertson Act.

10. In 1972 an 11 percent tax was placed on _____ equipment.

11. It is estimated that taxes paid by sport hunters have provided _____ for wildlife conservation since the program began.

12. Pittman-Robertson monies are typically used to purchase land for wildlife habitat, to manage and maintain existing habitat, and to _____ .

13. Hunters must purchase _____ and tags for the wildlife they intend to harvest.

14. License and tag fees provide _____ dollars to the wildlife departments in most states.

15. Waterfowl hunters must purchase _____ stamps regardless of which state they live in.

16. Sport hunters spend millions of dollars on such things as hotel rooms, _____ , gas, clothing, and other _____ .

17. _____ fees encourage farmers and ranchers to leave some areas for wildlife.

18. Hunting-generated money is important to wildlife and to the _____ in general.

19. There are _____ of privately funded and managed wildlife conservation organizations.

20. Some conservation organizations are designed to help a _____ species.

21. Hundreds of _____ species benefit from the management of a few game species.

22. _____ are some of the most intensely managed species in the world.

23. _____ , one of the oldest conservation organizations in the United States, deals exclusively with waterfowl.

24. Ducks Unlimited was incorporated in _____ .

25. By 1943 DU had 103 projects over _____ acres.

26. Today DU has _____ members and has raised at least _____ for wetland conservation.

27. _____ is another private conservation organization dedicated to waterfowl.

28. One of Delta Waterfowl's key programs is the Adopt a _____ program.

29. List four private conservation organizations other than DU and the Delta Waterfowl Foundation.

    (a) _____

    (b) _____

    (c) _____

    (d) _____

30. Regulated _____ has never been the cause of a single threatened or endangered species.

31. _____ and competition from introduced species are responsible for the bulk of our endangered and threatened species.

Short Answer

1. How can nonhunters help to support wildlife and wildlife management?

2. How important have sport hunters been to the success of wildlife management in America? Why?

Discussion

1. In your opinion, how important is continued funding for wildlife? Why is it important?

2. How can we work to ensure continued funding for wildlife, wildlife habitat, and wildlife management?

Learning Activities

1. Contact your state wildlife and fisheries management agency (Appendix B) and determine what percentage of its total budget comes from Pittman-Robertson funds. How are these funds used? This activity is similar to learning activity 1 in Chapter 4. The information you need to complete this activity should be in the material you received to complete activity 1 in Chapter 4.

2. Contact one of the private conservation organizations listed in Appendix C. Request general information about the organization, what it does to help wildlife, and how it accomplishes its goals. Present a report on your findings to your class.

Useful Web Sites

DELTA WATERFOWL
<http://www.deltawaterfowl.org>

DUCKS UNLIMITED
<http://www.ducks.org>

BOONE AND CROCKETT CLUB
<http://www.boone-crockett.org>

QUAIL UNLIMITED
<http://www.qu.org>

NATIONAL SHOOTING SPORTS FOUNDATION
<http://www.nssf.org>

Appendix C contains additional private conservation organizations and their Web addresses.

CHAPTER 8

Modern Wildlife Management

TERMS TO KNOW

adequate

contagious

decimate

disastrous

eliminate

exterminate

fluctuate

food plots

guzzlers

moderating

monitoring

nonnative species

nutrition

overpopulate

potential

poults

regulating

road-killed

supplemental

techniques

OBJECTIVES

After completing this chapter, you should be able to

- Identify the practices used in modern wildlife management.
- Describe the factors that limit a wildlife population.
- List several species whose numbers have increased due to proper wildlife management.
- Identify private conservation organizations that are dedicated to wildlife conservation and management.

INTRODUCTION

In its simplest form, modern wildlife management is the care and administration of a wildlife population for the purpose of ensuring survival of the species. In reality wildlife management is seldom simple. What we do to control a given species almost always affects several other species. The factors that affect a given species are usually complex. Wildlife management has long since gone past the question, "Should we attempt to manage a species or not?" We have altered the environment to such an extent that failure to manage wildlife would be **disastrous** for many species. In the United States alone dozens of species have been **exterminated**. There are hundreds of species of animals and plants on the Endangered Species List and we show little sign of slowing the process. With dwindling wildlife habitat, we must do our best to manage what we have left.

FACTORS THAT AFFECT WILDLIFE POPULATIONS

Numerous factors or conditions affect the population of a given species. A species' population is determined by four main factors:

1. The ability to reproduce itself
2. The quantity and quality of habitat
3. Weather factors such as drought and severity of winter
4. Losses due to disease, predators, accidents, and other causes.

If a drastic change in a species' environment occurs or a natural or human-made disaster causes the deaths of a sizeable portion of their population, that species must be able to reproduce at a rate that will allow it to recover. Throughout history species that are slow to reach reproductive age and/or have relatively low reproductive rates are more likely to become extinct. The productivity of a species is the rate at which mature breeding stock can produce other mature breeding stock (Figure 8-1). The following are two examples of the reproductive potential of two species under ideal environmental conditions. Even though we realize that such an ideal environment can never exist in the wild, it is important to know the maximum reproductive **potential** of a species. By comparing these two examples you will be able to see that reproductive potential varies greatly between species.

FIGURE 8-1 Species such as pheasants have a higher reproductive rate than larger species, such as deer. Generally, smaller mammals and birds have higher death losses and shorter life spans than larger mammals. Higher birthrates are needed to offset higher death losses.

(Source: Photo courtesy of Texas Parks & Wildlife Department © 2002.)

EXAMPLE 1—BOBWHITE QUAIL

Certain traits help determine the reproductive rates of each species. Bobwhite quail, for example, normally raise only one brood per year, with an average of 12 eggs per hen. It is also known that they will not raise their first brood until they are about a year old. For our purposes we will assume that there are six males and six females in each brood.

| | Number at start of year | | Number at end of year |
	Adult	Young	Total
First year	2	12	14
Second year	14	84	98
Third year	98	588	686

The figures for the second and third years are obtained by taking the number of adults at the start of the year, dividing this number by 2 (assuming 50 percent males and 50 percent females), and multiplying by the average brood size of 12. The result is the maximum reproductive potential of bobwhite quail under ideal conditions during a 3-year period. Each species has a fixed maximum reproductive rate. In poor reproductive years a bobwhite pair might not nest successfully at all. However, it is now known that the hen bobwhite, in good reproductive years, may leave her first clutch to be brooded and raised by the male. She then finds another unattached male and lays another clutch of eggs. It is quite rare for conditions to be right for this to occur, but it does happen. Let's look at a species with a much lower reproductive potential.

EXAMPLE 2—PRONGHORN ANTELOPE

Pronghorn antelope reproduce once per year, after reaching age 2, and normally have twin fawns. Knowing this, let's see what their maximum reproductive potential is over the same 3-year period.

| | Number at start of year | | Number at end of year | |
	Adult	Young	Yearling	Total
First year	2	2	0	4
Second year	2	2	2	6
Third year	4	4	2	10

(Source: © Jeff Banke, 2010. Used under license from Shutterstock.com)

FIGURE 8-2 This group of pronghorns should be capable of adding seven to 14 young antelope to the local population each year.

It is apparent that the reproductive potential of the bobwhite quail is much greater than that of the pronghorn. However, small birds and mammals generally have a much higher death rate, due to a variety of factors, than do larger mammals. Therefore, they must have a much higher reproductive rate. Of course, neither of these species can reach their reproductive potential in these examples because ideal conditions do not exist in the wild. The reproductive capability of a species is a factor over which wildlife managers have little or no control (Figure 8-2).

The quantity and quality of an animal's habitat is the next factor that has a direct impact on its production. (The habitat requirements of wildlife were dealt with in detail in Chapter 5.) Wildlife must have food, water, and cover, all suitably arranged, in order to survive. A lack of any one of these factors is not necessarily fatal to wildlife. Starvation is an extreme case of lack of the proper quantity or quality of food, and drought is an extreme shortage of water. Both of these can be fatal.

The quantity and quality of an animal's habitat is a factor over which wildlife managers can exert some control. Quality in particular can be improved. The food supply available to a species may be improved through **food plots**, **supplemental** feeders, or simply by mowing or plowing an area to promote weed and insect production. In areas with limited water supplies, **guzzlers** may be constructed—livestock water troughs built low enough for small animals to reach them. Livestock water troughs may also be allowed to overflow so that all wildlife have access to the water. Cover for wildlife can be enhanced by mowing or not mowing some areas, depending on the needs of the species. Additional shrubs, grasses, or trees may be planted to benefit wildlife.

Of course, quality means very little if there is not adequate quantity. We must leave some habitat for wildlife, and it needs to be quality habitat, not just places where we choose not to build a shopping mall, subdivision, or factory. These are factors over which people exert a great deal of control. Weather

FIGURE 8-3 Heavy snow and bitter cold have an adverse effect on most wildlife populations, particularly small species, such as songbirds, upland game birds, and rodents.

(Source: Photo courtesy of Texas Parks & Wildlife Department © 2002.)

conditions are factors over which we have no control. We cannot control floods, drought, ice storms, and heavy snowfall. A severe winter, particularly in the northern states, can devastate a wildlife population (Figure 8-3). Freezing rain can cover all available food supplies, resulting in starvation, or it may cause some small birds and mammals to freeze to death. Ice storms in north Texas can **decimate** the bobwhite quail population. These storms are equally devastating to other species of wildlife. Young birds or mammals that may be only hours or days old can drown in a May or June thunderstorm. Deep snows can be hard on wildlife, particularly if they occur late in the year, when most of the available food supplies have been consumed. Wildlife managers can do little to alter the impact of adverse weather on wildlife.

Predators, whether human or animal, are not normally a serious problem for wildlife, provided that **adequate** escape cover is available. Regulated sport hunting has never endangered a single species of wildlife. Natural predators, such as coyotes, foxes, birds of prey, raccoons, and bobcats, seldom limit a species' population to the point that it becomes a serious problem. There are exceptions to this, however, particularly in areas with an abundance of predators. Fur trapping used to keep predator numbers in check. However, with the decreased demand for items made with fur, predators such as the raccoon and coyote are endemic in parts of North America. Ground-nesting birds are the most adversely affected by the population explosion of the egg-eating, nest-destroying raccoons. In some years, more than 80 percent of all the duck nests in a study area are destroyed, primarily by raccoons. Similarly, there is considerable evidence that coyotes may kill 80 to 90 percent of the fawns in an area where they are especially abundant. These predators are particularly effective in times of poor habitat conditions, such as lack of available ground cover caused by drought.

Predator numbers are generally held in check by the amount of available prey. Young and old predators alike starve to death when prey species are scarce. Predator species also tend to produce fewer young in years when prey species are less abundant. In areas where large predators no longer exist, it is necessary for humans to fill this role. Such species as white-tailed deer quickly **overpopulate** their habitat. When a species becomes too numerous it can cause serious, often permanent, damage to habitat. Mass starvation and disease outbreaks are the normal result. By setting proper bag limits and seasons, wildlife managers help prevent species overpopulation. Wildlife managers can control the natural predators of wildlife, but most do not. It is generally better to allow predator and prey numbers to **fluctuate** naturally, even if it means that more than half of a year's fawn crop is lost to coyotes.

Other factors that can limit a species' population are disease, old age, pollution, and accidents. Old age is a natural process over which wildlife managers have no control. However, by setting hunting regulations and bag limits managers can influence the harvest of some species to increase the number of older animals taken. In the wild few animals ever reach old age. Most succumb to predators, disease, or starvation before they die of old age.

Pollution is a factor over which humans can exert a great deal of control. Pollution takes many forms, from trash thrown out of vehicle windows to misuse of lawn chemicals. By being good stewards of the earth and properly disposing of trash, waste oil, pesticides, and other pollutants, we can limit our impact on the environment. Accidents occur in the wild on a more regular basis than most people realize. Animals make mistakes, just as humans do. Of course, humans present wildlife with chances to have accidents, in the form of fences and highways to name just two. As human populations increase, and therefore vehicles and roads expand, **road-killed** wildlife have become much more rampant. In some states the number of white-tailed deer killed on the highway exceeds the number harvested by sport hunters. Accidents are a limiting factor on wildlife populations over which wildlife managers have little control.

Wildlife diseases are a factor over which wildlife managers can exert some degree of control (Figure 8-4). It may not be possible to **eliminate** disease, but

FIGURE 8-4 Many species become diseased or malnourished during the winter months. For those that do, such as this white-tailed deer, a slow and painful death is the normal result.

(Source: Photo courtesy of Texas Parks & Wildlife Department © 2002.)

it is possible to exert some control over it. Disease is often the product of over-crowding or poor **nutrition**. By helping to ensure that these conditions do not arise or by **moderating** them, wildlife managers can limit disease outbreaks and their effects.

BIOLOGICAL SURPLUS

Animals reproduce each year. The number of offspring produced varies greatly between species. Larger animals, such as deer and elk, generally produce fewer young but live longer. Smaller animals, such as cottontail rabbits and quail, generally live fairly short lives but produce more offspring. In many smaller animals, as much as 80 percent of the population may die in a given year. However, every year animals produce more offspring than is necessary to continue the species. This excess is known as the *biological surplus*. This is nature's way of ensuring that enough breeding stock is left each year to continue the species. In a healthy population, it is this biological surplus that predators harvest and that disease, starvation, and old age overcome.

THE WILDLIFE MANAGER'S JOB

It is the wildlife manager's job to ensure that there are adequate numbers of a species to maintain it over a long period. This is accomplished by providing adequate habitat and by regulating harvests of game animals. It is the wildlife manager's job to prevent overpopulation of a species. Overpopulation will cause the species to damage its habitat, often permanently. This will cause the species to exist in lower numbers in the future. Wildlife managers help to prevent over- or under population of game animals by **regulating** the harvest of those animals. Wildlife managers work to keep habitat healthy enough to support wildlife and to prevent or control wildlife diseases.

Wildlife managers see to it that we do not allow a species to become endangered or extinct. These things can be accomplished by **monitoring** populations and conserving their habitat (Figure 8-5). Wildlife managers try to ensure that we purchase or develop new habitat to replace habitat destroyed or altered by human activities. They work to stock animals in habitats that are healthy but do not have an adequate population of a particular species. It is also the wildlife manager's job to regulate hunters and control the harvest of game animals to ensure healthy populations for future generations.

(Source: Photo courtesy of Texas Parks & Wildlife Department © 2002.)

FIGURE 8-5 Wildlife managers gain valuable insight into a species' habits through the use of radio transmitters.

CARRYING CAPACITY

A given area of wildlife habitat can support only a certain number of animals. The number of a species that can occupy a given habitat without overpopulating it is known as the habitat's carrying capacity. If more animals live in the habitat than the habitat can support, the wildlife will damage the plants that make up the habitat. If the species' numbers are not reduced, permanent damage to the ecosystem can result. There will also be a die-off of animals due to starvation or stress and malnutrition-related diseases. A good example of overpopulation and the resulting destruction of habitat is the mid-continent snow goose population. Because of their wary nature, excellent migration and wintering areas, and estimated 20-year life span, snow geese have greatly increased in number. They have done extensive damage to their fragile arctic nesting areas, which has also adversely affected shorebirds, other geese, ducks, and other species that use the same nesting habitat. Wildlife managers have greatly expanded bag limits and liberalized harvest methods in an effort to reduce the mid-continent snow goose population. It remains to be seen if these efforts will work. Proper wildlife management is vital if such situations are to be avoided.

Similarly, for our enjoyment and the benefit of wildlife, it is in our best interest to maintain wildlife populations near the carrying capacity of the habitat. For the past 60 years wildlife managers have been doing an outstanding job of caring for wildlife resources. Because of wildlife management, most areas have now reached their carrying capacity.

WILDLIFE MANAGEMENT SUCCESSES

Many species of wildlife are much more numerous today than they were 100 years ago. Species such as elk, white-tailed deer, pronghorn antelope, and wild turkeys have been successfully restocked in areas where there were none. These areas had proper habitat to support the species but lacked enough breeding stock to maintain a healthy population of the species. Many areas that have been restocked were traditional habitat for the species, but for reasons ranging from over-hunting to changes in land use practices, there were no longer healthy populations in the area. The species that have been managed back to healthy populations are great successes for wildlife management. Animals such as the pronghorn antelope, which numbered around 25,000 at the turn of the century, now exceed 750,000 in population. There were once fewer than 40,000 elk in the United States. Today there are more than 500,000. More than 10,000 beavers have been restocked into suitable habitat, and today they are found in every state. Canada geese numbers were once around 1 million. Due to management and restocking, there are now more than 4 million. These and many other "unendangered species" are the proof of what proper wildlife management can do.

WILDLIFE MANAGEMENT PRACTICES

Numerous wildlife management practices benefit wildlife. Generally, wildlife management practices that protect, maintain, or develop wildlife habitat are considered most important.

Habitat management is the most important practice for the future of America's wildlife. As the habitat available for wildlife shrinks each year, it becomes more and more important to properly manage what is left. Some of the habitat management practices that directly benefit wildlife include these:

Planting food plots and plants known to be used by wildlife. This is beneficial for most wildlife.

Half-cutting brush or building brush piles. This benefits species such as rabbits, which use such structures for shelter.

Artificial nesting structures. Species that benefit include wood ducks, geese, and bluebirds.

Controlled burns. This practice promotes the growth of new grasses and forbs used by such species as white-tailed deer.

Timber cutting, if done as smaller clear-cuts or selective cutting. It opens up the forest floor, allowing grasses, scrubs, and other forage plants to grow. It benefits such species as turkeys, elk, and white-tailed deer.

Grazing by livestock. If done in moderation, it cuts back or slows plant growth, allowing growth of browse plants and easier travel for ground-dwelling wildlife such as bobwhite quail and meadowlarks.

Providing water sources. Farm tanks, ponds, and guzzlers can provide water in desert or dry-land environments. This practice is beneficial to most forms of wildlife (Figure 8-6).

FIGURE 8-6 Artificial watering structures, such as this guzzler, give wildlife a reliable source of essential water in places with inadequate natural water supplies.

(Source: Photo courtesy of Texas Parks & Wildlife Department © 2002.)

Mechanical alteration of habitat. This generally takes the form of mowing or plowing strips or blocks of ground. It promotes weed growth that attracts insects, an important source of protein to such species as turkey **poults** and quail chicks. In late summer and fall the seeds produced by the weeds will be used by such species as doves, quail, and songbirds.

Farming. Land planted in crops provides food and cover for many ground-dwelling species, such as field mice, pheasants, and meadowlarks.

Controlling or preventing diseases and their spread can be an important management practice when dealing with species such as waterfowl that tend to concentrate in large numbers when in their wintering areas. Among ducks and geese, an outbreak of avian cholera, a very **contagious** disease, can be a serious threat to the local population.

(Source: Photo courtesy of Texas Parks & Wildlife Department © 2002.)

FIGURE 8-7 Hunters, such as this successful goose hunter, help to control populations and provide millions of dollars to fund various wildlife projects.

Hunting regulations are designed to protect animal populations and habitat. Some examples of hunting regulations that benefit wildlife are bag limits, limits on the legal methods to take game, and daily and seasonal time limits.

Artificial stocking has benefited species such as pronghorn antelope, wild turkeys, beavers, and white-tailed deer by establishing healthy populations where there were none. **Nonnative species** such as pheasants and aoudad sheep have also been helped by the stocking efforts of wildlife managers.

Hunting assists wildlife managers to reach harvest goals for many species, such as white-tailed deer, to keep the species' population compatible with the available habitat. Failure to maintain the proper number of a species can result in serious habitat damage (Figure 8-7).

Management of funds is a very important aspect of wildlife management. Funds are not unlimited. The Pittman-Robertson Act has provided vital funds for wildlife management, and many states are also generating funds through other programs, such as special stamps. Many private organizations, such as Ducks Unlimited, the Rocky Mountain Elk Foundation, Bighorn Sheep Society, Pheasants Forever, the National Wild Turkey Federation, and Quail Unlimited, provide additional funds and support for wildlife programs. The management or allocation of funds is one of the greatest challenges facing state wildlife and natural resource agencies today. There are many areas—habitat protection, research, habitat management, game law enforcement, fisheries management, and endangered species protection and management, just to name a few—competing for the available money. In most states, state directors, with input from professional wildlife managers, must decide how much money each area receives. Priorities often vary from area to area each year as needs arise. To further complicate the situation, many state wildlife and natural resources agencies receive no funding from general state tax revenue. They rely solely on license and tag fees and excise taxes generated by sportsmen and women.

SUMMARY

Successful wildlife management practices and **techniques** are well established. Wildlife managers and biologists are professional caretakers of our wildlife resources. The past 60 years have seen a remarkable recovery for much of America's wildlife. If we leave wildlife some habitat in which to live, allow our wildlife professionals to do their jobs, and give them the funds with which to do their jobs, our wildlife can have a bright future. As human populations and the associated urban sprawl continue to expand, the pressure on wildlife and wildlife habitat will also increase. We must be aware of the effect this has on wildlife and its effect on the wildlife manager's ability to do his or her job.

REVIEW QUESTIONS

Fill in the Blank

Fill in the blanks to complete these statements.

1. In its simplest form, modern wildlife management is _____ .
2. In reality wildlife management is seldom _____ .
3. List the four major factors that affect a wildlife population.
   (a) _____
   (b) _____
   (c) _____
   (d) _____
4. Species with low reproductive rates are more likely to become _____.
5. Certain traits help determine the _____ of each species.
6. Reproductive rates vary greatly among _____ .
7. Smaller birds and mammals tend to have _____ reproductive rates.
8. The _____ of a species is a factor over which wildlife managers have little or no control.
9. The _____ and _____ of an animal's habitat has a direct impact on its production.

10. List three ways in which the food supply available to a species may be improved.

    (a) _____

    (b) _____

    (c) _____

11. In areas with limited water supplies, _____ may be built to provide water for wildlife.

12. _____ and _____ of habitat are factors over which humans can exert a great deal of control.

13. _____ conditions are factors over which we have no control.

14. _____ are not normally a serious problem for wildlife.

15. _____ has never endangered a single species of wildlife.

16. By setting proper bag limits and seasons, wildlife managers help prevent species _____ .

17. Factors such as _____, old age, pollution, and _____ also limit a species' population.

18. In many smaller animals, as much as _____ the population may die in a given year.

19. Each year animals produce more offspring than is necessary to continue the species. This is known as the _____ .

20. It is the wildlife manager's job to ensure that _____ .

21. A given area of wildlife habitat can support only a certain number of animals. This is known as the habitat's _____ .

22. Many species of _____ are much more numerous today than they were 100 years ago.

23. List three species that are more numerous today than they were 100 years ago.

    (a) _____

    (b) _____

    (c) _____

24. Wildlife management practices that _____ , _____ , or _____ a wildlife habitat are considered most important.

25. _____ is the single most important practice for the future of America's wildlife.

Short Answer

1. List six common wildlife management practices and explain how each benefits wildlife.

2. Why are the birthrates of smaller mammals generally much higher than those of large mammals? How is this important to a species' long-term population and survival?

Discussion

1. In your opinion, what needs to be done to ensure the future of wildlife in America?

2. How has modern wildlife management helped America's wildlife? Why has it been effective?

Learning Activities

1. Select a species of wildlife, preferably a game species native to your area. Using Appendix B, contact your state wildlife and fisheries management agency. Through contact with the agency's professional wildlife managers, determine what factor(s) have the greatest impact on the selected species' population. What,

if anything, can be done from a management standpoint to limit the impact these factors have on the species. Present your findings to your class. Be sure to include specific management tactics if possible.

2. Select a game species not native to your state. Choose a state where the species is relatively abundant and contact that state's wildlife agency. Request information on the agency's management plan for the species. What factors most affect the species' population? Why? How do these factors differ from those within your state? How does each wildlife agency's management plan differ? How are they similar? Present your findings to your class.

Useful Web Site

U.S. FISH AND WILDLIFE SERVICE

http://www.fws.gov

The U.S. Fish and Wildlife Service Web site may also be used to access all 50 state wildlife and natural resource management agency Web sites. Just use the site index and click on the state of your choice.

CHAPTER 9

Modern Waterfowl Management

OBJECTIVES

After completing this chapter, you should be able to

- Describe waterfowl management practices.
- Explain the migratory nature of waterfowl and how it affects their management.
- Describe waterfowl migration routes.
- List factors that limit waterfowl numbers.
- Explain why international cooperation and coordination are necessary to manage waterfowl successfully.

INTRODUCTION

North America's waterfowl benefit from the most **complex** wildlife management program in the world. Due to their migratory nature, waterfowl are a difficult group of species to manage. Wildlife and waterfowl in particular, do not recognize state or **international** borders. Most species of waterfowl nest in the northern United States and Canada but spend the winter in the southern United States, Mexico, and South and Central America. For example, green-winged teal, which may nest in Saskatchewan, Canada, may spend the winter months in South America, traveling several thousand miles and crossing several borders in the process. Thus it is vital that waterfowl management be an international affair.

Early in the twentieth century, wildlife managers recognized the importance of international **cooperation** in managing North America's waterfowl. In 1916 the United States and Canada signed the Migratory Bird Treaty Act, which called for joint management of waterfowl and their habitat. To ensure protection of wintering waterfowl and their habitat, a similar treaty was signed with Mexico in 1936. Treaties were also signed with Japan in 1972 and with the former Soviet Union in 1976. These agreements help to ensure the protection and proper management of waterfowl, regardless of what borders they cross or what country they are in at a given time of the year.

WATERFOWL MIGRATION

Let's begin our study of waterfowl migration with an examination of why waterfowl migrate. Simply put, waterfowl migrate because they must. The prairie potholes, marshes, arctic tundra, and other habitats that are such wonderfully productive wildlife factories during the spring and summer are frozen and **barren** in the long months of winter. Yet when spring arrives they become lush waterfowl breeding grounds (Figure 9-1). Due to a relatively short spring and summer, vegetation appears quickly and matures rapidly, providing an excellent food supply for adult birds. The abundant water supply in potholes, marshes, and bogs not only provides brood-rearing sites for waterfowl, it also produces tremendous clouds of insects. These insects provide a generous, high-protein diet for young waterfowl, which allows young birds to grow rapidly during the short northern summer.

With the coming of fall, ice begins to form on the marshes, and the insects are gone. These climatic changes make it necessary for waterfowl to migrate to areas where they can secure food and water. The areas that can supply waterfowl's needs during the winter become their wintering grounds. Southern

FIGURE 9-1 Although most geese nest in the Arctic, Canada geese are one of the many species that use prairie potholes as nesting sites.

(Source: Photo courtesy of PhotoDisc.)

hardwood swamps and bottomland hardwood forests; the Gulf Coast of Texas and Louisiana; inland marshes of Mexico, the Atlantic coast, the coast of California, and Baja California; and the inland marshes around the Great Salt Lake have **traditionally** been prime wintering grounds for waterfowl. These areas contain open water and generally abundant food supplies.

Waterfowl must store fat for both the long flight south in the fall and the return trip in the spring. Consequently, they must have plentiful food supplies in all seasons. It is not enough for them to store adequate fat to survive the winter, like many mammals do; they must have energy reserves left in the spring to make the long trip to their nesting grounds. By being mobile, waterfowl best utilize the entire habitat available to them.

Waterfowl migration routes and times vary tremendously among the many different species of waterfowl. Most species of geese and swans, for example, nest farther north than do most species of ducks. However, Canada geese are often the first arrivals in spring (Figure 9-2). Often they sit amid snow and ice waiting for the arrival of spring. Mallards and pintails are usually not too far behind the Canada geese. After the arrival of the mallards and pintails, a multitude of species appear—redhead, gadwall, canvasback, green-winged teal, scaup, and wigeon. Generally the last to arrive at the northern nesting grounds are the ruddy duck, shoveler, and blue-winged teal. The earliest migrates are usually mated pairs, ready to begin housekeeping chores. Most waterfowl select a mate while they are still on their wintering grounds. Almost all would have selected a mate by the time they arrive at the chosen nesting site. The female typically leads the male to the area where she was raised or has nested in the past. Fall migrations, at least among some species, are triggered by colder weather. However, some species, such as blue-winged teal, leave the northern nesting grounds in late summer, when food is still abundant and the weather relatively mild. No one knows for sure what causes their departure. Conversely, mallards and the larger species of Canada geese stay in the north as long as the weather permits. And

FIGURE 9-2 Many waterfowl, such as these Canada and snow geese, migrate in huge flocks, often resting in shallow waters.

(Source: Photo courtesy of PhotoDisc.)

though some species travel thousands of miles, others go only far enough south to find food and open water. The fall migration includes greater concentrations of waterfowl than are normally seen in the spring, which may be due in part to smaller areas of available water and more restricted food supplies.

MIGRATION ROUTES

The migration routes of waterfowl have been well documented for many years. Waterfowl utilize the same routes, staging areas, resting sites, and wintering grounds year after year. This was learned from the recovery of **banded birds**, the use of radar, and observation of migrations over many years. Waterfowl use four major migration corridors or flyways—the Atlantic, Mississippi, Central, and Pacific (Figure 9-3). As the names suggest, the Atlantic and Pacific flyways generally follow the east and west coasts, respectively. The Mississippi flyway generally follows the Mississippi River valley, and the Central flyway goes through the plains states. Waterfowl also use many smaller migration paths. Migration routes generally run north and south, but many species travel great distances east and west as they make their way south.

WATERFOWL NAVIGATION

How do waterfowl, particularly ducks and geese, know where they are going and how to get there? The urge to migrate is **instinctive** in waterfowl, but is their **navigation** also instinctive? Although some migratory behavior may be instinctive, it is believed that most routes are learned from parents or other adult birds during a young bird's first migration. Ducks do not normally migrate as a family unit. The bond between a female and her brood appears to lessen as the young birds mature. When the birds congregate in staging areas, just before the fall migration, family groups mix freely and it is possible for brood mates to winter on opposite coasts. Geese, on the other hand, have much stronger family bonds. Young geese learn their migration routes, stopping points, and wintering sites from their parents and other close family members as they migrate together.

Ducks Unlimited has conducted several studies of waterfowl that reveal their use of the sun and stars as navigational aids. Waterfowl also seem to use the same landmarks from year to year. However, when skies are overcast, blocking out the sun, stars, and landmarks, waterfowl are still able to navigate accurately. It is thought that they may use directional cues from the Earth's magnetic field. Much is still to be learned about waterfowl navigation.

WATERFOWL MANAGEMENT PRACTICES

Waterfowl must have adequate habitat in order to survive. A major problem for waterfowl has been habitat loss and **degradation**. The loss of productive wetlands has been staggering (Figure 9-4). In its report to Congress in 1990, the U.S. Fish and Wildlife Service estimated that more than half of all the wetlands in the lower 48 states had been destroyed.

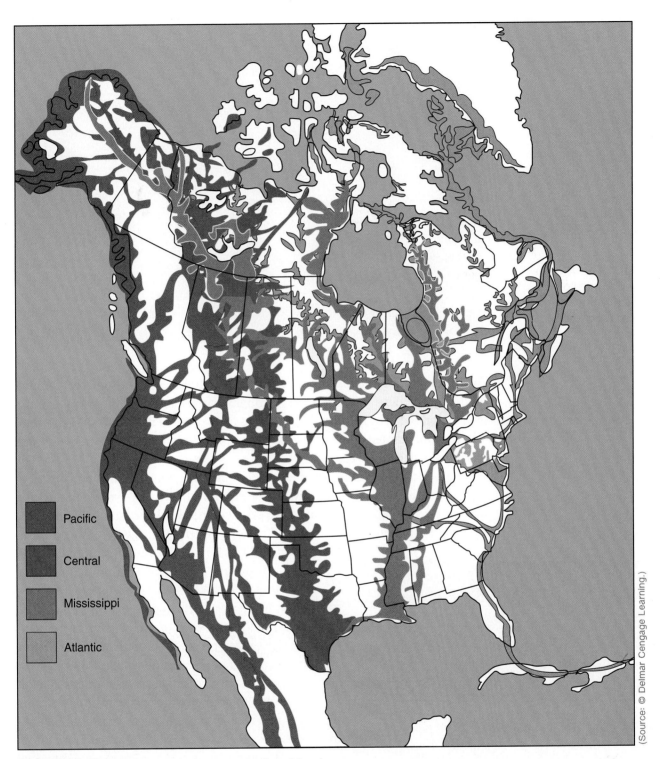

FIGURE 9-3 Fall duck migration routes (corridors).

Some areas are worse off than others. For example, it is estimated that Maryland has lost 73 percent of its original wetlands, Missouri 87 percent, Kentucky 81 percent, Iowa 89 percent, Indiana 87 percent, Illinois 85 percent, and California 91 percent. Twenty-two states have lost 50 percent or more of their original wetlands. Three hundred years ago there were nearly 25 million acres of bottomland hardwood forests. Now less than 4 million acres remain.

(Source: Courtesy of U.S. Fish and Wildlife.)

FIGURE 9-4 Human development and pollution have destroyed many of America's wetlands.

Waterfowl by the millions nest in the prairie pothole region of southern Canada and the northern United States. Drought causes many potholes to dry up. Normal snowfall or rain would have filled most of them back up, but many were plowed and planted. Once this occurs, the pothole is usually lost unless serious **reclamation** efforts are made.

The primary waterfowl management practice must be habitat protection. Available waterfowl habitat may also be enhanced with water-retaining **dikes, terraces,** and dams, ensuring adequate water levels for nesting waterfowl. It is generally beneficial for a wetland area to undergo periods of reduced water levels, but it is important that levels remain fairly constant during periods of nesting activity. It is also vital that the hens have adequate vegetative cover. If the area surrounding a pothole is farmed or mowed to the edge of the water, the remaining area gives predators an advantage. Narrow fringes of vegetation around wetlands let the predator search a much smaller area for its prey. Nest losses due to predation have been estimated at up to 90 percent in some areas of marginal habitat. These losses are far too high for a species to maintain its numbers, much less increase. Waterfowl prefer to nest in undisturbed upland areas with dense vegetative cover (Figure 9-5). Artificial nest sites such as nest boxes, human-made islands, and nesting baskets can be used if adequate natural nesting sites are not available. However, natural nesting sites are the most cost effective.

The primary management practice used for wintering waterfowl is habitat protection. Many millions of acres of coastal marshes and bottomland hardwood forests have been drained. As the available wintering habitat decreases,

(Source: Courtesy of U.S. Fish and Wildlife Service. Photo by Robert Gill.)

FIGURE 9-5 Good nesting habitat is becoming increasingly rare, but the federal Conservation Reserve Program has helped in many areas.

waterfowl concentrations in the remaining areas rise. This increases the chance of disease outbreaks, such as avian cholera and **duck plague**, which cause large numbers of deaths. Should a disease outbreak occur, wildlife managers must do what they can to disperse waterfowl concentrations and destroy as many dead birds as possible to help prevent the disease's spread.

Agriculture, which can be so damaging to nesting grounds, can be very beneficial to wintering waterfowl. Unharvested crops and waste grain from harvest provide excellent feed for many species of waterfowl. Another important practice for wildlife managers is to regulate the harvest of waterfowl to ensure healthy populations. In response to declining duck numbers, harvests have become more restricted, whereas limits on geese have been increased in response to their population increases.

FACTORS THAT LIMIT WATERFOWL POPULATIONS

Lack of habitat is a limiting factor for waterfowl populations. Similarly, a lack of high-quality habitat can be very damaging to a waterfowl population. Inadequate moisture in the nesting grounds can be very **detrimental** to nesting waterfowl. The drought of the 1980s in the prairie pothole region had a tremendous negative impact on duck populations.

Because waterfowl are often found in large concentrations, diseases can cause serious losses. In 1980 an outbreak of avian cholera in Nebraska's Rainwater Basin killed an estimated 70,000 to 100,000 waterfowl. Although diseases cannot be eliminated, wildlife managers try to limit their spread. Predators can be a serious problem, particularly on the nesting grounds, when enough dense vegetation is not available for nesting hens. If a skunk destroys a nest by eating the eggs, the hen will often attempt to renest (Figure 9-6). However, a predator such as a red fox may kill the hen. In this case that year's brood is lost along with the productive female that otherwise might have raised several more broods in the years to come. In poor or inadequate nesting habitat, predators can seriously affect brood-rearing success.

If waterfowl have enough resting areas to stop at on their long flight south and adequate wintering areas to prevent overcrowding and the related stress, they generally survive winter pretty well. Because of their mobility, food supplies seldom affect wintering waterfowl; they simply move to a better area. In this regard agriculture has been a big boon to waterfowl. Spring and winter crops, especially winter wheat, unharvested grain, and grain spilled by harvesting equipment, are excellent, high-energy waterfowl foods. Such food helps to ensure good winter survival as well as a healthy population when the birds reach the nesting grounds in the spring.

(Source: © Geoffrey Kuchera, 2010. Used under license from Shutterstock.com)

FIGURE 9-6 The striped skunk is a serious nest predator of all ground-nesting species.

SUMMARY

It is essential that all the countries where waterfowl spend any portion of their lives cooperate in wildlife management. It does little good for Canada to protect nesting grounds if, when birds migrate south for the winter, they do not have adequate wintering habitat in the United States and Mexico. If Canada and the United States place harvest limits on a species but Mexico allows excessive harvest, that species cannot be effectively managed. As with most species of wildlife, if they are left enough high-quality, unpolluted habitat, waterfowl will continue to thrive as they have for centuries.

REVIEW QUESTIONS

Fill in the Blank

Fill in the blanks to complete these statements.

1. The management of North America's waterfowl is the most _____ wildlife management program in the world.

2. Wildlife, especially _____ , do not recognize international borders.

3. In 1916 the United States and Canada signed the _____ , which called for joint management of waterfowl and their habitats.

4. _____ provide a generous, high-protein diet for young waterfowl.

5. Climatic changes make it necessary for waterfowl to migrate to areas where they can secure _____ and _____ .

6. By being _____ , waterfowl best utilize all the habitat available to them.

7. Waterfowl migrations and migration routes and times vary tremendously among the many _____ of waterfowl.

8. The earliest migrates are usually _____ , ready to begin housekeeping chores.

9. Fall migrations, at least among some species, are triggered by _____ .

10. The fall migration also sees _____ concentrations of waterfowl than those normally seen in the spring.

11. List the four major waterfowl migration corridors.

    (a) _____

    (b) _____

    (c) _____

    (d) _____

12. Waterfowl use many _____ migration paths.

13. The urge to migrate is _____ in waterfowl.
14. Ducks do not normally migrate as a _____.
15. Young geese learn their migration routes, stopping points, and wintering sites from their _____ and other _____ as they migrate.
16. Waterfowl may use the _____ and _____ as navigational aids.
17. Waterfowl must have adequate _____ in order to survive.
18. The loss of _____ has been staggering.
19. _____ states have lost 50 percent or more of their original wetlands.
20. The primary waterfowl management practice must be _____.
21. It is vital that hens have adequate _____ in which to nest.
22. Nest losses due to predation have been estimated at up to _____ percent in some areas.
23. Artificial nest sites such as _____ , _____ , and _____ can be used if adequate natural nesting sites are unavailable.
24. The primary management practice used for wintering waterfowl is _____.
25. Unharvested crops and _____ provide excellent feed for many species of waterfowl.
26. Lack of habitat is a _____ for waterfowl populations.
27. The _____ of the 1980s in the prairie pothole region had a tremendous negative impact on duck populations.
28. Because waterfowl are often found in large concentrations, _____ can cause serious losses.
29. In poor nesting habitat, _____ can seriously affect brood-rearing success.
30. Because of waterfowl's _____ , food supplies are seldom a factor for wintering waterfowl.
31. Due to the _____ nature of waterfowl, it is essential that all the countries where waterfowl spend any part of their lives cooperate in wildlife management.

Short Answer

1. What methods have the United States and other countries developed for managing migratory waterfowl?
2. Why are treaties between Russia, Japan, Mexico, Canada, and the United States so important to waterfowl management? How has cooperation between governments helped waterfowl?

Discussion

1. How does migration affect waterfowl populations? Why?
2. In your opinion, how does the mobility of waterfowl challenge wildlife managers? Why?

Learning Activities

1. Draw a map of Canada and the United States. Color in the main breeding areas for ducks and geese. Draw the major and significant minor migration flyways and show the main wintering grounds in the United States.
2. Using your local library, your state wildlife agency, and any other appropriate sources determine the greatest factor limiting waterfowl populations in your state. Decide what can be done to help the situation and compile a report listing specific tactics and recommendations. Present your report to your class.

Useful Web Sites

U.S. FISH AND WILDLIFE SERVICE
<http://www.fws.gov>

DUCKS UNLIMITED
<http://www.ducks.org>

DELTA WATERFOWL
<http://www.deltawaterfowl.org>

CHAPTER 10

Endangered Species

OBJECTIVES

After completing this chapter, you should be able to

- List some common endangered species.
- Explain why species become endangered.
- List management practices for endangered species.
- List several species that have been successfully managed to prevent their extinction.
- Explain how we have driven several species in the United States to extinction.
- Identify several endangered species in the United States.
- Describe the characteristics of each of these species.
- Identify the type of habitat where you might find each species.
- List the main reasons for each species being listed as endangered.

INTRODUCTION

An endangered species is any species in danger of extinction throughout all or a large portion of its range. More than 1,000 species of plants and animals are listed as endangered. If you include species listed as threatened, there are over 1,300. Animal and plant species have been going extinct since life on Earth began. However, the *rate* at which species are becoming extinct is far from natural. Human **alteration** of the Earth has greatly accelerated the rate at which species are going extinct.

EXTINCTION FACTORS

The main reason a species declines is the destruction or alteration of its habitat. An animal's habitat is its home, without which it cannot survive. However, many factors other than habitat destruction can cause a species to decline (Figure 10-1).

Habitat degradation can be as detrimental to a species as habitat destruction. For example, a species' habitat may become so polluted that it poisons the animals. Although this is not the problem it once was in the United States, it is still a serious problem in developing countries of the world. Habitat may also become so **fragmented** that normal activities, such as breeding, become impossible, or animals may begin to inbreed.

The introduction of nonnative species has caused the demise of more than one species. Introductions such as the English sparrow, European starling, and Norway rat, just to name three nonnative species, compete with native wildlife species for habitat. Often our native species lose out to more aggressive nonnative ones. Such losses to introduced species are even more severe in fragile environments such as the Hawaiian Islands, where dozens of native species of plants and animals have lost out to competitors. Native species in island habitats tend to evolve in a "vacuum," so to speak, and are generally very poorly prepared to compete with a sudden introduction of a new species into their environment.

Some of the species that were listed as endangered were once controlled as pests. Species such as the gray wolf and grizzly bear were serious predators of domestic livestock during the human expansion across America. Domestic livestock is generally easier to capture and kill than native wildlife. With severe decreases in the numbers of natural prey, such as bison, large predators turned to domestic livestock. Seeking to protect their livelihoods, farmers and ranchers fought back, often with help from the government. As is always the case when wildlife and humans collide, the wildlife lost.

Another factor is the fact that some species have always been rare. Some species are very specialized in their needs and/or have very low reproductive

What Has Caused Threatened and Endangered Species?

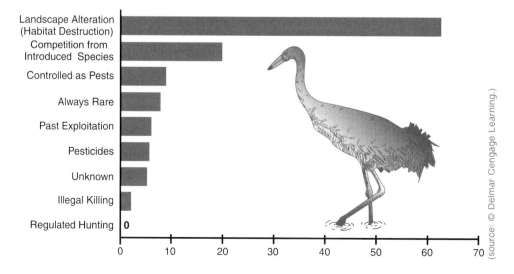

FIGURE 10-1 This graph shows the relative influence of the factors that have endangered wildlife species, from most to least influential.

(source: © Delmar Cengage Learning.)

FIGURE 10-2 The bald eagle, our national symbol, was all but eliminated from the lower 48 states.

(Source: Photo courtesy of Texas Parks & Wildlife Department.)

rates. These factors combine to keep their numbers low. Past exploitation has been a factor in several species' fortunes. Market hunting drove the passenger pigeon to extinction and nearly did the same for the bison. The whooping crane, although never thought to be especially abundant, was driven to the brink of extinction due to demand for its feathers to decorate women's hats during the early part of the twentieth century. Pesticides, in particular DDT, caused the decline of several species. The peregrine falcon and bald eagle were both driven to the edge of extinction in the lower 48 states, primarily due to DDT (Figure 10-2). DDT does not break down readily once it is sprayed. Consequently, it became concentrated in lakes and streams after being washed out of fields. This caused fish and waterfowl to carry major concentrations of DDT in their bodies. Because the bald eagle and peregrine falcon preyed on the fish and waterfowl, they collected even greater concentrations of the pesticide in their own systems. This caused numerous problems, the most serious of which was the thinning of their egg shells. It became increasingly difficult for them to raise their young because their eggs would crack and break when they tried to incubate them. The use of DDT was banned in the United States over 30 years ago. This has helped the bald eagle and peregrine falcon make a comeback in the lower 48 states. Some species become endangered for reasons we cannot determine. Extinction is a natural process of **evolution** that has been going on since there has been life on Earth. Species that are highly specialized or that cannot adapt to a changing environment become endangered or extinct.

The last factor that is known to have contributed to the endangerment of wildlife species is illegal killing. Sadly, when there is money to be made there will always be someone willing to break the law to get it. It is estimated that illegal killing has been a factor in less than 3 percent of the endangered species in America, but it is a significant problem in developing countries around the world.

ADAPTABILITY

Some plants and animals may **flourish** for hundreds or even thousands of years, whereas other species may come and go in a relatively short period. On the North American continent many great mammals, such as camels, mammoths, and sabre-toothed tigers, flourished until the end of the last ice age. Also present during this period were such species as mule deer, pronghorn antelope, and grizzly bears. For a variety of reasons that no one completely understands, only the latter species survived into our times. However, most species that survive through the years have a common trait—adaptability. An animal or plant species that is capable of changing as its environment changes has a better chance of survival. Species that are highly specialized or require specialized habitats have a poorer chance of survival. For example, species such as the white-tailed deer, cottontail rabbit, and common field mouse are highly adaptable. They adjust to a wide variety of habitats and habitat situations, and so they are able to survive drastic changes in their environments.

The opposite of highly adaptable species are species that are specialized, or not very adaptable to changes in their environment. These species are particularly vulnerable to extinction. The everglade kite is a good example of a highly specialized species (Figure 10-3). It feeds exclusively on the apple snail. Therefore, as apple snail populations have decreased, so have everglade kites. Another example is the black-footed ferret of the Great Plains, which feeds almost entirely on the prairie dog. With the **demise** of many of the great prairie dog colonies, the black-footed ferret population plummeted. Animals or plants that are highly specialized are comparable to humans with only one job skill in a changing job market. Should the job market shift or change quickly, these are the people most often left unemployed.

FIGURE 10-3 Everglade kites, such as this one, are becoming an increasingly rare sight over much of their former habitat.

Plant and animal species get into serious trouble when humans alter their environment at a faster rate than the plant or animal can adapt. Most species are capable of adapting to changes in their habitats if those changes take place over an extended period. The Earth's environment has been changing, and will continue to change, throughout evolution. However, with bulldozers and other modern construction equipment, changes occur much faster than the rate at which most species can adapt. Of course, if we pave, plow, or pollute a species' entire habitat, it has little chance of survival, regardless of how quickly it can adapt.

SPECIES THAT WE HAVE ENDANGERED

More than 1,000 native plants and animals are listed as endangered at this time. Some 80 species of mammals, 90 North American birds, 40 reptiles, 23 amphibians, 130 species of native fish, and more than 700 native plants are listed as endangered. A few species that have been endangered by human encroachment are the California condor, red wolf, ocelot, Florida panther, gray wolf, grizzly bear, whooping crane, and the Kemp's Ridley sea turtle (Figure 10-4). These are some of the species that receive the most public attention, but there are dozens more that are less appealing. Species such as the Lower Keys rabbit, Fresno kangaroo rat, northern swift fox, Ozark big-eared bat, and the Virginia big-eared bat are just as endangered, yet they receive little public attention. These are just a few of the hundreds of endangered and threatened species.

Human activities in North America have resulted in several species becoming extinct. The passenger pigeon and the heath hen are two of the more familiar species that we have driven to extinction. Destruction of habitat and unregulated market hunting were the main causes of the passenger pigeon's demise. They were also shot by the thousands by farmers protecting their fields. Passenger pigeons were communal nesters, that is, they nested together in large colonies. When the oak and beech forests on which they depended were cleared for farm fields and lumber for homes, furniture, and so on, they could not adapt. This habitat destruction, coupled with market hunters killing millions to be shipped to the eastern cities, doomed the passenger pigeon.

The heath hen was an eastern subspecies of the prairie chicken. Habitat destruction and unregulated subsistence hunting helped push the heath hen to extinction. Once commonly found in the eastern coastal plains, the heath hen ceased to exist early in the 20th century.

The dusky seaside sparrow is another species pushed to extinction by human activity. Its Florida coastal marsh habitat was altered and destroyed by humans to such an extent that the bird could not recover. Many dozens more species have become extinct, primarily due to human population increases, which foster habitat destruction. As people occupy more of our planet and demand more land and other resources that wildlife use, we will continue to exterminate species.

FIGURE 10-4 Many species, such as the Kemp's Ridley sea turtle, are endangered for various reasons, ranging from habitat destruction to pesticide poisoning.

(Source: Photo courtesy of U.S. Fish and Wildlife Service.)

WHY SAVE ENDANGERED SPECIES?

If extinction is a natural process, why should we worry about species becoming extinct during our time? For one thing, the rate at which species are going extinct is far from natural. When the Pilgrims first arrived in America we had approximately 500 more species and **subspecies** of plants and animals than we have today. In other words, some 500 species, subspecies, and varieties of our nation's plants, invertebrates, and animals have been lost forever. In contrast, all of North America lost only about three species every 100 years or so during the 3,000 years of the Pleistocene Ice Age. Most species are lost as a result of human activities, such as habitat destruction, habitat degradation, introduction of non-native species, environmental pollution, and exploitation.

Each species of plant, wildlife, and fish is of scientific, ecological, recreational, aesthetic, and educational value to our country. They can also be of economic value. No matter how small and unknown a species is, it could be of great value to us one day. For example, **penicillin** was discovered in a fungus and since its discovery has saved millions of lives. The anticancer drug Taxol comes from trees, as does rubber. Plants provide us with sources for the drugs **quinine**, **morphine**, and **curare**. Although there are some 80,000 species of edible plants, 90 percent of the world's food comes from fewer than 20 species. This means that there are many species that may prove useful in the future. If these species are destroyed before we have a chance to study them, imagine what we stand to lose. This is exactly what is happening around the world. Species are becoming extinct before we can study them, in some cases before we have even identified them. Thus there are many good reasons to save our great diversity of plants and animals.

THE ENDANGERED SPECIES ACT

The Endangered Species Act is America's attempt to save endangered and threatened species. The Act, passed by Congress in 1973, stated that its purpose in part, "was to provide a means whereby the **ecosystems** upon which endangered species and threatened species depend may be conserved and to provide a program for the conservation of such endangered species and threatened species." Efforts undertaken since the Act became law have been successful in restoring many species and preventing many others from becoming extinct.

The bald eagle is probably the most easily recognizable former endangered species. Although the bald eagle was never endangered in Canada or Alaska, its numbers fell to an estimated 400 nesting pairs in the lower 48 states in the late 1960s. The banning of the pesticide DDT, protection of habitat, public education, reintroduction, and stricter law enforcement have combined to greatly increase the number of bald eagles in the contiguous United States. It is estimated that there are now more than 3,000 nesting pairs in the lower 48 states alone.

Whooping cranes once flourished in bogs and grassy marshes, but they were nearly lost forever when croplands and cities replaced much of their habitat. Once numbering fewer than 20 birds in the wild, there are now more than 200 whooping cranes. Although still not numerous, they are in much better shape than they were 40 years ago. It is hoped that through captive breeding programs and habitat protection, this magnificent bird will continue to increase in number.

The Aleutian Canada goose was once thought to be beyond saving. In the early 1900s commercial fox farmers introduced nonnative foxes to many of the Aleutian Islands where the birds nested (Figure 10-5). The ground-nesting birds proved easy prey for the foxes, and their numbers plunged. At one point in the late 1960s fewer than 200 were thought to exist. Wildlife professionals went to work to ensure fox-free nesting grounds and to protect vital wintering grounds in Oregon and California. These measures resulted in an impressive comeback by the species. They have made such a wonderful recovery that they are now delisted, the ultimate goal for any endangered species.

These are just a few examples of endangered species success stories. The American alligator has recovered to the point that it no longer requires endangered species protection. Surveys of the endangered Lange's metalmark butterfly

(Source: Photo courtesy of Texas Parks & Wildlife Department.)

FIGURE 10-5 The Aleutian Canada goose is one of many species that human activities endangered. Scientific wildlife management has saved the species from extinction.

once revealed only 20 to 30 specimens. While over 1,200 were counted in 1990, they are still listed as endangered. The greenback cutthroat trout population has shown improvement and is now listed as threatened. Many more lesser-known species have also made improvements.

The Endangered Species Act has saved dozens of species from almost certain extinction. When professional wildlife scientists identify a species as threatened, they can use the Endangered Species Act to protect the species and its critical habitat. Through captive breeding programs, animals such as the red wolf, California condor, and black-footed ferret have a chance at being restored to their former habitats. In the lower 48 states, the gray wolf, bald eagle, grizzly bear, and peregrine falcon have increased in number and expanded their ranges primarily due to the protection offered by the Endangered Species Act. The habitat protection provisions of the Act ensure that these animals had habitat to return to.

Unfortunately, protection for some species came too late. The dusky seaside sparrow suffered extensive loss of its coastal salt marsh habitat and is now extinct. Protection of species is not cheap, and it will require some sacrifices. If we are to preserve the rich diversity of flora and fauna in the United States, we must make these sacrifices.

MANAGEMENT PRACTICES FOR ENDANGERED SPECIES

When a species is listed as endangered, a recovery plan is written by an expert or team of experts. The recovery plan may call for captive breeding and reintroduction, habitat protection or improvement, new research, or special wildlife and habitat management techniques (Figure 10-6). A plan may use each of these practices or any combination of them. Some recovery plans are more complex

(Source: Photo courtesy of Texas Parks & Wildlife Department.)

FIGURE 10-6 Captive breeding programs are the only hope for many endangered species.

than others. For any recovery plan to be successful it must stop and reverse the decline of the species and provide the means with which to ensure its long-term survival. Listed below are the major factors involved in protecting a species. (They are not listed in any particular order.)

- Habitat protection or improvement
- Scientific research
- Captive breeding
- Reintroduction into the wild
- Special habitat or wildlife management techniques

WILDLIFE IDENTIFICATION: ENDANGERED SPECIES

There are currently more than 1,000 species of plants and animals native to the United States on the list of threatened and endangered species. We will discuss a few of these species, many of which may already be familiar to you. Several less "glamorous" species (those that generally receive less press) are also included. It is extremely important to realize that all endangered species are of equal importance, whether invertebrate, mammal, or plant. By law each endangered life form must be treated equally. The Endangered Species Act gives no preference to a species just because it is popular or receives more coverage from news media. There are more than 230 invertebrates and 700 plants on the Endangered Species List. All life forms are important, not just mammals, birds, and fish. The Peregrine falcon is another example of a species that was once endangered but has recovered to the point of being delisted, or removed, from the Endangered Species List.

WEST INDIAN MANATEE (*TRICHECHUS MANATUS*)

The manatee is a large marine mammal found in the rivers and along the coast of Florida (Figure 10-7). They prefer large, slow-moving rivers; areas where rivers flow into the sea; and shallow areas along the coast, such as inlets, bays, and coves. Manatees spend most of their time traveling from saltwater to brackish water, to freshwater habitats, and back again. They may cover large distances during their travels, especially when traveling between winter and

(Source: Courtesy of USFW/Gaylen Rathburn.)

FIGURE 10-7 This West Indian manatee cow and her calf are endangered primarily by human activities in their habitat, from collisions with motor boats to development.

summer habitats. Florida manatees may grow to 12 feet in length and weigh 3,500 pounds or more, but the average adult is about 10 feet long and weighs 800 to 1,200 pounds. Manatees resemble a walrus without tusks. They have stiff facial whiskers and large, flattened tails that they use to propel themselves through the water. Two flippers located toward the front of the manatee are used for steering. Manatees are vegetarian, feeding on a variety of aquatic plants, such as hydrilla and water hyacinth. Although manatees are mammals and must surface to breathe air, they feed underwater and may remain submerged for as long as 10 to 12 minutes. The average underwater time is 4 to 5 minutes. Because manatees are such large animals, they must eat a large amount each day. It is estimated that they need to consume from 4 to 9 percent of their body weight each day. If a manatee weighs 1,000 pounds, it must eat 40 pounds of vegetation every day! When it is not feeding, which takes 6 to 8 hours each day, it rests just below the surface of the water, with only its nose above the surface. Manatees are also playful at times, and they are quite graceful and athletic in the water.

Manatees have a very low reproductive rate. They typically do not breed until 7 to 9 years of age, and then only one calf is born every 2 to 5 years. Twins do occur but are rare. The newborn calf is about 4 feet long and weighs 60 to 70 pounds at birth. The mother nurses her young underwater for about 3 minutes at a time. Although the young are born with teeth and begin to eat plants within a few weeks of birth, they may remain with the mother for up to 2 years.

Why the Manatee Is Endangered

Manatees are endangered for a variety of reasons, not the least of which is their low rate of reproduction. The reproductive rate makes the manatee vulnerable because it may not be able to reproduce fast enough to replace animals that die.

Manatees face natural hazards such as diseases and accidents, but their greatest problems are human-made hazards. Floodgates and canal locks may crush or drown manatees, but the biggest threat is speeding boats. Because the manatees rest just below the surface of the water, they are vulnerable to boat traffic. As human population has increased, so have the number of boats speeding around Florida's coasts and rivers. Manatees are killed either by the impact of the boat striking them or when their backs are gashed open by the propellers. Prohibiting boats from areas with heavy populations of manatees and enforcing speed limits in other areas have reduced manatee deaths from boating incidents. These efforts may not be enough. Manatees are difficult to count; however, census results from 1990 to 2007 have shown a low of around 1,200 to a high of around 3,000. In any case, the West Indian manatee is struggling to survive. It is feared that losses resulting from manatee–boat collisions are greater than the manatee birthrate. If these losses do not decrease, the manatee will not survive.

(Source: Photo courtesy of U.S. Fish and Wildlife Service.)

FIGURE 10-8 California condors are carrion eaters.

CALIFORNIA CONDOR (*GYMNOGYPS CALIFORNIANUS*)

California condors are the largest birds in North America, with a wingspan of up to 9.5 feet and a weight of up to 25 pounds. Although California condors were probably never found in large numbers, they originally inhabited the entire Pacific coast, from Baja California to British Columbia. California condor numbers have been in decline since the late 1800s, and in recent years they have been found only in a small area north of Los Angeles. As this is being written there are about 50 California condors in the wild in California and Arizona. While the Condor restoration program still has a long way to go, this is an improvement over 30 or so condors in the captive breeding program in the 1980s.

California condors are in the vulture family and as such are carrion eaters (Figure 10-8). They tend to prefer larger animals, such as deer, sheep, and cattle, but will also consume fish, birds, rodents, and other smaller carrion. Condors have very keen eyesight and locate their meals by soaring over the wilderness and rangeland. They can soar for hours on warm thermal updrafts, reaching altitudes of 15,000 feet. Condors are gray-black, with bare heads and necks. They look magnificent as they soar high above the ground on massive wings.

California condors prefer to nest in caves and on ledges of cliffs, particularly those with clear approaches. An approach clear of obstructions is essential for a bird as large as a condor to make clean landings and takeoffs. Condors have a low rate of reproduction, with each adult pair laying only one egg every other year. The young mature slowly and usually remain with their parents until their second year.

Why California Condors Are Endangered

Although scientists have yet to discover the exact cause of the condor's decline, several factors have contributed to the condor's **perilous** position. Their low reproductive rate and the high number of years (6 to 8) it takes the young to reach sexual maturity are certainly a factor. Of course, human-made problems are the most important in the condor's decline. For many years they were shot by ranchers, who mistakenly thought they killed livestock, and their eggs were stolen for private collections. However, the biggest factor is a decline in the quality and quantity of habitat available to condors. The condor's historic range, the Pacific coast, has an ever-increasing human population. Southern California in particular has lost much of its condor habitat. Shopping malls, roads, and subdivisions have replaced much of the ranchland and wilderness that condors need to survive. Isolated nest sites are fast becoming impossible to find.

All is not lost, however. Scientists have been working overtime on the California condor problem for many years. Using the closely related Andean condor as a model, scientists have learned a great deal about condor behavior and reproduction. As a last-ditch effort to save the birds, fish and wildlife officials captured the last wild California condor in 1987. The first captive-bred condor chick arrived in 1988. By 1992 the population of captive condors had risen and scientists were comfortable enough in their knowledge of condor behavior to begin the release of captive-bred condors. This release occurred several years sooner than most wildlife officials had anticipated. With education, cooperative effort, and public support the California condor is once again sailing the skies over the Pacific coast.

(Photo courtesy of Texas Parks & Wildlife Department.)

FIGURE 10-9 Peregrine falcon.

PEREGRINE FALCON (*FALCO PEREGRINUS*)

The peregrine falcon is a bird of prey. It demonstrates awesome skill and grace while airborne, as well as great speed. Scientists have estimated the peregrine's speed during a dive on its prey to exceed 200 miles per hour. It is a medium-sized bird, about the size of a large crow, with a long tail and, long pointed wings. Adults are blue-gray with black barring and have large, dark eyes and a white face, with a black stripe on each side. The females are usually larger than the males, a common trait among birds of prey (Figure 10-9).

Peregrine nests usually consist of a shallow depression on a cliff ledge. On rare occasions peregrines use an old eagle's nest or tree cavity. Peregrines are also known to nest in cities, utilizing skyscrapers for nesting sites and the usually overly abundant Rock Dove or common feral pigeon as a primary food source. Normally the male selects a nesting site and performs spectacular aerial displays to attract a mate. The female lays an average clutch of four eggs, which are incubated for 32 to 34 days. Peregrines have to be severely harassed for them to abandon the nest site, which they generally defend with considerable determination.

Peregrine falcons feed almost exclusively on other birds. Pigeons, starlings, jays, gulls, and ducks are favorite targets. Soaring high above their victims, peregrines suddenly dive, or "stoop," on their intended prey. Descending on their victims at over 200 miles per hour, their needle-sharp talons often kill their prey instantly.

Although peregrines have probably never been very numerous, they were once found over large portions of the United States. In the past they were most common in the valleys of the Appalachian Mountains, from Georgia north to New England. They were also present in the Rocky Mountains, the upper Mississippi River valley, the Arctic tundra, the Pacific coast from Alaska south to Mexico, and other parts of the world. Peregrines that nest in Alaska, Canada, and Greenland migrate south in the fall, often spending the winter in South and Central America. Peregrines that nest in the lower 48 states generally do not migrate as far as those nesting in the north and some may not migrate at all.

Why Peregrine Falcons Became Endangered

Peregrine numbers, particularly in the eastern part of the United States, began to decline drastically in the late 1940s. By the mid-1960s, peregrines were nearly impossible to find east of the Rockies. As with most problems with endangered species, the reasons for their decline can be traced to human activities. The U.S. Fish and Wildlife Service had scientists at the Pawtuxet Wildlife Research Center investigate the decline. In studying peregrine falcons and other birds of prey, the researchers found high concentrations of DDT and a product formed by the partial breakdown of DDT, DDE, in the birds. It was discovered that the birds **accumulated** these concentrations by eating birds that had fed on DDT-contaminated seeds and insects. Because of their location near the top of the food chain, birds of prey accumulated large amounts of DDT and DDE from feeding on contaminated prey. DDT and DDE are toxic chemicals, and they had a direct effect on the reproduction of birds of prey, causing the birds to produce **abnormally** thin egg shells. The fragile eggs broke as the female tried to incubate them. It quickly became such a problem that many birds of prey, including peregrine falcons, could not produce enough young to replace the adults who died.

In 1972 DDT was banned in the United States. Although DDE residues may still be found in the United States, the banning of DDT has been essential to the recovery of many birds. Wildlife scientists began raising and releasing peregrines in the early 1970s. The case of the Peregrine Falcon is a good example of the successful enactment of the Endangered Species Act.

RED WOLF (*CANIS NIGER*)

The red wolf is a prime example of an animal that humans, through **persecution** and habitat destruction and degradation, have all but exterminated. The red wolf has been misunderstood since it was first identified in 1851. First, it is often not red; many are gray or black. Red wolves are smaller than gray wolves, weighing on average 40 to 80 pounds (Figure 10-10). They are larger than coyotes, but they are not as aggressive as gray wolves and tolerate coyotes within their territory. Unfortunately, all species of Canis can interbreed, and this happened between red wolves and coyotes. Red wolves once roamed from central Texas to Pennsylvania. However, as the forests they inhabited were cleared for farming, their numbers declined. The more open areas attracted coyotes, which expanded eastward. When red wolves were killed as potential livestock predators, their numbers continued to dwindle and they began to interbreed with coyotes. By 1973, when the red wolf was listed as endangered, this was mainly because true red wolves could be found only in small, isolated areas of the southeastern Texas Gulf Coast and the southwestern corner of Louisiana. With the coyote continuing to expand its range, it was feared that the species would be lost forever, so all the remaining purebred red wolves were captured and placed in a captive breeding program. The wolves were sent to the Point Defiance Zoological Park in Tacoma, Washington.

Free-roaming red wolves normally breed in February or March. Males and females appear to mate for life and begin to breed in their second or third year. Dens are usually dug out along the sides of gullies, ditches, or canals. They may also den in hollow logs on occasion. After a gestation period of approximately 63 days, the female gives birth to two to six pups. The pups are born in much the same manner as domestic dogs, with their eyes closed and dependent on their mother for milk. The pups grow rapidly and are out and about with their parents by 2 months of age. The young remain with their parents, hunting in small packs.

(Source: Photo courtesy of U.S. Fish and Wildlife Service.)

FIGURE 10-10 Adult red wolf.

Why Red Wolves Became Endangered

The red wolf has become endangered because of persecution by humans, loss and degradation of habitat, and interbreeding with coyotes. Parasites such as hookworms and heartworms also took a certain number of animals each year. The captive breeding program at Point Defiance has been producing captive-bred red wolves since 1977. Scientists have gone to great lengths to ensure that the wild instincts of these wolves are not lost and that they do not develop a dependence on humans. Captive-bred wolves have been reintroduced to suitable habitat since 1987, when four pairs were released in Alligator River National Wildlife Refuge in North Carolina. These wolves were equipped with radio transmitters and were monitored carefully. Unfortunately, it is difficult to locate large tracts of habitat that are free of coyotes, thus avoiding the possibility of interbreeding. Since the Alligator River release, several islands off the Southeast coast have been used as release sites. Islands are ideal because there is no opportunity for interbreeding with coyotes, the animals are easier to monitor, and the offspring are raised wild and are easier to capture for release at other sites. There are now an estimated 200 Red Wolves in the wild in North Carolina. The Red Wolf recovery plan appears to be working and the species is in considerably better shape than 20 years ago.

BLACK-FOOTED FERRET (*MUSTELA NIGRIPES*)

The black-footed ferret came as close to becoming extinct, without actually doing so, as a species can get. In fact by 1972 black-footed ferrets were thought to be extinct. Although many ferret sightings were reported, none were actually found until 1981. A rancher in Wyoming discovered a strange animal that a dog on his ranch had killed. Unable to identify it, he took it to a taxidermist, who recognized it as a black-footed ferret. Further investigation revealed a tiny ferret population near Meeteetse, Wyoming (Figure 10-11).

Black-footed ferrets are not the same species as ferrets commonly sold as pets, which developed in Europe. The black-footed ferret evolved in North America. However, both kinds are members of the weasel family, as are skunks, wolverines, badgers, and minks. The black-footed ferret is about mink-sized, with a long, slim body similar to a weasel's, but larger. It is a buff-yellow overall, with a distinctive

(Source: Photo courtesy of Texas Parks & Wildlife Department.)

FIGURE 10-11
Black-footed ferret.

black or dark brown facial mask. It has a black-tipped tail, black legs and feet, and a brownish head. The black-footed ferret feeds on prairie dogs and lives in the burrows that make up a prairie dog town. Black-footed ferrets were once found wherever prairie dog towns were found, from southern Canada to Texas, and from the Rocky Mountains eastward across much of the Great Plains.

Why Black-Footed Ferrets Became Endangered

Black-footed ferrets have a serious weakness: they are very specialized, living only in prairie dog towns and feeding almost exclusively on prairie dogs. As our vast prairies were settled, efforts were made to eliminate the "pest" prairie dogs. They competed with livestock for pasture, and their burrows were a danger to live-stock, particularly horses. So without regard to the damage done to other species, prairie dogs were poisoned, gassed, and shot by the millions. With the destruction of their prey, which also provided them with burrows in which to live, black-footed ferrets struggled to survive. The destruction of the prairie dog towns was the main cause of the severe population decline of the black-footed ferret.

Their future is looking brighter today. The population discovered near Meeteetse, Wyoming, in 1981 grew to an estimated 130 ferrets in 1984. However, from that point the population began to decline, and the last 18 ferrets known to exist were captured in 1987. Captive breeding and reintroduction began almost immediately. The captive ferrets increased to over 300 by 1991, when 49 were released to the wild. Although only 7 of the original 49 survived to reproduce, progress was made and an additional 90 were released in 1992. This program will continue, with the goal of establishing 10 separate populations of at least 30 breeding adults in the wild. In 2006 there were an estimated 1,000 black-footed ferrets in Arizona, Colorado, Montana, South Dakota, Utah, and Wyoming. The goal is to have 1,500 ferrets in the wild by the year 2010. With continued efforts on the part of wildlife professionals, and if we leave them some habitat to live in, the black-footed ferret should again roam the Great Plains.

WHOOPING CRANE (GRUS AMERICANA)

The whooping crane stands 5 feet tall and has a wingspan exceeding 7 feet. They are found nowhere else in the world but North America. Whooping cranes once nested from southern Canada to Illinois and spent their winters from Mexico to the Carolinas. However, in 1941 only 16 birds were counted on their wintering grounds in Texas. Whooping cranes are snow-white with a black and red head and black wing tips (Figure 10-12).

Whooping cranes nest in wet areas, such as marshes, and build their nests in cattails, bulrushes, or sedges. These areas provide food as well as protection from predators. Whooping cranes mate for life and have an elaborate courtship display that includes head-bowing, wing-flapping, strutting, and tremendous leaps in the air. The female lays two eggs, but only one young crane is raised per year. Young cranes grow quickly and are soon eating frogs, crabs, crayfish, and other small aquatic animals. Whooping cranes rarely eat fish. The young

(Source: Photo courtesy of Texas Parks & Wildlife Department.)

FIGURE 10-12
Whooping crane.

cranes follow their parents on a perilous 2,600-mile journey from the Wood Buffalo National Park in Canada to the Aransas National Wildlife Refuge on the Texas Gulf Coast.

Why Whooping Cranes Became Endangered

The primary factor that endangered the whooping crane was the conversion of its habitat into farmland and cities. Both their nesting habitat in the northern United States and their wintering habitat along the Gulf Coast were converted to cities, farm fields, and oil refineries. The first step in helping the cranes was to protect their remaining habitat—their nesting grounds in Canada, their wintering grounds in Texas, and important staging areas in between. A program to breed whooping cranes in captivity was begun in 1967. Realizing that the cranes raised only one chick of their two eggs, wildlife officials removed the extra egg from several nests. These eggs were sent to the Pawtuxet Wildlife Research Center in Laurel, Maryland, where they were placed in **incubators**. A captive flock was established from these first 12 eggs. Using modern techniques such as artificial lighting and **artificial insemination** scientists were able to increase egg production in the captive birds. By 1989 the captive flock numbered 41 birds. With the entire population of whooping cranes spending the winter at the Aransas National Wildlife Refuge on the Gulf Coast of Texas, there is the potential for a disaster such as disease or a hurricane that wipes out the entire population. With that in mind, scientists began to work toward establishing a population with a different nesting area, migration path, and winter home. It was decided to use sandhill cranes, a species closely related to whooping cranes, as foster parents. Whooping crane eggs were placed in sandhill nests at the Grays Lake National Wildlife Refuge in Idaho. The sandhill cranes hatched the eggs and raised the young, teaching them an entirely different migration route in the process. This new flock of whooper's, 10 to 12 strong, now flies 850 miles to the Bosque Del Apache National Wildlife Refuge in New Mexico. However, this new flock has yet to reproduce.

Even with the great amount of work, time, and expense over many years, the whooping crane population continues to fluctuate. With their low reproductive rate, whooping cranes have a difficult time replacing members that die from disease or, as often happen, from flying into obstacles, such as power lines. However, the number of birds in wild flocks now exceeds 250, and there are another 200 or so in captivity.

SUMMARY

The six species discussed in detail represent a cross section of species that have had problems for a variety of reasons. It is obviously impossible to cover all 1,300 and more species of plants, birds, fishes, mammals, reptiles, mollusks, and other life forms that are currently listed as endangered or threatened. You can obtain a current and complete list of endangered and threatened species by visiting the U.S. Fish and Wildlife Service Web site at <http://www.fws.gov>.

One common factor is present in each endangered species we covered, as well as most others on the list—the influence of humans. Our destruction and alteration of habitat without regard for wildlife survival must stop. If we continue to dam, build on, and bulldoze wildlife habitat, despite the best efforts of wildlife professionals, we will continue to lose species at an alarming rate.

We have made great progress in endangered species identification and management and have seen many species, from the Bald Eagle to the Eastern Gray Kangaroo Rat removed from the Endangered Species List. However, there is still plenty of work to do and many species to save. The Endangered Species Act of 1973 has been a valuable tool in helping to save many species.

REVIEW QUESTIONS

Fill in the Blank

Fill in the blanks to complete these statements.

1. An _____ is any species that is in danger of extinction throughout all or a large portion of its range.

2. _____ of the Earth has greatly accelerated the rate at which species are going extinct.

3. List the five most common factors that cause a species to decline.

(a) _____

(b) _____

(c) _____

(d) _____

(e) _____

4. Most species that survive for long periods have a common trait, _____ .

5. Species that are highly _____ have a much poorer chance of survival.

6. An example of a large mammal that is very adaptable is the _____ .

7. A good example of a highly specialized species is the _____ , which feeds exclusively on the apple snail.

8. Animals that are highly specialized are comparable to people who have only one _____ .

9. Most _____ are capable of adapting to changes in their _____ if those changes take place over an extended period.

10. More than _____ native plants and animals are currently listed as endangered.

11. More than 80 species of _____ , 90 North American _____ , _____ reptiles, _____ amphibians, and _____ species of native fish are listed as endangered.

12. List four of the better-known species that humans have endangered.

(a) _____

(b) _____

(c) _____

(d) _____

13. _____ activities in North America have resulted in several species becoming extinct.

14. Some _____ species, subspecies, and varieties of American plants and animals have been lost forever.

15. Each species is of scientific, _____ , _____ , _____ , and educational value.

16. _____ was discovered in a fungus and since its discovery has saved millions of lives.

17. The anticancer drug Taxol comes from trees, as does _____ .

18. There are _____ species of edible plants, but 90 percent of the world's food comes from fewer than _____ species.

19. List three species whose numbers have increased as a direct result of the Endangered Species Act.

(a) _____

(b) _____

(c) _____

20. List the five major factors involved in protecting a species.

(a) _____

(b) _____

(c) _____

(d) _____

(e) _____

Short Answer

1. What is the purpose of the Endangered Species Act?
2. An animal's adaptability is an important factor in its ability to survive. How does human alteration affect this ability? Why?
3. For each of the species discussed in this chapter, give at least one factor that contributed to the species becoming endangered. Why did this occur? What is being done to save the species?

Discussion

1. In your opinion, is it important that we try to save threatened and endangered species? Why?
2. Select one endangered species. What would you do to save this species?

Learning Activities

1. Choose an endangered species not discussed in this chapter. Using information from your library, state natural resources agency, the U.S. Fish and Wildlife Service, and other sources, write a detailed account of your chosen species. Include such information as preferred habitat, feeding habits, reproduction, and the reasons the species became endangered. Explain what is being done to save the species. Present your paper to the class.
2. Write a report discussing the major factor(s) that have caused species to become endangered in your state. Why has this occurred? What is being done or can be done to slow, stop, or reverse the process? Present your report to the class.

CHAPTER 11

Wildlife Parks and Zoos

OBJECTIVES

After completing this chapter, you should be able to

- Describe the history of zoos and wildlife parks.
- Understand the importance of zoos and wildlife parks in the preservation of endangered species.
- List specific events that led to the development of modern zoos and wildlife parks.
- Identify methods used to protect and propagate endangered species in zoos and wildlife parks.
- Understand the role zoological societies played in the development of zoos in America.

INTRODUCTION

Zoological gardens, more commonly known as zoos, have been in existence in one form or another for many years. Thousands of years ago the rulers of Egypt, China, and Rome kept wild animals. As far back as 1500 B.C. the Egyptian **pharaoh**, Thutmose III, reportedly kept large numbers of wild and exotic animals for his personal pleasure. The Chinese emperor, Wen Wang of the Zhou Dynasty (1027–221 B.C.) called the garden where his animals were kept the Garden of Intelligence. These early zoos were signs of power among world rulers and were often used for the entertainment of special guests. They were also places to study wildlife and nature.

During the Roman Empire (27 B.C.–A.D. 476) the grandiose use of animals for entertainment reached a peak. Roman leaders used wild animals to demonstrate their power. The Romans built enormous stadiums, called coliseums (Figure 11-1), in which animal fought animal, **gladiator** fought

FIGURE 11-1 The Coliseum in Rome, Italy, where gladiators once fought animals.

animal, or unarmed people were thrown to the lions as punishment. Large numbers of animals, such as bears and lions, were killed in fights with each other or against gladiators. Most of these animals were captured from the wild, but some were bred in captivity so that there would always be a ready supply for Roman blood sports.

Not all early rulers used their animal collections without regard for the animal's welfare. The ruler of Macedonia, Alexander the Great, kept large numbers of animals brought to him by the men of his **conquering** armies. Alexander is known to have kept elephants, monkeys, bears, and a large variety of other animals. During the period 336 to 323 B.C., Alexander's armies conquered most of the known world, including India, Persia, and Egypt. Alexander is said to have taken great care of the animals in his collection.

As a result of Alexander the Great's work, the first organized zoo came into being. When Alexander left his collection to King Ptolemy I of Egypt, Ptolemy established what became known as the first organized zoo. This was a culturally rich period in the world's history, and many people became interested in the study of animals. One such individual was the famous Greek **philosopher**, and Alexander's tutor, Aristotle. Aristotle wrote a book based on his observations, called *History of Animals*. In his encyclopedia of **zoology** Aristotle described 300 species of **vertebrates**.

After the fall of the Roman Empire in the fifth century, little attention was paid to nature. During the Middle Ages, collections of animals appear to have been unimportant to people. By the thirteenth century, however, animal collections had again become fashionable. Emperors, kings, and queens often exchanged gifts of animals. Animals were once again kept for the enjoyment of people, although often only for royalty, the rich, and scientific research. As the nobility lost power over the common people in various parts of the world, animal collections became more accessible to all people.

During the early part of the 1700s, organized collections of wild animals became known as menageries. *Menageries* were defined simply as organized collections of wild animals kept in cages. There were many famous and large menageries during the later part of the eighteenth century. Henry I of England had a large collection of animals, as did Louis IX of France. There was a large and famous menagerie in Florence, Italy, in the late 1400s, and Akbar (1542–1605), the third Mogul emperor of India, kept more than 5,000 elephants and a thousand camels.

Early in the nineteenth century increased wealth and the development of large cities led people to begin to preserve natural areas. These factors also gave rise to the development of parkland for recreation and concern for the survival of the natural world. People began to protect and exhibit animals and plants together in parks called zoological gardens or parks. (The word *zoo* comes from the ancient Greek word, *zoion*, which means "living being." From this comes the word *zoology*, or the study of living beings.) People's strong interest in wildlife and nature during this period created an atmosphere in which the Zoological Society of London could be formed.

Zoological societies were formed to plan, raise money for, and build zoological gardens (zoos) in which to display wild animals for the public. Zoological societies are made up of individuals with an interest in animals and their welfare. Members of these early zoological societies raised the money necessary to buy land, construct buildings and cages, collect animals, and hire a staff. Without the help of the Zoological Society of London, it is doubtful that one of the world's more famous zoos, Regent's Park, in London, would have opened in 1828. The Regent's Park Zoo became the model for the zoological gardens that were later developed in the United States.

During the 1800s, as cities grew more crowded, zoological gardens exploded in popularity. For the average city dweller, zoos provided the only relief from the ugly, dirty cities that characterized the period. At some point during the mid-1800s, *zoological garden* was shortened to *zoo*. In all parts of the world, zoological gardens, collections, and menageries have displayed animals in captivity. Today zoos and wildlife parks continue to evolve, along with our knowledge of animals and their needs (Figure 11-2).

FIGURE 11-2 The natural surroundings this rhinoceros enjoys is an example of the continual development of our understanding of how wild animals live.

(Source: © Albo, 2010. Used under license from Shutterstock.com)

EARLY ZOOS IN THE UNITED STATES

Early animal collections in this country differed little from their European counterparts. Traveling menageries were common in the 1700s and early 1800s; they differed little from early circuses. Animals were exhibited in cages, typically made of wood and iron bars, which offered the public a clear view but little comfort for the animals. Like early menageries around the world, deaths were frequent, with poor diet, inadequate housing, and stress being the likely cause of most deaths. As larger cities developed, many of these traveling menageries found permanent homes. New York had menageries as early as 1781. The Central Park Wildlife Center has had a collection of animals since 1861.

In March of 1859 civic leaders and leading members of Philadelphia society formed the Zoological Society of Philadelphia. Many of these citizens had seen the London zoo and felt that a similar facility would be beneficial to Philadelphia. Dr. William Camac was the leading **proponent** of a zoo in Philadelphia. He had traveled throughout Europe and visited several of the leading zoological gardens. It took years of planning and hard work before Dr. Camac's vision could be realized. Finally, in 1872, land was acquired and an architect was hired. Unlike earlier menageries, the Philadelphia Zoo was well planned and organized. On July 1, 1874, the gates to the Philadelphia Zoo opened and approximately 3,000 people strolled through. In 1900 the Philadelphia Zoo again set a trend by opening a research facility for animals associated with the zoo.

The city of Chicago opened its Lincoln Park Zoo to the public in 1868; by 1873 it had more than 50 animals and birds on display. Washington, D.C., had a world-**renowned** natural history museum, the Smithsonian Institution, and a zoological society as early as 1870. The Smithsonian maintained a menagerie of donated animals, which its **taxidermists** used as models. However, it took nearly 20 years for the society to gather support for the National Zoo. Finally the collection at the Smithsonian grew so large that Congress was pressured to provide funding for a National Zoological Garden. In 1889 a site was selected and in 1891 the National Zoo opened to the public for exhibition and study.

Although New York had various menageries for decades, including the one in Central Park, the New York Zoological Society was not formed until the late 1800s. This organization, forerunner of today's Wildlife Conservation Society, went to work immediately gathering support and raising money for a zoological garden outside the city. On November 8, 1899 the Bronx Zoo opened.

Most of the early zoos were a part of a city's park system. By the early 1900s America had numerous large cities. The park system of a major city was considered incomplete without a botanical or zoological garden. By the early 1900s, more than 20 cities had zoological gardens, including the Bronx Zoo, Lincoln Park Zoo, the San Diego Zoo, Zoo Atlanta, and the National Zoo. All these zoos exhibited animals primarily in bare, concrete enclosures with iron bars. This arrangement allowed the public an unobstructed, close-range view of the animals, but conditions for the animals were terrible. These types of enclosures are far from any **semblance** of an animal's natural habitat. Animals were most often exhibited individually, regardless of their natural social groups. Animals that normally lived in herds, troops, or prides, such as antelopes, gazelles, apes, chimpanzees, and lions, did not do well alone. Although the conditions in these zoos were far better than those of the early menageries, they were still far from ideal.

A MORE NATURAL SETTING

A German named Carl Hagenbeck changed the nature and philosophy of zoo exhibits around the world. In 1907, Hagenbeck opened his own zoo, with a radical new approach to the display of animals. Hagenbeck had started as a trapper of wild animals for circuses and zoos and moved on to training animals for circuses. Never satisfied with zoos and circuses of the day, Hagenbeck created his own exotic animal show. He brought together various people, dressed in traditional native clothing and displayed them with the native fauna of their country or region. For example, his European shows might include Laplanders with reindeer, Native Americans with bison and Inuits with polar bears. Hagenbeck's shows were so popular that some 60,000 people visited his traveling exhibit on a single day in 1878, when it was on display at the Berlin Zoo. Because of his extensive travels and vast experience with animals, Hagenbeck developed extensive knowledge of animal behavior. His continued pursuit of a better way to exhibit animals led him to open his own zoo. It was not to be an ordinary zoo, with animals exhibited in rows of cages. Hagenbeck wrote, "I wished to exhibit them not as captives, confined to narrow spaces and looked at between bars, but as free to wander from place to place within as large limits as possible and with no bars to obstruct the view and serve as a reminder of captivity." He originated the concept of using moats around animal exhibits. Because of his experience training animals, he knew the maximum height and distance each species was comfortable jumping. He began to build the Hagenbeck Tierpark in 1890, hiding the moats and walls that confined the animals behind creative landscaping. He designed his zoo to look as if predator and prey were in the same exhibit. Hagenbeck's **innovative** design allowed the public to view a pride of lions, rather than just one lion, with a herd of zebra or gazelles in the background. Although the animals seemed to be sharing the same habitat, they were actually separated by a hidden moat. The public could now see animals in a much more natural setting and could picture what the African landscape might actually look like. The animals had more freedom of movement and more to do, which created a much healthier habitat and atmosphere for the animals. Viewing animals in a more natural setting was a major step in changing the public view of how wildlife should be exhibited (Figure 11-3). There was some resistance to Hagenbeck's idea of **naturalistic** exhibits. Some zoo directors

(Source: © Sankar, 2010. Used under license from Shutterstock.com)

FIGURE 11-3 This polar bear is enclosed by a moat rather than a cage. The moat allows the public to view the polar bear in its natural environment.

FIGURE 11-4 This lion relaxes in the shade and comfort of his manmade naturalistic domain.

felt that the animals would be too far from the public to be seen and appreciated properly. However, most zoos, recognizing the value of Hagenbeck's naturalistic exhibits, rushed to change theirs. Carl Hagenbeck's sons were instrumental in designing the Detroit Zoo and they helped with the Cincinnati Zoo. A team from Chicago went to Hamburg, Germany, to study the Hagenbecks' ideas and used many of them in the construction of the Brookfield Zoo. Many other zoos around the world followed Hagenbeck's example. The result was an improved view for the public and a much better life for the animals. Animals were now free to move about and interact with others of their species. The trend of more freedom for the animals in zoos and a more natural habitat continues today (Figure 11-4).

MODERN ZOOS AND WILDLIFE PARKS

Beginning in the late 1950s the protection and conservation of wildlife began to be the focus of most professional zoos around the world. Wildlife populations in the wild were decreasing rapidly. Growing human populations placed increasing pressure on wildlife resources while available wildlife habitat was destroyed at a record pace. Wildlife was squeezed into smaller and smaller fragments of its former habitat. **Poaching** wildlife for profit became a widespread problem, particularly in Africa and Central and South America. In Africa, the black rhino has been poached nearly to extinction. Thousands of elephants have been slaughtered for their **ivory** tusks (Figure 11-5).

FIGURE 11-5 Ivory tusks intact, this elephant is free to roam in a wildlife park where he will be protected from poachers.

Zoos and wildlife parks have recognized the significant role they could play in helping to preserve many species. In many cases the very survival of a species is at stake due to the very small numbers that were surviving in the wild. Zoos and wildlife parks began to cooperate with each other in the breeding and care of endangered species. Crucial to the success of the effort to save these endangered species has been the leadership role taken by the American Zoo and Aquarium Association (AZAA).

In 1924 the American Association of Zoological Parks and Aquariums (AAZPA) was organized as an affiliate of the American Institute of Park Executives. Forty-two years later, when the National Recreation and Park Association (NRPA) was formed, the AAZPA became a professional branch of the NRPA. In the fall of 1971, recognizing the changing needs of captive wildlife and the professionals who cared for them, the AAZPA separated from the NRPA. In 1972 the AAZPA was incorporated and opened its first office in Wheeling, West Virginia. The AAZPA immediately set the goal of increasing the professionalism of its member **institutions** and of the employees in wildlife parks and zoos. By 1985 all member institutions had to meet the AAZPA's accreditation standards. Members also pledged to follow the Code of Professional Ethics adopted by the AAZPA in 1976. In January 1994 the AAZPA changed its name to the American Zoo and Aquarium Association (AZAA). Today the AZAA represents nearly all the professional zoos, wildlife parks, and aquariums in North America, as well as most of the professional employees in the zoological field. More than 180 AZAA-accredited zoos, wildlife parks, and aquariums are located throughout North America (Figure 11-6).

Beginning in 1981 the AZAA, working in cooperation with its members, developed the species survival plan (SSP). SSPs are cooperative conservation and population management plans for selected species at North American zoos and

FIGURE 11-6 Aquariums are a good means to educate the public about marine and other aquatic life.

(Source: © Tororo Reaction, 2010. Used under license from Shutterstock.com)

aquariums. To maintain a healthy population of a captive species, it is essential to maintain its genetic **diversity**. Genetic diversity can be lost within a closed population, such as that in a zoo, where closely related animals are mated. It also occurs in the wild where fragmented habitat keeps animals from dispersing naturally and small populations result in close breeding. By keeping a studbook for each species, the AZAA and its members maintain a species' genetic diversity by carefully recording an individual animal's history. This practice ensures that each female is mated to the male who is most likely to produce healthy offspring. Today more than 90 SSPs covering more than 120 species are administered by the AZAA. More than 134 million people visited AZAA member institutions in 1998. (For more information on the AZAA, SSPs, or a list of affiliated zoos, aquariums, or wildlife parks in your area, visit the AZAA's Web site <http://www.AZA.org>.)

SUMMARY

The zoos and wildlife parks we enjoy today are far better places for the animals than they were even 50 years ago. In the late 1950s and early 1960s Americans began to understand the value and importance of wildlife and wildlife habitat. This educational process continues today. Hands-on laboratories, computer simulations, interactive technology, and immersion exhibits are some of the means modern zoos use to educate the public. The use of interactive exhibits and auditoriums extends the educational reach of many zoos. When the first television wildlife program, *Wild Kingdom*, aired in 1963, host Marlon Perkins, a zoo director, did not think that they would have enough material for a year of shows. However, the show continued more or less in its original format, until 1991, six years after his death, for a total of 329 episodes. Today there are dozens of animal and wildlife shows on a variety of cable and network stations, many CD-ROM computer programs, and hundreds, perhaps thousands, of Web sites (see Appendix A, B, and C). There is no reason for the average citizen to be uneducated or undereducated about wildlife.

For the past 30 years or so, wildlife parks and zoos have led the effort to breed endangered species in captivity. They have been largely successful, with the captive populations of almost every species increasing. As we move into the twenty-first century, zoos and parks are leading the fight to preserve what remains of many species' habitats. If we are to save most species we must reintroduce them into their former native habitats. As the human population continues to expand, we place tremendous pressure on wildlife and wildlife habitat. In all cases in which wildlife and human life struggle for survival, wildlife loses.

REVIEW QUESTIONS

Fill in the Blank

Fill in the blanks to complete these statements.

1. _____ have been in existence in one form or another for many years.
2. The rulers of Egypt, _____ , and _____ kept wild animals thousands of years ago.
3. Roman leaders used _____ as a demonstration of their power.
4. Not all early rulers used animals without regard for their welfare. _____ is said to have taken good care of his animals.
5. King _____ of Egypt established what became known as the first organized zoo.
6. The early Greek philosopher, _____ , wrote a book called the *History of Animals* based on his observation of animals.
7. During the period known as the _____ , little attention was paid to nature.
8. In the early 1700s, organized collections of animals became known as _____ .
9. The world's first zoological society was the _____ Zoological Society.
10. _____ were formed to plan, raise money for, and build zoos in which to display wild animals for the public.
11. The _____ became the model for zoos that later developed in the United States.
12. _____ had menageries as early as 1781.
13. In March of 1859 the Zoological Society of _____ was formed.
14. Unlike early menageries, the Philadelphia Zoo was well _____ and _____ .
15. In 1891 the _____ Zoo opened to the public, and in 1899 the _____ Zoo opened.
16. _____ changed the nature and philosophy of zoo exhibits around the world.
17. Viewing animals in a more _____ setting was a major step in changing the public view of how wildlife should be exhibited.
18. Beginning in the late 1950s the _____ and _____ of wildlife began to be the focus of most professional zoos around the world.
19. AZAA stands for _____ and _____ .
20. Beginning in 1981 the AZAA, in cooperation with its members, developed the _____ , or SSP.
21. More than _____ million people visited AZAA member institutions in 1998.
22. The nation's first television program about wildlife, _____ , first aired in _____ .

Short Answer

1. How have zoos evolved, particularly in the past 100 years?
2. What factors led to the development of modern zoos?

Discussion

1. Explain how zoological societies contribute to zoo development.
2. In your opinion, what have zoos and wildlife parks done to help save endangered wildlife?

Learning Activities

1. Using additional references as needed, make a chronological list of the dates and events that were important in the development of zoos in America.
2. Select an animal species commonly kept in zoos. How is this species doing in its native country or habitat? If it is endangered, what have zoos done to help preserve it?

Section
TWO

Wildlife and Fish Identification

CHAPTER 12

Large Mammals

OBJECTIVES

After completing this chapter, you should be able to

- Identify the common large mammals found in the United States.
- Describe the characteristics of each of the common large mammals found in the United States.
- Identify the type of habitat where you might find each of these species.
- List at least one major food source for each species.
- Describe some of the behavior traits of each species.

White-Tailed Deer
(Odocoileus virginianus)

DESCRIPTION

White-tailed deer vary greatly in size from state to state and even within different areas of a state (Figure 12-1). The body is usually 4 to 6 feet long with a 7- to 12-inch tail. Body color varies from reddish brown to tan to gray. The underside of the tail is white, and the animals "flag," or raise the tail, when alarmed. Males, or bucks, have antlers that consist of a main beam and tines or points that branch off these main beams. Antlers are shed in late winter each year, and a new set is grown during the spring and summer. Females, or does, do not normally have antlers. Weights also vary greatly, with bucks weighing from 80 to 300 pounds and does weighing from 60 to 150 pounds, with individuals occasionally exceeding these weights. As a general rule the farther north you go the larger the deer are.

Our most numerous hoofed mammal can be found throughout the eastern, the Midwestern, and southern United States, as well as in many river drainages in the west where appropriate habitat can be found. The conservation of white-tailed deer is a tremendous wildlife management success story. Numbering fewer than 500,000 at the turn of the twentieth century, they now exceed 15 million. Habitat destruction and unrestricted market and **subsistence** hunting had greatly reduced their population. Today, they are often regarded as pests in many suburban areas, where they make themselves at home feeding on shrubbery. Deer–vehicle collisions are also increasing at an alarming rate, resulting in many injuries and millions of dollars of damage to vehicles each year.

HABITAT: White-tailed deer are very **adaptable** creatures. They can be found in swamps, forested areas, river bottoms, or the edges of open plains. They have adapted well to the presence of humans, often living in close proximity to settled areas. Whitetails have expanded their range into almost all suitable areas.

FEEDING HABITS: White-tailed deer are **herbivores**. They eat forbs, **browse**, shrubs, mast (acorns), fungi, and grasses during various times of the year. Browse and mast are very important to their diet during the fall and winter months; grasses and forbs are more important during spring and summer.

LIFE CYCLE: The whitetail **rut**, or breeding season, usually occurs in October and November, with **dominant** bucks mating with several does. Bucks establish a **territory** and defend it **vigorously** from other bucks. These fights can be vicious and are, on rare occasions, fatal to one or both of the participants. Does give birth to one or two fawns about 7 months after being bred, normally in May or June. Fawns are spotted at birth, with almost no odor. This helps them avoid **detection** by predators. Within a few weeks the young are able to follow their mothers and escape most predators by running. Mature whitetails have an extraordinary sense of hearing and smell and see movement well. White-tailed deer usually spend their entire lives within a 1-square-mile area, although bucks may wander widely during the rut. Average life expectancy is around 8 to 10 years, though whitetails have been known to live to 16 years. Mortality is often high, especially during the first 6 months of their lives. Coyotes are the major predator, often taking over 50 percent of the fawn crop.

(Source: Photo courtesy of Texas Parks & Wildlife Department.)

FIGURE 12-1 White-tailed deer—doe with fawn.

Mule Deer
(Odocoileus hemionus)

DESCRIPTION

Mule deer are a medium- to large-hoofed mammal, with large ears and a black-tipped tail (Figure 12-2). Their body length is from 4.5 to 6.5 feet, with a 5- to 9-inch tail. Mule deer typically stand around 3 to 3.5 feet tall, with does weighing 150 to 200 pounds and bucks weighing 200 to 300 pounds and up. Mule deer are reddish brown, with a tendency to turn grayish in the winter. Bucks sport antlers that fork into nearly equal branches (unlike the whitetail, which has a main beam and secondary points), which they shed during late winter and re-grow during spring and summer.

Mule deer, named for their large, mule-like ears, tend to be found in less populated areas than whitetails. Mule deer are found throughout the western United States and Canada. Some can be found in the western Plains states, using **draws, coulees**, and river bottoms for habitat. Many are found at higher **elevations** in the western mountain states.

HABITAT: Mule deer are somewhat less adaptable than whitetails, preferring mountain forests, brushy desert areas, wooded hills, and surrounding areas. They are also often seen in open plains areas. Mule deer habitat is more restricted today than when European settlers first arrived.

FEEDING HABITS: Mule deer are herbivores, eating browse, forbs, shrubs, and some grasses. They migrate from higher to lower elevations during the winter to have better access to food.

LIFE CYCLE: Mule deer experience a rut or breeding season similar to whitetails, although their rut may begin a little earlier, especially at the higher elevations. Bucks are polygamous, breeding with as many does as possible. Does normally carry their fawns for 210 days, giving birth to one or two fawns. Fawns are spotted at birth, and like white-tailed fawns, have little odor. They are capable of following their mothers within a short period of time. Coyotes are major predators of the young, as are black bears over much of their range. Cougars, or mountain lions, are major predators of both the young and adult mule deer. Mule deer may live to be 16 years old in the wild, but this is rare, with 8 to 10 probably being about average. Mule deer generally roam over a much larger territory than whitetails.

(Source: Photo courtesy of Texas Parks & Wildlife Department.)

FIGURE 12-2 Mule deer—buck and doe.

Black-Tailed Deer
(Odocoileus hemionus columbianus)

DESCRIPTION

Black-tailed deer are similar to mule deer but with smaller, black tails. They are small- to medium-size hoofed mammals. Their body size is similar to a large whitetail, but bucks have the characteristic branched antlers of the mule deer. Their ears are also larger than those of the average whitetail.

Black-tailed deer were once thought to be a separate species from other mule deer but are now known to be a major subspecies. Blacktail's are primarily found in the western coastal states of California, Oregon, Washington, coastal Canada, and southern coastal Alaska.

HABITAT: Blacktails can be found from the scrub oak canyons of northern California to the coastal forests of Oregon and Washington. They are also found on many of the islands and in coastal areas of Alaska and British Columbia.

FEEDING HABITS: Blacktails are herbivores and heavy browsers, normally feeding on scrubs, bushes, and weeds. In winter they are often forced to eat dead grasses and browse, which has little nutritional value.

LIFE CYCLE: With the exception that they are somewhat more likely to have twins, the blacktail life cycle is very similar to that of mule deer.

Elk
(Cervus canadensis)

DESCRIPTION

Elk are large, hoofed mammals, with bulls having a massive set of antlers, often six or more points to a side (Figure 12-3). Bulls shed their antlers in late winter or early spring each year. Elk have a reddish brown body, which becomes dark brown in the neck area, and a cream-colored rump patch. Bulls sport an impressive set of antlers; cows do not normally have antlers. Body length is usually from 7.5 to 9.5 feet, with a 4- to 8-inch tail. Height at the shoulder is 4 to 5 feet. Bulls may weigh 700 pounds or more but probably average around 500 pounds. Cows are considerably smaller.

The elk is a creature of our western mountains. Formerly found throughout most of the eastern United States and the Plains states, elk are now found in remote, generally mountainous areas of the United States and Canada. Elk have also been reintroduced in some eastern states, such as Pennsylvania, Michigan, and Arkansas, where they were formerly found. They are also known as wapiti, Native American for "white," no doubt referring to the cream-colored rump patch.

HABITAT: Elk prefer semi-open forested areas in mountains, mountain meadows, and foothills. They are found at higher elevations during the summer and generally migrate to lower elevations with the first heavy snowfall. For a large mammal, elk are quite **elusive**.

FEEDING HABITS: Elk are herbivores, eating grasses, forbs, and some brush and bark. They graze more than do whitetails or mule deer. Cows and calves form herds and graze together during the summer.

LIFE CYCLE: Elk have a rutting period that usually begins in September. Bulls "bugle" or call to attract cows and to advertise their presence. Bulls are polygamous and gather a herd or "harem" of cows, which they defend from all other bulls. Cows have a **gestation** period of approximately 8.5 months. Birth of a single calf normally occurs in May or June; twins are rare. Elk calves remain hidden, except when nursing, for about 2 weeks, until they are able to follow their mothers. Once this occurs the cow and her calf rejoin a group of cows and young. Coyotes, black bears, grizzly bears, and cougars prey on young elk. Cougars, black bears, and grizzly bears also prey on sick, weak, wounded, or unwary adults. Elk have been known to live to 14 years of age in the wild, but 8 to 10 years is probably average.

(Source: Courtesy of U.S. Fish and Wildlife Service. Photo by Gary Zahm.)

FIGURE 12-3 Elk—bull.

Moose
(Alces alces)

DESCRIPTION

Moose are very large, hoofed mammals (Figure 12-4). Bulls have huge, **palmate** antlers with multiple points. Cows do not normally have antlers. Medium brown, with upper lip overhanging the lower lip, these are long-legged, massive animals. Body length is 8 to 10 feet, with a short tail of 3 to 4 inches. Height is 5 to 6.5 feet at the shoulder, with bulls weighing in excess of 1,000 pounds in the fall.

Moose are the largest members of the deer family. There are three main varieties or subspecies of moose: the Shiras, Canada, and Alaska-Yukon. The main differences between the varieties are range and size. The Shiras moose is found in the mountains of the western United States, and the Canada is found across much of southern Canada. The Alaska-Yukon is found in Alaska and the Yukon Territory of northern Canada. The Alaska-Yukon is the largest variety, with the Shiras the smallest and the Canada somewhere in between.

HABITAT: Prefers bogs, swamps, river bottoms, and other areas along watercourses or around freshwater lakes. Often found in dense alder or willow thickets and are usually solitary.

FEEDING HABITS: Moose are herbivores and feed on many kinds of **aquatic** plants in water or along lakes or rivers. They consume bark, twigs, and other browse during the winter.

LIFE CYCLE: Moose have a breeding period or rut during the early fall. Bulls are very aggressive and mate with as many cows as possible. Combat between bulls is common. Calves, often twins, are born in the spring, and do not have spots. They are **gangling** and weak at birth and remain hidden for several days. As soon as they are strong enough, calves follow their mothers during her feeding periods. Wolves and bears prey on the young, the sick, and the weak, as well as on the adults. Adults are particularly vulnerable during the winter, when deep snow limits their mobility somewhat.

(Source: Courtesy of U.S. Fish and Wildlife Service.)

FIGURE 12-4 Moose—cow.

Caribou

DESCRIPTION

Caribou are medium-size members of the deer family (Figure 12-5). Unlike other deer, both the males and females have antlers, although those of the females are considerably smaller. They are a medium brown, with a white rump and neck. Body length is 5.5 to 7.5 feet, with a 4- to 5-inch tail. Height at the shoulder is 3.5 to 4 feet, and bulls may weigh 300 pounds or more. Females are usually 20 to 30 percent smaller. The much larger antlers of the males usually have three branches, with the antlers being flattened or palmated to a lesser degree than those of the moose.

Caribou are northern-dwelling species with two major varieties: woodland (*Rangifer caribou*) and barren-ground (*Rangifer arcticus*). The primary difference between the varieties is range and size. Barren-ground caribou are found in the **tundra** regions of Canada and Alaska; woodland caribou live in the forested regions of southern Canada.

HABITAT: The barren-ground variety inhabits the northern tundra of Canada and Alaska. They migrate south in the fall and spend the winter months in the lightly forested areas just south of the tundra. The woodland variety is found in the southern, forested areas of Canada, particularly in Alberta and Newfoundland.

FEEDING HABITS: Caribou are herbivores whose main diet consists of grasses, **lichens**, and browse from a variety of shrubs. They are also known to eat **fungi**.

LIFE CYCLE: Caribou have a rutting period, usually in September, during which bulls breed with as many cows as possible. Bulls gather up a harem of a dozen or so cows and guard them jealously from all challengers. A single calf (occasionally twins) is born in late May or June. Unlike most other members of the deer family, caribou calves are not spotted at birth but solid brown. They are capable of following their mother shortly after they are dry. Wolves are the major predator of both adults and young; however, wolverines and black and grizzly bears take a large number of newborn calves each year. In the wild, caribou rarely exceed 6 to 8 years of age.

(Source: Courtesy of U.S. Fish and Wildlife Service. Photo by Dean Biggins.)

FIGURE 12-5 Caribou.

Musk Ox
(Ovibos moschatus)

DESCRIPTION

The musk ox looks similar to a bison but is much smaller and has considerably longer hair (Figure 12-6). The hair ranges from 6 inches long on the back to 2 to 3 feet along their flank, neck, and chest. Both males and females have horns that curve down, out, and then up. They have a pair of glands, one just below each eye, that give off a strong musky odor when they are disturbed. Body length is from 7 to 8 feet, and height at the shoulder is 4.5 to 5.5 feet. Musk ox bulls weigh from 500 to 900 pounds, with females 20 to 30 percent smaller.

The musk ox is perhaps the **hardiest**-hoofed mammal in the world, certainly the hardiest in North America. They exist on the Arctic tundra, far above tree line. They do not migrate but spend the winters in a habitat that often has temperatures of −50°F.

HABITAT: Musk oxen inhabit the arctic tundra of far northern Canada, Alaska, Greenland, and a few Arctic islands. The landscape is bleak, treeless, and barren. Their home is very **inhospitable**; rarely does the temperature rise above freezing, even during the summer.

FEEDING HABITS: The bulk of the musk ox diet is various grasses and sedge, although dwarf alder, birch, and willow are also important food sources.

LIFE CYCLE: The breeding season typically falls in July and August, with dominant bulls within the group protecting the cows from all interlopers. Fights between bulls can be vicious. Cows give birth to a single calf in April or May. It is a cold, bleak world into which the calf is born. Mortality is quite high. Calves may freeze before they dry and gain their feet. They often huddle under their mother's belly, warmed by her body and long hair. Due to the tough arctic conditions, a cow musk ox may produce a calf only every other year.

The primary predator of the musk ox is the wolf. Adults are seldom attacked, at least not successfully, but calves are vulnerable. Musk oxen have a unique defense system. When danger approaches they form a rough circle, with calves in the center and bulls and cows facing outward. This presents a formidable array of horns to a potential predator, but wolves are noted for their patience and are often successful in snatching a calf or two.

(Source: Courtesy of U.S. Fish and Wildlife Service. Photo by Jerry L. Hout.)

FIGURE 12-6 Musk oxen.

Bison
(Bison bison)

DESCRIPTION

The bison is a very large, hoofed mammal (Figure 12-7). Both cows and bulls have a set of black, upturned horns, although the bulls' horns are usually larger. They are a uniform chocolate brown, with a darker, longer mane from behind the front shoulders forward. Bison have a distinct hump over the front shoulder that is massive and thick. Bulls may be up to 10 feet long and stand 6 feet tall at the shoulder. Bulls weigh 2,000 to 3,000 pounds. Cows are considerably smaller, weighing 700 to 900 pounds and standing 5 feet high at the shoulder.

The bison, or buffalo, as it is more commonly but improperly called, once numbered in the millions. They were a mainstay of Native Americans of the Great Plains for hundreds of years. They were a plains animal, ranging from what is now central Texas to Canada and from the Rocky Mountains to the Mississippi River and beyond.

HABITAT: Bison used to inhabit much of North America, but they were found in largest numbers on the open plains. Today, bison are found on private ranches and various national parks in the western United States and Canada.

FEEDING HABITS: Herbivores, primarily grazers, they feed on a variety of grasses, such as bluestem, gramma, and buffalo grass.

LIFE CYCLE: The breeding season is typically in July and August. Bulls fight furiously over cows and breed as many as possible. A single calf (twins are rare) is born in the spring, usually in May. The calf is able to follow its mother in 3 to 4 days and will not be weaned until 6 to 7 months of age. Bison may live 10 to 20 years in the wild, 30 in captivity. Their only natural predators, before they were exterminated from the bison's range, were grizzly bears and wolves. Wolf packs migrated with, and preyed on, the young, old, and weak bison.

(Source: Courtesy of U.S. Fish and Wildlife Service. Photo by Jesse Achentenberg.)

FIGURE 12-7 Bison—bull.

Pronghorn
(Antilocapra americana)

DESCRIPTION

Pronghorns are a medium-size, hoofed mammal (Figure 12-8). Their basic color is tan, with a white rump patch, belly, and throat. Pronghorns raise or "flare" their rump patches when alarmed, thus warning other pronghorns of danger. Pronghorns are the fastest land mammal in North America. They have been known to run at speeds exceeding 50 miles per hour! Body length is 4 to 4.5 feet and height at the shoulder is 3 to 3.5 feet. Bucks weigh up to 125 pounds. Does are about 20 percent smaller. Both bucks and does have horns, but the does' horns are seldom longer than her ears. The horns are black, with a single, forward-protruding prong. Pronghorns are unique among North American animals in that they shed the outer sheath of their horns each year. The hairs on a pronghorn are hollow and can be raised to allow the body to cool in the summer or held close to insulate against the cold.

The pronghorn is truly an American original. It is found nowhere else in the world. The swiftest land mammal in North America, pronghorns once roamed our prairies by the millions. Though nearly exterminated by settlers and loss of habitat, pronghorns have made a great comeback.

HABITAT: Pronghorns are creatures of the prairies and plains. They have exceptional eyesight to go along with their outstanding speed and hence are well adapted to this habitat.

FEEDING HABITS: Pronghorns are herbivores. They browse and graze on sagebrush and a variety of grasses, such as gramma and wheat. They normally feed in mixed groups of does, fawns, and bucks.

LIFE CYCLE: The breeding season usually occurs during October and November, with the dominant bucks gathering a harem of does, which they vigorously defend. Fights among bucks are common. The doe gives birth to one to three young, often twins, in May or June. Fawns are able to follow their mothers shortly after birth and are able to outrun a coyote within 3 or 4 weeks. The average life span of a pronghorn is around 5 years, but they may live 8 or more years. The only serious predator is the coyote, which has little chance of catching a healthy adult. However, coyotes and golden eagles are known to take fawns.

(Source: Photo courtesy of Texas Parks & Wildlife Department.)

FIGURE 12-8 Pronghorn.

Bighorn Sheep
(Ovis canadensis)

DESCRIPTION

A medium- to large-hoofed mammal (Figure 12-9). Both the rams and ewes have horns; however, the ewes' are short and fairly straight. The horns on a mature ram are large, curving upward from the skull, around behind the ear, generally dropping below the jawline and curving back out and up toward the animal's nose, in a spiraling pattern. Body color is brown to grayish, with a cream to white rump. Body length is 5 to 6 feet, and height at the shoulder is usually 2.5 to 3.5 feet. Rams weigh as much as 250 pounds. Ewes are much smaller.

The most common (and it is not very common) of the bighorn sheep is the Rocky Mountain variety. A smaller subspecies, desert bighorns, inhabit the desert mountains of western North America and Mexico. Competition from more-aggressive introduced species, such as the burro and the wild horse, contributed to a dramatic decline in desert bighorn sheep populations in the western United States.

HABITAT: Bighorn sheep inhabit the western mountains of North America, from Alberta southward to Colorado. The desert variety inhabits the desert mountains of the southwest United States and Mexico.

FEEDING HABITS: Bighorn sheep are herbivores. They graze high mountain pastures during the summer months and move to lower elevations during the winter.

LIFE CYCLE: The breeding season for bighorns is usually in November and December. Rams fight for dominance, running at and crashing head-on into each other. The resulting crash can be heard for great distances in the mountains. In this manner a ram gathers together a small band of ewes and defends them from all challengers. Except during the breeding season rams are normally solitary, but they may form bachelor groups during the late spring and summer. Ewes give birth to a single lamb, occasionally to twins, in May or June. The lambs are precocious and soon follow their mothers all around the mountain. Their normal life span in the wild is 12 to 14 years. Golden eagles and coyotes prey on the young. Mountain lions are a serious predator of young and adults both throughout the bighorn's range. In its northernmost range, wolves are another potential predator.

(Source: Photo courtesy of Texas Parks & Wildlife Department.)

FIGURE 12-9 Desert Bighorn ram.

DALL SHEEP (Ovis dalli)

A pure-white form of mountain sheep, the dall has habits that are very similar to the bighorn. Found in the mountain ranges of Alaska, these sheep are exceptionally beautiful.

Mountain Goat
(Oreamnos americanus)

DESCRIPTION

Mountain goats are stocky, almost pure white, and difficult to view if you are not an eagle (Figure 12-10). Both males and females have short, black horns. The horns on the males are considerably heavier and tend to curve slightly back. Distinguishing characteristics are their long, shaggy white coat and the "beard" under their jaws. Body length is about 5 feet; height at the shoulder is around 3 to 3.5 feet. A big billy may weigh 300 pounds, but is usually lighter. Females are generally 15 to 20 percent smaller.

The mountain goat is not a true goat at all; it is an antelope. Its appearance is quite goat-like, except for its heavy front shoulders and non-spiraling horns, and it behaves much like a goat.

HABITAT: The highest, most **inaccessible** mountain tops in North America. They can be found in the mountains of western Montana, Idaho, and eastern Washington state, northward into Alaska.

FEEDING HABITS: Being herbivores, they graze and browse on scrubs, grasses, sedges, and lichens well above the timberline. Winter snows may force them to lower elevations to forage.

LIFE CYCLE: Breeding season is generally in November, with fights among Billies common and occasionally serious. One to two kids are born in late April, May, or June. One is the normal number of young, but twins are not uncommon. The nanny finds a secluded spot away from the herd to give birth. The young are capable of running within an hour of their birth and generally follow their mothers back to the herd in 3 to 4 days. Avalanches and rockslides are the principal hazards for mountain goats. Their summer homes are inaccessible to most predators, although golden eagles take a kid on occasion. Mountain lions, lynxes, and wolves can take mountain goats when they descend to the more accessible lower elevations during the winter. Life span in the wild is about 10 years.

(Source: Courtesy of U.S. Fish and Wildlife Service. Photo by Dave Grickson.)

FIGURE 12-10 Mountain goat.

Black Bear
(Ursus americanus)

DESCRIPTION

Black bears are not necessarily black in color (Figure 12-11). They may range from jet black to a cinnamon or blonde color and most shades in between. They often have a white blaze or spot on their chests. They have a rolling gait and no hump over their front shoulders. Body length is from 5 to 6 feet; height at the shoulder is 2 to 3 feet. Weight is highly variable, depending on such factors as location, time of year, and sex. Average weight is 300 to 600 pounds for mature boars, although it is not unheard of for a black bear to weigh 800 pounds. Sows and immature bears generally weigh much less.

Our most common and widely **distributed** bear, the species became famous when a black bear was rescued from a New Mexico forest fire in 1950. The bear was called "Smokey," and his picture was used on National Forest Service posters and public service announcements to educate the public about the dangers of forest fires. In 1902 another black bear cub **catapulted** to fame when President Theodore Roosevelt, being the sportsman that he was, refused to shoot it. Instead he adopted it as a pet. A Brooklyn doll maker after receiving permission from the White House, created a stuffed bear modeled after this young cub and named it "Teddy."

HABITAT: Black bears prefer swamps, bogs, forested areas, and mountains and are seldom seen in the open. Once widespread throughout North America, they are now restricted to swamps, deep woods areas, western forested mountains, and remote areas in several eastern and southeastern states. They have been expanding their range in recent years.

FEEDING HABITS: **Omnivorous** and very **opportunistic**, they feed on anything from insects to large mammals and berries to green grass. In the western states they can be a formidable predator of mule deer fawns and elk calves during fawning and calving seasons.

LIFE CYCLE: Bears are generally solitary animals except when a female is in estrus. Black bears are not true hibernators during the winter, as people once thought. Their body temperatures do not drop dramatically and they may awaken and wander around occasionally. The young are generally born in January, when the sow is still in her den. Compared to their 300-pound mothers, they are tiny at birth, weighing around 8 ounces. One to three is the normal number of young. On rare occasions four may be born, and twins are common. The young bears stay with their mothers for 2 years. Natural predators are few, although a boar sometimes kills and eats young and immature cubs. Life span in the wild is about 12 to 15 years.

(Source: Courtesy of U.S. Fish and Wildlife Service. Photo by Mike Bender.)

FIGURE 12-11 Black bear.

Grizzly Bear
(Ursus arctos horribilis)

DESCRIPTION

A massive bear with a pronounced hump over its front shoulders (Figure 12-12). There are many subspecies, ranging from the great brown or kodiak bears (Ursus arctos middendorffi) of coastal Alaska to the smaller grizzly bears of interior Alaska, western Canada, and limited areas of the lower 48 states, primarily in and around Yellowstone National Park. They are generally a uniform brown color, with some older animals having white or silver guard hairs over the shoulder hump, hence the nickname "silvertip." They have a direct, lumbering gait and are as fast as or faster than a fast horse for short distances. Body length is from 6 to 9 feet; height at the shoulder is usually 3 to 4 feet. Weight ranges from 500 to 1,000 pounds, with many of the coastal brown bears weighing well in excess of 1,000 pounds. Sows are generally considerably smaller than the boars.

The grizzly bear was once fairly common throughout the western half of North America. It is one of the largest predators on the continent and has little to fear except humans. When the first settlers arrived, their domestic livestock proved to be easy kills for the bears. Consequently, the bears were hunted to near extinction in the lower 48 states.

HABITAT: Can be found in the remote northwestern mountains of Montana, Wyoming, and Idaho, particularly around Yellowstone National Park and throughout western Canada and Alaska. The brown bear is found in the coastal areas of Alaska and on nearby islands, particularly Kodiak Island. Grizzly bears require remote areas, basically devoid of humans.

FEEDING HABITS: Omnivorous, eating everything from berries to fish to large mammals, such as elk and moose. Very opportunistic, eating whatever is available at the time.

LIFE CYCLE: Grizzly bears are solitary animals, except when a female is in estrus and will therefore tolerate a male. These big bears fight over females. The fights are usually ferocious affairs, with much roaring and noise, but are seldom fatal. The young are born in January or February while the female is still in her den. They are tiny at birth and stay with their mothers at least until their second June and often longer. Twins are the rule, rarely one or three and very rarely four. The only natural predator is other grizzly bears. Boars kill and eat cubs, even their own. Sows put up a tremendous fight to protect their young and are usually successful. However, big boars have been known on rare occasions to kill and eat the sow and her cubs. Life span in the wild is probably around 20 years. They have been known to live 25 to 30 years in zoos.

(Source: Courtesy of U.S. Fish and Wildlife Service. Photo by LuRay Parker.)

FIGURE 12-12 Grizzly bear.

Polar Bear
(Thalarctos maritimus)

DESCRIPTION ■

A large, white to yellowish bear (Figure 12-13). Has a lumbering gait, similar to the grizzlies, but both bears are amazingly quick and agile. Polar bears have large, webbed paws; thick, oily fur; and large, well-muscled shoulders. These **attributes** make them excellent swimmers. They come close to rivaling brown bears in size, being 6.5 to 8 feet long and standing 3 to 4 feet at the shoulder. Average weight is around 600 to 800 pounds, with the females being considerably smaller than the males.

The polar bear is found in the far north. Eskimos have long feared this arctic bear, as it has little or no fear of them.

HABITAT: Rocky shores, ice floes, and islands of the Arctic.

FEEDING HABITS: Polar bears are more **carnivorous** than other bears, feeding almost exclusively on meat. They feed heavily on seals and the washed-up carcasses of whales and walrus.

The diminishing ice pack, due to warmer temperatures, is having an adverse effect on polar bears. These great bears need the ice pack to successfully hunt seals, which constitute the majority of their diet.

LIFE CYCLE: Like most bears, polar bears are loners, accepting companionship only for mating purposes. Most females are bred in June or July, and the young are born in the den in January. The sow may weigh 600 pounds or more, but the cubs rarely weigh more than 2 pounds. The sow usually gives birth to two cubs in a den carved from drift snow or upturned ice floes. The cubs are totally dependent on their mother for the first 2 years of their lives. As in the case of most bears, this allows the females to breed only every other year. Their only natural predator is other polar bears. The males kill and eat the young if possible and have been known to dig a sow and her offspring out of their den to kill and eat them. In the wild polar bears may live to be 20 years old, but this is probably the exception.

(Source: Courtesy of U.S. Fish and Wildlife Service. Photo by LuRay Parker.)

FIGURE 12-13 Polar bear.

Mountain Lion
(Felis concolor)

DESCRIPTION

A large, long-tailed, secretive, and largely nocturnal cat (Figure 12-14). Tan to grayish; only the young have spots. Mountain lions, like most cats, have claws that are at least partially retractable. They are 5 to 8 feet long; of which 2 to 3 feet is tail. The body is long and slender. Mature weight varies from 100 to 200 pounds, with males sometimes exceeding 200 pounds. The cats are fast for short distances and stalk within striking distance of their prey before dashing in to attempt a kill.

Also known by the names catamount, panther, puma, and cougar, among others, the mountain lion is the only large cat found in North America.

HABITAT: Forests, mountains, swamps, and deserts. As a general rule the mountain lion does not do well in close proximity to human settlements. The mountain lion once inhabited much of the United States, but encroaching human civilization reduced its range. In some areas the mountain lion has begun to expand its range, and in areas of encroaching subdivisions human and pet contact with mountain lions are increasing dramatically.

FEEDING HABITS: A **carnivore** that feeds heavily on both white-tailed and mule deer; it also feeds on smaller mammals, such as dogs, and rarely on domestic livestock. Cougars are capable of killing young elk or even adults if the snow is deep.

LIFE CYCLE: These big cats are solitary creatures except when females are receptive to the males' advances. They require vast areas of largely uninhabited space in which to survive. Most young lions are born in April or May, although the adults appear capable of breeding any time of the year. Two to five spotted kittens are born in a den selected by the female. The kittens are at least partially dependent on their mothers for 2 years. Thus, females raise a litter only every other year. Males sometimes kill the kittens, just as male house cats do, so females are very protective of their young. Mountain lions have been known to live for up to 20 years in captivity, but 10 to 12 years is probably more realistic in the wild.

(Source: Photo courtesy of Texas Parks & Wildlife Department.)

FIGURE 12-14 Mountain lion.

Gray Wolf
(Canis lupus)

DESCRIPTION

A large, dog-like predator, with a thick coat of fur and a bushy tail (Figure 12-15). There are a number of subspecies, most notably the endangered red and Mexican wolves. The red wolf formerly inhabited the southeast United States and has been reintroduced in some areas, after a successful captive breeding program. (See Chapter 10 for additional information on red wolves.) Generally found in groups, known as packs, they may be black, tan, gray, white, or any shade in between. A big male might weigh 200 pounds, though the average is 100 to 150 pounds; females are smaller, 3.5 to 5 feet long, with tails of about 16 inches.

One of the most misunderstood and feared predators; the wolf has been exterminated from most of its former range. Wolves have recently begun to expand their range, repopulating small portions of their former ranges in the northern and western United States. This has occurred primarily due to migration from Canada and reintroduction efforts.

HABITAT: Prefers open forests and tundra; needs remote areas with little or no human disturbance. The gray wolf once ranged over most of North America but is now found in northern Minnesota, Michigan, and the northern Rocky Mountain states. Canada and Alaska are its strongholds, where it is still abundant.

FEEDING HABITS: Largely carnivorous, they eat anything from a mouse to a moose, as well as ripe fruits and berries on occasion. They generally hunt as a pack and are capable of attacking and killing an animal as large as an adult moose, particularly in deep snow.

LIFE CYCLE: Very social animals; they may mate for life. The dominant male and female in the pack produce all the offspring. Raising the young appears to be a community affair. Four to six pups are born in a den selected by the female. Like most **canid** young, they are born helpless, with their eyes closed. They have no natural wild predators, although a grizzly bear may take the young if they are left unguarded. As soon as they can keep up, the young follow their parents and other members of the pack as they hunt and so learn to hunt for themselves. Wolf pups are generally on their own at about 1 year of age. Though they may live 15 years, 10 years is more common in the wild.

(Source: Photo courtesy of Texas Parks & Wildlife Department.)

FIGURE 12-15 Gray wolf.

SUMMARY

Large mammals are a vital part of most ecosystems in North America. While some, such as the white-tailed deer, are very adaptable and have adapted very well to human changes in their habitat, many other large mammals have not adapted as well. Large predators in particular, gray wolf, grizzly bear, and mountain lion to name three, require large territories that are relatively free of human activity. If we are to continue to enjoy these wonderful animals we must work hard at preserving their habitat.

REVIEW QUESTIONS

Fill in the Blank

Fill in the blanks to complete these statements.

1. _____ are our most numerous hoofed mammal.
2. White-tailed deer are _____, feeding entirely on plant material.
3. Numbering fewer than 500,000 at the turn of the century, the whitetail population now exceeds_____.
4. White-tailed deer are very _____, being found in a great variety of habitats.
5. _____ are named for their large, mule-like ears.
6. Mule deer experience a _____, or breeding season, similar to whitetails.
7. _____ are a major subspecies of mule deer.
8. The _____ is a creature of the western mountains.
9. Bull elk may weigh _____ or more pounds.
10. Elk tend to do more _____ than do mule deer or whitetails.
11. _____ are the largest members of the deer family.
12. List the two major varieties of caribou.
    (a) _____
    (b) _____
13. The _____ is probably the hardiest hoofed animal in the world.
14. The proper name for the buffalo is _____.
15. The _____ is truly an American original, being found nowhere else in the world.
16. _____ inhabit the highest, most inaccessible mountain tops in North America.
17. Black bears are our most _____ and widely _____ bear.
18. Black bears are _____ in diet and very opportunistic.
19. _____ require large, remote territories that are basically devoid of humans.
20. _____ are large, white to yellowish bears.
21. The _____ is our most common and widespread large cat.
22. Gray wolves are largely _____ and can eat anything from a mouse to a moose.

Short Answers

1. Why have white-tailed deer numbers increased greatly since the early 1900s? How was this increase accomplished?

2. Why are some species, such as white-tailed deer, more widespread than others, such as the grizzly bear?

Discussion

1. In your opinion, what can be done to ensure that a species like the gray wolf, which requires large, uninhabited areas, has a future in America?

2. What large mammals are found in your area? How has human development affected their population and distribution?

Learning Activities

1. Draw a map of your state. Use various colors to illustrate areas known to hold a population of any of the species discussed in this chapter.

2. Select a species from your map and list all the factors, such as weather, available habitat, food, and predators, which might limit the species' population in your area.

CHAPTER 13

Small Mammals

TERMS TO KNOW

camouflaged

conspicuous

ferocious

marsupial

nocturnal

pelage

polygamous

preference

promiscuous

ungulates

vegetarian

OBJECTIVES

After completing this chapter, you should be able to

- Identify the common small mammals found in the United States.
- Describe the characteristics of each common small mammal in the United States.
- Identify the type of habitat where you might find each of these species.
- List common food sources for each species.
- Describe some of the behavior traits of each species.

Cottontail Rabbit
(Sylvilagus floridanus)

DESCRIPTION

Cottontails are, for the most part, a uniform brown, except for their white "cottonball" tail, which is **conspicuous** when they run (Figure 13-1). Body length is 13 to 17 inches, with another 2 inches of tail. Weight averages 2 to 4 pounds.

Our most common and numerous rabbit, the cottontail is found throughout most of the United States, particularly east of the Rocky Mountains. A number of subspecies are found in a wide variety of habitats across North America.

HABITAT: Abandoned farmsteads, field edges, brushy areas; nearly anywhere that provides brushy, dense escape cover.

FEEDING HABITS: Cottontails feed on grasses, forbs, vegetables, herbs, and berries during the summer. They primarily survive on bark and dried vegetation during winter.

LIFE CYCLE: Breeding season may begin as early as February and typically runs through September. Rabbits are quite **promiscuous.** Five to seven young are born after a gestation period of 28 days. The young are born blind and naked but are generally completely independent by 3 weeks of age. Female cottontails may produce four or five litters per year. This prolific rate of reproduction is necessary due to the cottontail's many predators. All meat-eating mammals, particularly foxes, coyotes, and bobcats, prey on cottontails. Various birds of prey and snakes also take a great many cottontails. A cottontail may live to be 3 to 5 years of age, but very few survive their first year. More than half of all cottontails are killed by predators or die from a variety of causes each year.

(Source: Photo courtesy of Texas Parks & Wildlife Department.)

FIGURE 13-1 Cottontail rabbit.

Jackrabbit

DESCRIPTION

Jackrabbits got their name from their very long ears, which resemble those of a jackass (Figure 13-2). They are 17 to 21 inches long, not counting those big ears. Jackrabbits have tails about 4 inches long and weigh from 4 to 10 pounds. The major differences between blacktails and whitetails are their tail color and ranges. Except for a black fringe on their ears, jackrabbits are a uniform brown. Their rear legs are much longer than their front legs. Because of this they have a loping gait and are capable of leaping 20 or more feet in a single bound.

There are two major species of jackrabbits in the United States, the blacktail (*Lepus californicus*) and the whitetail (*Lepus townsendi*). However, neither is actually a rabbit; they are hares.

HABITAT: Blacktails seem to prefer southwestern deserts, prairies, and areas around cultivated fields. Whitetails are mainly found in the north central Plains states, whereas blacktails prefer the dry, short grass prairies of the American southwest.

FEEDING HABITS: Jackrabbits feed on grasses and brush, such as mesquite and sagebrush, as well as alfalfa and other farm crops at various times of the year.

LIFE CYCLE: Jackrabbits are solitary creatures except when mating occurs. After a gestation period of about 45 days, three or four fully furred young are born, with their eyes open. Jackrabbits may live 4 to 6 years in the wild. Coyotes are their greatest enemy, but birds of prey, badgers, bobcats, and foxes eat them if they can catch them. A jackrabbit's main defense is staying hidden. Should this fail, they possess great speed, having been clocked at 45 miles per hour.

(Source: Photo courtesy of Texas Parks & Wildlife Department.)

FIGURE 13-2 Jackrabbit.

Swamp Rabbit
(Sylvilagus aquaticus)

DESCRIPTION ■

Somewhat larger than the cottontail, the swamp rabbit is also a darker brown, with less of a gray tinge to its fur. A close relative of the cottontail, but generally a little larger, it weighs 3 to 6 pounds and is 16 to 18 inches long (Figure 13-3).

HABITAT: As the name implies, swamp rabbits prefer low-lying, poorly drained areas. Swamps, marshes, creeks, and river bottoms throughout the southeastern states are areas where a swamp rabbit might be found. They prefer dense thickets and briar tangles.

FEEDING HABITS: Swamp rabbits eat many types of green grasses and browse during spring and summer. Winter diet appears to consist mainly of forbs, bark, and twigs.

LIFE CYCLE: Peak breeding season is in February and March, but the season extends until at least September. Two or more litters are usually born per season, each containing two to four young. After a gestation period of 39 to 40 days, the young are born in a fur-lined depression, usually under heavy undergrowth. At birth the young are fully furred, but their eyes and ears are not yet open. As with most small mammals that are preyed on heavily, life expectancy is short. Foxes and owls are common predators, and bobcats undoubtedly account for some. Because of the area they inhabit and their tendency to swim streams and rivers, alligators are also important predators. However, floods are probably the swamp rabbit's worst natural enemy.

(Source: Photo courtesy of Texas Parks & Wildlife Department.)

FIGURE 13-3 Swamp rabbit.

Snowshoe Hare
(Lepus americanus)

DESCRIPTION ■

The snowshoe hare has the ability to change color (Figure 13-4). In the summer it is a uniform brown color, but as the days become shorter in the fall, it turns white. This provides the hare with excellent camouflage throughout the year. However, should a sudden early snowfall occur or an early spring melt, it may be caught in the wrong **pelage.** When this occurs, it is very vulnerable to predators because it is more visible.

Although known as the snowshoe hare, the *varying hare* is its more proper name.

HABITAT: Preferring thickets and dense vegetation, it is found in forested areas of the northeastern United States, Canada, and the mountains of the western United States.

FEEDING HABITS: Snowshoe hares eat a variety of grasses, herbs, and green vegetation during spring and summer. They feed heavily on bark during the winter.

LIFE CYCLE: Like all rabbits and hares, they are quite promiscuous and generally have two to three litters each year. Four to six young are normally born after a gestation period of 30 to 38 days. The young are furred at birth and are capable of running on their very first day. The young hares are weaned and usually on their own at 1 month of age. Most hares do not live out the year, instead becoming food for a wide variety of predators. However, a rare few may live to be 3 years of age. Common predators include the lynx, bobcat, foxes, and great horned owls.

(Source: © Nialat, 2010. Used under license from Shutterstock.com)

FIGURE 13-4 Snowshoe hare in winter pelage.

Red Squirrel
(Tamiasciurus hudsonicus)

DESCRIPTION

The red squirrel is generally a rusty red color, except during the winter, when its coat turns a more subdued brown (Figure 13-5). It is a small creature, weighing only 7 to 9 ounces and measuring 14 inches long, including a 6-inch tail. A very active squirrel, it gives an almost birdlike chirp when alarmed. It is quite vocal, scolding intruders in its territory.

HABITAT: Evergreen forests from Newfoundland to Alaska and southward into the northern Rocky Mountains and Appalachian Mountains in the east. Two western species, the Douglas squirrel of the northwest and the pine squirrel of the Rocky Mountains, do not have as extensive a range.

FEEDING HABITS: They eat a wide variety of nuts, such as acorns, walnuts, hickory, and beechnuts. They also feed on mushrooms and pine cones.

LIFE CYCLE: These squirrels usually have a single litter of young in April or May. The gestation period is approximately 40 days and the young are normally born in a hollow or abandoned woodpecker hole. A leaf nest may be built but is not preferred. Three to five young are born blind and naked, but within 6 to 8 weeks are weaned and venturing out on their own. Nearly every predator enjoys squirrel for dinner—hawks, foxes, owls, snakes, bobcats, lynxes, martins, and fishers all prey on red squirrels. The martin and the fisher are probably its most deadly enemies, being good climbers and fast enough to run a squirrel down. Red squirrels that avoid becoming dinner may live up to 10 years.

(Source: Courtesy of U.S. Fish and Wildlife Service. Photo by Donna Dewhurst.)

FIGURE 13-5 Red squirrel.

Eastern Gray Squirrel
(Sciurus carolinensis)

DESCRIPTION ▣▬

One of our best known and most visible small mammals (Figure 13-6). Gray squirrels are common in many city parks and neighborhoods. Most gray squirrels are gray, as the name suggests, but all-white and all-black color variations are fairly common. Gray squirrels are less nervous than reds and can become quite tame around humans. They weigh from 0.75 to 1.75 pounds and are about 18 inches long, including an 8- to 9-inch tail.

HABITAT: Gray squirrels prefer a mixture of hardwood and evergreen forests. They are particularly fond of oak trees throughout their range, which extends from the Dakotas to Maine and from the Gulf of Mexico to the St. Lawrence River.

FEEDING HABITS: They feed on acorns, hickory, and chestnuts; they also eat fruits, birds, if they can catch them, and baby birds during spring and summer.

LIFE CYCLE: These squirrels may breed year-round, especially in their southern range. They are promiscuous, with several males usually attempting to breed each receptive female. After a gestation period of 40 days or so, two to four blind and naked young are born. The young remain in the nest for about 6 weeks, until they are able to eat solid foods. Although some may live to be 15 years old, the majority do not. Foxes, bobcats, house cats, dogs, martins, hawks, owls, fishers, and snakes all eat gray squirrels whenever they can. This helps keep the squirrels from overpopulating their habitat.

(Source: Courtesy of Photodisc.)

FIGURE 13-6 Gray squirrel.

Eastern Fox Squirrel
(Sciurus niger)

DESCRIPTION

The eastern fox squirrel varies from rusty red to gray to black (Figure 13-7). They are the largest member of the squirrel family in the United States, weighing from 2 to 3 pounds and measuring up to 28 inches in length. Fox squirrels range throughout the eastern half of the United States. They tend to use large leafy nests more than other squirrels.

HABITAT: Has definite **preference** for oak forests. They prefer more open woodlands to dense forest.

FEEDING HABITS: The fox squirrel is a true acorn lover, but it eats other nuts and seeds, as well as baby birds eggs and berries in season. Fox squirrels spend more time on the ground foraging than do most other squirrels.

LIFE CYCLE: Fox squirrels are somewhat more solitary than other squirrels, preferring to live alone except during the breeding season, which usually starts in January. About 45 days after mating occurs, three or four blind and naked young are born. One litter per year is the norm, but in its southern range it may raise two. The young may rely on their mother for up to 10 weeks. Although it is possible for a fox squirrel to live 10 or more years in the wild, few actually do. Snakes are serious predators of young squirrels, as are hawks and owls. Foxes and bobcats eat young and adult alike. Raccoons have also been known to take young squirrels from their nests.

(Source: Courtesy of U.S. Fish and Wildlife Service. Photo by W. H. Julian.)

FIGURE 13-7 Fox squirrel.

Red Fox
(Vulpes fulva)

DESCRIPTION ■—

Most numerous and widespread fox in America (Figure 13-8). Red Foxes are characterized by light to very dark red coat and long bushy tail. The body and head may be 30 inches long, with the tail adding another 14 inches. They typically weigh 8 to 10 pounds, with females being somewhat smaller than males. Red foxes have bright yellow eyes, pointed ears, and a long narrow muzzle. They range across most of North America, being absent only from the far Arctic, the deserts of the southwest, and a few plains areas.

HABITAT: Open woodlands, brushy areas, suburbs, and treeless areas.

FEEDING HABITS: Red foxes eat nearly anything they can catch and kill, but they prefer pheasants, quail, grouse, ducks, rabbits, frogs, snakes, mice, and rats. They also eat domestic poultry if the opportunity arises.

LIFE CYCLE: Foxes typically breed in February, and six to eight pups are born approximately 56 days later. The male remains with the female and assists in raising the pups. Pups are normally born in an old woodchuck burrow, or in a burrow dug by the female, or vixen, if the soil is soft and loose enough. The pups are born fully furred, but their eyes do not open until 9 to 10 days after birth. The pups grow rapidly and begin to venture out of the den at about 3 weeks of age. As the pups grow and mature the parents bring some prey back that is still alive. In this manner the young learn to kill for themselves. In late summer or early fall the family breaks up and each individual fends for itself. Foxes may live to be 15 years old, but this is rare. Distemper, encephalitis, and rabies take a number of foxes each year, and periodic outbreaks can kill large numbers. Young foxes are also preyed on by owls, hawks, and bobcats. The coyote, lynx, wolf, and cougar may take young or adult.

(Source: Photo courtesy of Texas Parks & Wildlife Department.)

FIGURE 13-8 Red fox.

Gray Fox
(Urocyon cinereoargenteus)

DESCRIPTION

The gray fox is often mistaken for the red because it may have quite a bit of red hair (Figure 13-9). However, its normal color pattern is salt-and-pepper gray. The gray fox is more compact, shorter legged, and slightly shorter bodied than the red. The gray fox has a black-tipped tail that is about 12 inches long. They weigh from 8 to 10 pounds, with a body length of about 40 inches including the tail. Females are usually slightly smaller than males. They range from Central America up through Mexico and into the southern and northeastern United States.

HABITAT: Open woodlands, fields, stream and river bottoms, and western plains. Gray foxes climb extremely well and often take refuge in trees.

FEEDING HABITS: They eat many kinds of small birds, mice, snakes, rabbits, frogs, most small mammals, and insects. They are also known to eat carrion, fruits, berries, and grains.

LIFE CYCLE: Breeding normally occurs in January, with three to five pups being born about 2 months later. Dens are usually in a hollow log, tree, or rocky area. Pups are blind but fully furred at birth and are weaned at about 60 days of age. Both parents participate in the rearing of the young, although the male tends to wander off in late summer, while the female remains with the young until fall. Natural enemies include coyotes, which kill young and adults if they can catch them. Young foxes are also preyed on by dogs, owls, and bobcats. Gray foxes may live to be 10 to 12 years of age in captivity, but their life span in the wild is probably considerably shorter.

(Source: Photo courtesy of Texas Parks & Wildlife Department.)

FIGURE 13-9 Gray fox.

Swift Fox
(Vulpes velox)

DESCRIPTION

This is the smallest of the American foxes, as well as the fastest, hence its name (Figure 13-10). It has unusually large ears and a slender grayish to buff yellow body. The swift fox is shy and more retiring than other foxes. They weigh from 3 to 6 pounds and are about 24 to 36 inches long, not much larger than an average house cat. They range across western North America, from Mexico to southern Canada.

HABITAT: Prefers short grasslands and prairies. Utilizes old burrows, but prefers areas with loose, sandy soils.

FEEDING HABITS: They largely feed on rabbits, mice, and other rodents. Insects, snakes, and lizards are also eaten, as well as ground-nesting birds, their eggs, and their young.

LIFE CYCLE: Swift foxes in the southern portion of their range may give birth as early as February, but birth occurs later in more northern areas. Three to five young are normal, but as few as two and as many as seven may be born. The young are weaned at about 10 weeks. Both the male and the female bring food for the pups, and they grow rapidly. Swift fox numbers have declined in recent years, probably because of the spread and proliferation of coyotes.

(Source: Photo courtesy of Texas Parks & Wildlife Department.)

FIGURE 13-10 Northern swift fox.

Spotted Skunk
(Spilogale putorius)

DESCRIPTION

The spotted skunk is the smallest and most agile of our skunks (Figure 13-11). Rather than actual spots, it has narrow, broken bands of white around a black body, which gives it a spotted appearance. It has a spot under each ear and another in the middle of the forehead. Spotted skunks weigh around 2 pounds and are 20 to 22 inches long, including a 9-inch tail. Females are generally a little smaller than males. They range across Central America, Mexico, and most of the southern two-thirds of the United States, except the northeast.

HABITAT: Dry desert areas, woodlands, river and stream bottoms, and prairies.

FEEDING HABITS: Spotted Skunks are omnivores, eating rats, mice, snakes, lizards, insects, fruits, berries, and chickens. Has a strong preference for the eggs of most ground-nesting birds, as well as chicken eggs.

LIFE CYCLE: Male spotted skunks are **polygamous** and mate with as many females as possible during a late winter breeding season. Five to eight young are born in early spring in a hollow log or other suitable den. They are born with their eyes closed and with hair so fine it is almost invisible. The young can follow their mother around on foraging trips at about 30 days of age. The young are weaned at about 2 months. Spotted skunks raise their tails and stamp their front feet if threatened. If pushed, they spray the characteristic strong musk, which may even temporarily blind a potential assailant. Owls, dogs, foxes, bobcats, and badgers have all been known to kill skunks. However, due to their nocturnal habits and poor sense of smell, great horned owls are probably the chief predator. Spotted skunks have been known to live up to 6 years.

(Source: Photo courtesy of Texas Parks & Wildlife Department.)

FIGURE 13-11 Spotted skunk.

Striped Skunk
(Mephitis mephitis)

DESCRIPTION

The striped skunk is the most common, widespread skunk found in North America (Figure 13-12). They are found from central Mexico to central Canada and throughout the lower 48 states. They are larger than spotted skunks, weighing 6 to 14 pounds and being around 30 inches long. Contrary to what their name implies, striped skunks may be almost solid black or largely white. However, the more common color pattern is two parallel white stripes, starting just behind the head and running the length of the black body. Another color pattern is stripes connected in a wide, white band down the center of the back.

HABITAT: Striped skunks can be found almost anywhere; they are well adapted to most habitats.

FEEDING HABITS: They feed on insects, the eggs and young of ground-nesting birds, mice, rats, and domestic poultry. Also known to dig up turtle nests and eat the eggs.

LIFE CYCLE: Like spotted skunks, striped skunks have a late winter breeding season. The males mate with as many females as possible. Three to eight young are born about 60 days later. The baby skunks' eyes open at about three weeks and by five weeks they are following their mother on her nightly forays for groceries. Captive skunks may live 10 to 12 years, but skunks in the wild rarely live that long. Coyotes, bobcats, badgers, eagles, and owls take a toll on the skunk population. However, most predatory mammals, with their keen sense of smell, are not overly interested in risking a spraying to get a skunk dinner. In contrast, great horned owls, with their apparently poor sense of smell, take a large number of skunks. Disease may also kill skunks; for example, in some years rabies infects and kills large numbers of skunks.

(Source: Courtesy of U.S. Fish and Wildlife Service.)

FIGURE 13-12 Striped skunk.

Coyote
(Canis latrans)

DESCRIPTION

A medium-size, slim, dog-like predator, the coyote has done extremely well, despite human efforts to control it (Figure 13-13). Color varies from reddish brown to dark gray. Coyotes have an excellent sense of smell and hearing and good eyesight. Length of body is around 48 inches and weights range from 25 to 40 pounds. Coyotes can breed with dogs, however, and the resulting crosses may be considerably larger. Coyotes were once a western species, but they have expanded into virtually all areas of the United States and northward into Canada and Alaska. Coyotes are extremely fast and may run at speeds of up to 40 miles per hour for short periods.

HABITAT: Virtually anywhere it can secure food—from deserts to woodlands to prairies. Coyotes can, and often do, live in suburbs, surviving on pet food and pets. The Coyote is one of the most adaptable species in North America.

FEEDING HABITS: Rodents, rabbits, deer (especially fawns), birds, domestic livestock (particularly sheep), fruits, and berries are all eaten by the coyote. They seem to relish domestic cats, where they are available.

LIFE CYCLE: Coyotes normally have one litter of 6 to 10 pups per year. The breeding season normally occurs in January, and the pups are born after a gestation period of 63 days or so. Coyotes, as well as most predators, have larger litters when food is abundant. The pups are born fully furred but with their eyes closed. They are weaned at about 6 weeks and begin to chase and kill rodents that the parents bring back to the den alive. They learn to hunt from their parents, and the family normally does not break up until early fall. Wolves, bobcats, lynxes, bears, and golden eagles may kill young coyotes, but humans are their greatest predator. Diseases, such as rabies and distemper, parasites, and mange probably kill more coyotes than anything else. Coyotes have lived up to 15 years in captivity.

FIGURE 13-13 Coyote.

(Source: Photo courtesy of Texas Parks & Wildlife Department.)

Wolverine
(Gulo gulo)

DESCRIPTION

The wolverine is the largest member of the weasel family and was once found throughout the northern United States (Figure 13-14). Its range is now restricted to northern Canada, Alaska, and the Rocky Mountain states. The wolverine has a rich dark brown fur with a light tan stripe down each side. For its size, it is one of the strongest animals alive. In the wild they weigh only 25 to 35 pounds but have sharp teeth and claws and can be aggressive. They have earned the nickname of "glutton" because they are voracious eaters, most probably an adaptation to frequent food shortages.

HABITAT: Inhabits forested areas, tundra, and northern woodlands. Wolverines prefer remote, uninhabited areas.

FEEDING HABITS: The wolverine will eat anything, but marmots, mice, rats, rabbits, gophers, birds, and bird eggs probably make up the bulk of their diet. They are aggressive in nature and have been known to run bears off their kills and claim them.

LIFE CYCLE: Wolverines are solitary creatures, getting together only during a short mating season in late summer or early fall. Due to delayed implantation, the young are not born until the following spring. One litter per year, consisting of two to three young, is the norm. The young are nursed by the mother until late summer and may remain with her for 2 years. Due to their strength and **ferocious** nature, wolverines have no natural enemies and may live up to 16 years.

(Source: © Geoffrey Kuchera, 2010. Used under license from Shutterstock.com)

FIGURE 13-14 Wolverine.

Badger
(Taxidea taxus)

DESCRIPTION

The badger's hair coat is coarse and sandy-brown, with darker head, feet, and lower legs (Figure 13-15). Badgers have a distinct white stripe that starts at the back of the nose and runs back to the front shoulders. A white cheek stripe runs up behind the eyes. Adults normally weigh 15 to 25 pounds and are about 30 inches long. Badgers are powerful members of the weasel family and are excellent diggers, with sharp claws and short, powerful legs. Badgers range from northern Mexico, throughout the western United States, eastward into the Midwest and southern Canada.

HABITAT: Open plains and up into the mountains of the west, preferably open to semi-open areas.

FEEDING HABITS: Eats rodents of all kinds, particularly those living in ground burrows, such as ground squirrels. Also eats snakes, insects, carrion, eggs, and birds.

LIFE CYCLE: Like wolverines, the badgers' reproductive cycle involves delayed implantation. They mate in late summer or early fall and give birth to two to four young, usually three, the following spring. Badgers are solitary animals outside of the breeding season. On rare occasions the male may help to rear the young. The young are born in a den deep in the ground and do not open their eyes for about 6 weeks. Sometime after 2 months of age the young are weaned and begin to follow their mother on nightly hunting trips. In early fall or late summer the young wander off on their own. Badgers are fierce fighters, but they are occasionally preyed on by wolves, cougars, or bears. They have been known to live up to 12 years in captivity.

(Source: Photo courtesy of Texas Parks & Wildlife Department.)

FIGURE 13-15 Badger.

Porcupine
(Erethizon dorsatum)

DESCRIPTION

One of the largest rodents in North America, the porcupine may weigh up to 40 pounds but averages 15 to 20 pounds (Figure 13-16). Mature adults are about 36 inches long; of which 6 inches is tail. Porcupines are covered with over 30,000 needle-sharp, barbed quills. The belly, face, and underside of the tail are the only quill-less areas. Porcupines cannot actually shoot their quills, as is commonly thought, but they do whip their tails around in an attempt to impale an aggressor. The quills are tapered where they enter the skin and thus are easy to detach. However, because they are barbed they are very difficult to remove once in a victim. Porcupines also have a dense undercoat, with a layer of longer guard hairs

HABITAT: Porcupines inhabit forested areas from Alaska south to New Mexico, from the eastern edge of the Rockies westward almost to the Pacific coast, and from the west coast of Canada to Newfoundland and the northeastern United States. They spend a great deal of time in trees.

FEEDING HABITS: They feed on leaves, bark, and twigs and are entirely **vegetarian.**

LIFE CYCLE: Breeding season is in November, with one offspring born in March. The young are born with a full set of soft quills that harden within 30 minutes or so of birth. Baby porcupines begin to feed themselves at one week and can climb by themselves when only a couple of days old. Although most predators avoid porcupines, with good reason, they are still killed and eaten by fishers, wolverines, wolves, lynxes, mountain lions, and coyotes. Most predators attempt to flip the porcupine onto its back to get at its soft, unprotected belly. However, the incautious predator may get a face or mouth full of quills, making it unable to catch prey or eat, and thus die of starvation. Average life span for porcupines is 6 to 8 years.

(Source: Photo courtesy of Texas Parks & Wildlife Department.)

FIGURE 13-16 Porcupine.

Opossum
(Didelphis virginiana)

DESCRIPTION

The opossum is the only **marsupial** found in North America (Figure 13-17). It has a darkish undercoat under long, silver guard hairs, which gives it a salt-and-pepper look. The face is white with a darker stripe running up the forehead from between the eyes to between the ears. The ears and 12- to 13-inch tail are hairless. The tail is scaly and capable of supporting its weight if necessary. Opossums may weigh as much as 12 pounds and may be up to 36 inches long, including the tail. Opossums climb well.

HABITAT: They prefer wooded areas, old farmsteads, and brushy areas. Found throughout the eastern half of the United States, along the west coast, and down into Mexico and Central America.

FEEDING HABITS: Fruits, nuts, berries, insects, bird eggs, and any meat it can catch or scavenge.

LIFE CYCLE: Female opossums give birth from April to September and may have up to three litters of 8 to 18 young each. The young are born only 13 to 14 days after conception. They are incubated in their mother's fur-lined pouch for about 6 weeks. The newborn opossums must make their way to the mother's pouch and attach themselves to a nipple. Once out of the pouch, young opossums often ride on the mother's back by clinging to her fur. Opossums are probably best known for feigning death or "playing possum" when threatened by an enemy. However, they are just as likely to run for the nearest tree or turn around and bite at the intruder. Coyotes, cats, and dogs all take a toll on opossums. In the wild an opossum rarely lives past 7 years of age.

FIGURE 13-17 Opossum.

(Source: Photo courtesy of Texas Parks & Wildlife Department.)

Raccoon
(Procyon lotor)

DESCRIPTION

Medium-size, robust predators, raccoons occupy most of the United States (Figure 13-18). They have a distinctive blackish facial mask that is outlined in white. The bushy tail has alternating black and buff (whitish) rings around it. The general body color is a grayish, dark-brown mix, often with light-tipped guard hairs. Raccoons normally weigh from 15 to 20 pounds but may weigh up to 60 pounds. They average 25 to 38 inches in length, including a 10-inch tail.

HABITAT: Raccoons are very adaptable. They can be found almost anywhere from Central America to southern Canada, with the exception being certain areas of the mountains in the western United States. They prefer to live in relatively close proximity to water, such as rivers, streams, lakes, or storm drains. They readily adapt to suburbia and are often found within the city limits of large towns.

FEEDING HABITS: Raccoons are true omnivores, eating frogs, berries, fish, small mammals, nuts, vegetables, shellfish, grain, birds, and bird eggs. Raccoons are serious predators of all ground-nesting birds, particularly waterfowl, and destroy many nests each year.

LIFE CYCLE: The breeding season typically runs from late January until March, with males mating with as many receptive females as they can find. Two to six young are born after a gestation period of around 64 days. The young are born in a den in a hollow log, hollow tree, or other site the female has selected. The young are born fully furred but with their eyes closed. Their eyes begin to open at about 18 to 20 days of age, and the young raccoons venture out of the den at about 6 weeks of age. They begin to follow their mother on foraging trips as soon as they are strong enough. In late summer or early fall the young raccoons become independent and wander off on their own. Raccoons may live for 14 years, but predators such as coyotes, bobcats, and great horned owls take a fair number of young raccoons. However, few predators will take on the adults, as they are fierce fighters.

(Source: Photo courtesy of Texas Parks & Wildlife Department.)

FIGURE 13-18 Raccoon.

Collared Peccary
(Tayassu tajacu)

DESCRIPTION

The collared peccary is a pig-like creature that is not actually related to pigs at all. Peccaries are of the family Dicotylidae, whereas pigs are of the family Suidae. Collared peccaries are native to the desert southwest and Mexico. Their range is more restricted today than in past years. Today they are primarily found in south and west Texas and southern New Mexico and Arizona. They have stiff, bristle-like hair that is gray to salt-and-pepper black in color. Adults have a distinct white collar or stripe around the neck. Peccaries also have musk glands along their backs. When excited they often raise the hair along their backs and give off a strong odor. They have a short, pig-like snout, crushing molars, and almost straight, dagger-like canines, similar to pig tusks. They are thought to have rather poor eyesight but

HABITAT: Javelinas are most often found in desert or desert-like habitats. They prefer habitat with an abundance of prickly pear cactus and are often found in thick stands of scrub oak, chaparral, guajillo, or prickly pear.

FEEDING HABITS: Collared peccaries are very opportunistic feeders, eating everything from snakes to cactus. They are mainly herbivorous, feeding on prickly pear, sotol, mesquite beans, and other succulent plants. However, they are known to eat snakes and the eggs of ground-nesting birds. Javelinas seldom need standing water because the succulent cactus they consume provides most of their water.

LIFE CYCLE: The males are known as boars and the females as sows. Sows generally have two young but may have as few as one or as many as five. Javelinas are one of the few wild ungulates to breed year-round. The gestation period is 142 to 145 days, or about 5 months. The young are reddish to yellowish brown, with a black stripe down the back. They are capable of running within hours of birth. This is a very important survival trait, as javelinas are killed and eaten by a wide variety of predators, including bobcats, coyotes, jaguars, and black bears, where their ranges overlap. Peccaries typically travel in groups ranging from a few animals to 20 or more. Sows leave the group to give birth but rejoin as soon as the young can keep up, usually within a few days. Sows, and the group as a whole, are very protective of the young and defend them from predators if at all possible.

an excellent sense of smell and very good hearing. They are most active during early morning hours and again in the late evening. They often bed down in dense thickets of brush during the heat of the day. Collared Peccaries are also known as javelinas.

Muskrat
(Ondatra zibethica)

DESCRIPTION

This relative of the beaver is found virtually wherever water is available, from Mexico to Alaska (Figure 13-19). Adults average 25 inches in length, including a 10-inch tail, and weigh about 3 pounds. The muskrat is covered with dense brown fur with lighter underparts and coarse guard hairs. Their tails are hairless and appear rat-like. Muskrats are strong swimmers and make lodges in the water, with underwater entrances, or burrow into stream and riverbanks.

HABITAT: Muskrats are always found near water. Their typical habitat is marshes, swamps, and riverbanks.

FEEDING HABITS: Muskrats eat a variety of aquatic plants, freshwater mussels, and fish. They have also been known to invade gardens for fresh vegetables.

LIFE CYCLE: Female muskrats bear two to four litters per year, with an average of three. Four to eight young are born blind and almost hairless. Their eyes open at about 2 weeks of age, and they are eating vegetation at 3 weeks. Shortly after they begin to eat vegetation they leave the family lodge and go off to find a home of their own. Young muskrats are preyed on by a wide variety of predators, from hawks and owls to alligators. Northern pike are also known to eat young muskrats, whereas adults are killed by bobcats, coyotes, foxes, and a variety of other predators. Normal life span is 5 to 6 years.

(Source: Courtesy of U.S. Fish and Wildlife Service.)

FIGURE 13-19 Muskrat.

Beaver
(Castor canadensis)

DESCRIPTION

The largest rodent in North America, the beaver is widespread, found throughout the lower 48 states, as well as subarctic Canada and Alaska (Figure 13-20). Noted for its dam-building expertise, the beaver uses its broad flat tail as a rudder in the water. They have a rich dark brown or chocolate coat of dense, thick fur. Adults weigh from 50 to 65 pounds and measure about 42 inches long, of which 12 inches is tail. Beavers are superbly adapted to their habitat, being excellent swimmers and divers and capable of staying under water for up to 15 minutes.

HABITAT: Rivers and streams in areas with woodlands. Most watercourses have at least a narrow band of trees along their banks, which the beaver can put to good use.

FEEDING HABITS: Beavers feed on a variety of aquatic plants, such as water lilies, as well as the inner bark of willow, aspen, and birch.

LIFE CYCLE: Beavers mate for life. The female gives birth, usually in April, to one to four young after a gestation period of about 4 months. An average beaver household consists of the parents, the previous year's young, and the most recent offspring. At about 2 years of age the young beavers leave to seek mates and their own lodge. Beaver young are born fully furred and open eyed and can swim when only hours old. The young beavers are weaned at about 1 month and begin to feed on vegetation as soon as they leave the lodge. Wolves, coyotes, bears, wolverines, and bobcats prey on beavers, should they catch them on land. However, adults are formidable, being equipped with a large set of teeth, and can be aggressive if cornered. Death by accidents has also been recorded, most commonly from falling trees the beaver has cut with its large incisors. Life span in the wild is around 15 years.

(Source: Courtesy of Photodisc.)

FIGURE 13-20 Beaver.

Long-Tailed Weasel
(Mustela frenata)

DESCRIPTION

Various species of weasels are found throughout North, South, and Central America (Figure 13-21). The most common and widely distributed is the long-tailed weasel. The long-tailed weasel is 12 to 20 inches long, of which 4 to 6 inches is tail, and weighs about 12 ounces. Females are considerably smaller than the males. Coat color is dark brown, with off-white underparts and a black-tipped tail. Except for the black-tipped tail, the coat of the long-tailed weasel in its northern range turns white during the winter. The long, low-slung body of the weasel gives the appearance of a snake with legs.

HABITAT: Woodlands, fields, abandoned farmsteads—almost anywhere their prey can be found.

FEEDING HABITS: Weasels eat mice, rats, and chipmunks and seem to have a special fondness for domestic poultry. Weasels are well known for killing more than they need, often killing every chicken in the pen.

LIFE CYCLE: Mating normally occurs in late summer or early fall; the following spring three to six young are born, often in an old mole burrow. The young do not open their eyes until they are around 5 weeks old, even though they have been weaned a week or two earlier. The male normally assists in raising the young. Sometime around 7 or 8 weeks of age, the young begin to hunt for themselves, but they remain with their mother until late summer. Although the long-tailed weasel may live to be 5 years old or so in the wild, this age is probably rarely obtained. Nearly all birds of prey, as well as house cats, mink, bobcats, lynxes, and fishers kill and eat weasels.

(Source: © Brian Sallee, 2010. Used under license from Shutterstock.com)

FIGURE 13-21 Long-tailed weasel.

Mink
(Mustela vison)

DESCRIPTION ■

The mink is a chocolate to almost black member of the weasel family (Figure 13-22). They inhabit the majority of the United States, with the exception of the southwest, and much of Canada, except the far Arctic. They average 20 to 30 inches in length, of which about 8 inches is tail. A large male may weigh 3.5 pounds. Females are usually only about half the size of the males. Like all members of the weasel family, mink have scent glands. The mink's glands are capable of omitting an acrid scent, should they be frightened, that is said to be even more offensive than the skunk's.

HABITAT: Preferred habitat is woodlands, near lakes, rivers, streams, or another water source. They prefer smaller streams to large rivers.

FEEDING HABITS: Mink prefer to hunt around water, eating frogs, salamanders, fish, waterfowl and their eggs, and snakes. They also eat small mammals such as rodents, rabbits, muskrats, as well as domestic poultry. Mink are known to kill in excess of their needs.

LIFE CYCLE: The breeding season for mink is in January, February, and March. Although they are promiscuous, a male normally stays with one of the females he has bred and assists in raising the young. Gestation period is normally around 45 days, although it has been known to be longer. Five to eight young are born in a den in rocks, under the roots of a tree, or in an old muskrat den. The young are born with their eyes closed and are nursed until about 5 weeks old, at which time they begin to eat solid food. Bobcats, foxes, owls, and lynxes are the mink's worst predators. Mink may live for up to 10 years in the wild.

FIGURE 13-22 Mink.

(Source: © Nialat, 2010. Used under license from Shutterstock.com)

Bobcat
(Lynx rufus)

DESCRIPTION

The bobcat is a medium-size, short-tailed predator and is widespread in the lower 48 states (Figure 13-23). It can also be found in Mexico and southern Canada. Bobcats are grayish to red-brown, with black streaks and spots. They have alert, pointed ears and long hair on the sides of their heads, often called a face ruff. They have sharp, pointed teeth and razor-sharp claws. They are well **camouflaged** regardless of their habitat. Those living in desert areas tend to be lighter in color than forest-dwelling bobcats. A large bobcat might weigh 40 pounds, but the average is 15 to 30 pounds. They average 2.5 to 3 feet in length, including their short tail.

HABITAT: Woodlands, marshes and swamps, deserts and mountains. Bobcats can be found almost anywhere their prey can be found, but they are largely **nocturnal** and are seldom seen.

FEEDING HABITS: Rodents, most small mammals, and birds, as well as fawns and deer weakened in winter or caught in deep snow.

LIFE CYCLE: The breeding season may begin as early as February. After a gestation period of about 50 days, two to seven young, usually four, are born. Caves, hollow logs, and rocky ledges are often chosen as den sites. The newborn kittens are fully furred, but their eyes are closed. At around 9 or 10 days the kittens' eyes open and they begin to crawl about. These cats appear to be promiscuous, with the female assuming sole responsibility for rearing the young. As the young approach weaning age the mother begins to bring back prey that is still alive. This allows the young to learn to kill for themselves. Bobcats in captivity have lived for up to 15 years. Life span in the wild is probably much shorter. Other than the occasional kitten killed and eaten by an owl, hawk, or coyote, bobcats appear to have few natural predators.

(Source: Photo courtesy of Texas Parks & Wildlife Department.)

FIGURE 13-23 Bobcat.

Lynx
(Lynx lynx)

DESCRIPTION ■

The lynx is a northern species, primarily found in Canada and Alaska (Figure 13-24). It has a short tail and is often confused with the bobcat. The lynx has prominent ear tufts and appears to be more solidly colored than the bobcat. The lynx is a light reddish brown, with a black tail tip. Its face ruff is longer and more prominent than the bobcat's and it has large, furred feet. Its feet allow it to move relatively freely and quickly in the deep snow common in its range during winter. The lynx is about 3 feet long, including the short tail, and weighs between 15 and 30 pounds. Lynx populations have plunged in recent years. Scientists are not really sure why, but the lynx may soon be on the Endangered Species List.

HABITAT: Northern woodlands and brushy areas. Generally found anywhere the varying or snowshoe hare can be found.

FEEDING HABITS: Eats grouse, squirrels, rodents, and other small mammals; however, main quarry is the snowshoe hare. The lynx depends so heavily on the hare that when the hare population crashes, as it does every few years, the lynx population also declines dramatically.

LIFE CYCLE: Female lynx carry their young for about 60 days. Favorite den sites are hollow logs, caves, and rocky crevasses. Two to four young are commonly born in late spring. The kittens are born with their eyes closed but fully furred. Their eyes open after about 10 days, and they get quite adventurous shortly thereafter. The kittens stay in or around the den until they are about 3 months old, at which time they begin to follow their mother on her hunting trips. At first about the only thing the young lynxes can catch are mice and other small rodents. However, they develop the skill and strength to catch and kill snowshoe hares fairly quickly. Few animals wish to tangle with an adult lynx's sharp claws and teeth; however, owls, eagles, wolves, and wolverines occasionally take a young lynx. Normal life span is 10 to 12 years.

(Source: Courtesy of U.S. Fish and Wildlife Service. Photo by Erwin and Peggy Bauer.)

FIGURE 13-24 Lynx.

Black-Tailed Prairie Dog
(Cynomys ludovicianus)

DESCRIPTION

The black-tailed prairie dog is a western prairie-dweller and our most common prairie dog (Figure 13-25). It can be found from northern Mexico northward to southern Canada. The prairie "dog" is not a dog, nor is it related to dogs. It does, however, give a bark-like call when alarmed and often sits on its haunches like a dog, no doubt getting its name from these two traits. They are a yellowish brown to buff-colored rodent. They were once much more numerous than they are today. The black-tailed prairie dog weighs about 3 pounds and is 15 inches or so in length, of which 3 inches is tail. They are communal animals, living in prairie dog "towns" or colonies. These colonies can be quite large.

HABITAT: Short grass and open prairies. Prairie dogs have large, often complex burrows that may be 10 to 15 feet deep, with lateral tunnels of 30 feet or more. The burrows contain sleeping quarters, toilets, and nesting areas.

FEEDING HABITS: Largely grasses and, occasionally, insects.

LIFE CYCLE: Prairie dogs have one litter per year, typically with three to five young. The gestation period is 35 to 40 days. The young are born in April, May, or June, with their eyes closed. They grow rapidly and are fully furred and able to crawl about at 3 weeks of age. The young are weaned at about 7 weeks and are eating grasses well at this age. The dominant male within a small area of the town appears to breed all the females within his area. A particularly strong or aggressive male may control several burrows, the females, and their young. After the young are weaned, the female leaves the nest burrow to her young and goes forth to dig a new burrow or occupy one whose previous owner was killed and eaten. Prairie dogs are preyed on by snakes, eagles, hawks, coyotes, foxes, and badgers. The black-footed ferret lives almost exclusively on prairie dogs, but it is very rare today. Prairie dogs may live to be 8 years old.

(Source: Photo courtesy of Texas Parks & Wildlife Department.)

FIGURE 13-25 Black-tailed prairie dog.

Woodchuck
(Marmota monax)

DESCRIPTION ▪

Woodchucks inhabit much of the northeastern United States and Canada (Figure 13-26). They are medium to light brown, burrowing rodents. Unlike the prairie dog, they live a solitary existence, except for the female and her young. Woodchucks are about 32 inches long, including a 6-inch tail, and may weigh up to 10 pounds. Although they live in burrows, they can climb and will take refuge in a tree if cut off from their burrow by a predator. The woodchuck likes to sit erect on its haunches while it watches for danger.

HABITAT: Open woodlands and field edges. Woodchucks like to have their burrow entrances under tree roots or rocks.

FEEDING HABITS: Eats many types of grasses, clover, garden vegetables where available, and alfalfa.

LIFE CYCLE: Woodchucks are promiscuous, with males breeding as many receptive females as they can find. Should two males meet during the breeding season, a fierce battle usually ensues. Old male woodchucks are often quite scarred. The female carries the young for about 28 days, after which she usually gives birth to four naked young. The young chuck's eyes open at about 3 weeks of age, and they are soon eating all types of vegetation. The young stay with their mother for a few months before venturing off to establish their own burrows. While a woodchuck may live to be 9 years old, it must dodge owls, dogs, foxes, coyotes and hawks to do so. The woodchuck is a genuine hibernator, in that all its bodily functions slow down during the winter period. When cold weather sets in, the woodchuck, which has put on multiple layers of fat, retires to its burrow. When it reaches a deep slumber, its heart beats only once per minute and it breathes once every 5 minutes. Its body temperature drops from a normal 96°F to about 37°F. Woodchucks normally come out of this deep slumber in late March.

(Source: © Mr Zap, 2010. Used under license from Shutterstock.com)

FIGURE 13-26 Woodchuck.

SUMMARY

Small mammals are as vital to an ecosystem as any other member, perhaps more so as they are food for many other species. For the most part, small mammals have adapted well to human encroachment in their habitat. However some, such as the Lynx, are struggling at this time. As with all wildlife, habitat both quality and quantity, must be maintained if we are to continue to have the great variety of small mammals we enjoy.

REVIEW QUESTIONS

Fill in the Blanks

Fill in the blanks to complete these statements.

1. The _____ is our most common and numerous rabbit.

2. Rabbits require a prolific rate of _____ due to their many predators.

3. List the two major species of jackrabbits found in the United States.

    (a) _____

    (b) _____

4. Jackrabbits are capable of leaping _____ or more feet in a single leap.

5. List three common predators of rabbits.

    (a) _____

    (b) _____

    (c) _____

6. List the three most common species of tree squirrels found in the United States.

    (a) _____

    (b) _____

    (c) _____

7. List three common squirrel foods.

    (a) _____

    (b) _____

    (c) _____

8. The _____ is the most common and widespread fox in America.

9. _____ foxes climb extremely well and often take refuge in trees.

10. Our two most common varieties of skunks are the _____ and _____.

11. The _____ is a medium-size, slim, doglike predator.

12. The _____ is the largest member of the weasel family.

13. _____ are powerful members of the weasel family and are excellent diggers.

14. Adult _____ are covered with over 30,000 needle-sharp, barbed quills.

15. The opossum is the only _____ found in North America.

16. A medium-size, robust predator, _____ are found in most of the United States.

17. The _____ is a pig-like creature that is not actually related to pigs at all.

18. Muskrats are always found near _____. Typical habitats are marshes, _____, and riverbanks.

19. The _____ is the largest rodent in North America.

20. The most common and widespread weasel is the _____.

21. Bobcats are widespread but largely _____ and are seldom seen.

22. The black-tailed prairie dog commonly inhabits _____ grass and open _____.

23. Unlike the prairie dog, _____ live a solitary existence, except for the female and her young.

Short Answer

1. Why must rabbits and many other small mammals be such prolific breeders?

2. Why does the wolverine have no serious natural predators?

Discussion

1. In your opinion, why are most small mammals highly adaptable? How do they achieve their adaptability?

2. Why do you think some small mammals, such as squirrels, tolerate the close presence of humans, but others do not?

Learning Activities

1. Draw a map of your state. Use various colors to illustrate areas that are known to hold a population of any of the species discussed in this chapter.

2. Select a species from your map and list all the factors, such as weather, available habitat, food, and predators, which might limit its population in your area.

3. Select a small mammal species not discussed in this chapter but native to your area, if possible. Use your local library and state fish and wildlife agency for information and write a detailed account of the species. Include a description, preferred habitat, life cycle, and feeding habits. Present your report to your class.

CHAPTER 14

Nonindigenous Species

OBJECTIVES

After completing this chapter, you should be able to

- List numerous nonindigenous species.
- Explain the detrimental effects of some invasive species.
- Describe how nonindigenous species have entered the United States.
- Identify common nonindigenous species that are not regarded as pests.
- Understand the economic costs of nonindigenous species.
- Understand the economic value of beneficial nonindigenous species.

INTRODUCTION

There are an estimated 50,000 **nonindigenous** species in the United States, including plants, fish, mammals, birds, amphibians, reptiles, microbes, and invertebrates. Many of these nonindigenous species are beneficial to the United States. Introduced crops, such as wheat (Figure 14-1) and rice, and introduced livestock, such as cattle, swine, and poultry, provide more than 90 percent of America's food. The value of these introduced food products exceeds $750 billion each year. However, many other introduced species are detrimental to the economy and harmful to the environment. An entire book could be written about the damage and environmental impact introduced species have had on the United States. This chapter can only give a brief overview of introduced species and the damage they cause.

(Source: Courtesy of NRCS. Photo by Jeff Vanuga.)

FIGURE 14-1 Wheat is an example of a nonindigenous plant.

SOME EFFECTS OF INVASIVE OR NONNATIVE SPECIES

- Invasive species compete both directly and indirectly with native species.
- In many cases invasive species lower the quality of vital habitats for native wildlife and fish species.
- Nonnative species can reduce or damage critical habitat needed by endangered and threatened native species.
- They often decrease the biological diversity of native ecosystems.
- Some invasive species pose direct human health risks.
- Many invasive species are a direct threat to agricultural crops.

Species have been introduced both **intentionally** and **unintentionally**. In addition to food crops and livestock, species have been introduced for use as landscape plants, for sport, as pets, and as **biological** pest control. As humans have evolved, travel has become both more frequent and much faster. We now trade with virtually every country on the planet, with thousands of ships and airplanes traveling throughout the world. Our travel habits have led to a greatly increased chance of accidental introduction. Many nonindigenous species arrive as hitchhikers aboard ships. Incompletely loaded or empty ships typically carry some form of ballast, which is necessary to stabilize the ship. One unintentional introduction is the red imported fire ant. These ants are native to South America, where their population is held in check by a variety of natural diseases and predators. They are thought to have arrived in the United States in soil used as ship ballast. With no natural predators, they spread very quickly and are now a scourge over most of the southern United States. In Texas alone, the red imported fire ant is estimated to cost agriculture millions of dollars in crop damage each year. Not included is the damage they cause to residential lawns and golf courses throughout the South or the medical costs of treating the thousands of people who are painfully stung each year. Almost half of the species listed as threatened or endangered (400 of the more than 900) are at risk primarily because of nonindigenous species. This is mainly due to predation or competition from introduced species.

PLANTS

Many of the estimated 5,000 introduced plants that have become established in the United States are displacing native plant species. **Invasive** plant species can reach the point where they dominate an area, which can completely change an ecosystem. For example, the European purple loosestrife (*Lythrum salicaria*), which was introduced in the early nineteenth century for use as an ornamental plant, is causing widespread damage to wetlands (Figure 14-2). This pest weed now grows in 48 states and is spreading at a rate of 115,000 hectares, or approximately 280,000 acres, per year. Loosestrife is changing the structure of most of the wetlands where it is now found, which adversely affects dozens of wildlife species that inhabit wetlands. The cost of attempting to control loosestrife—a single nonindigenous plant—is estimated to exceed $40 million per year.

Nonindigenous aquatic weeds are another grave problem, especially in the southern United States. In Florida, introduced aquatic plants such as water lettuce (*Pistia straiotes*) and hydrilla (*Hydrilla verticillata*) are causing tremendous damage to Florida's waterways. These aquatic weeds choke waterways, reduce the recreational use of many bodies of water, and alter nutrient cycles. They also change aquatic wildlife habitat, reducing and sometimes eliminating its usefulness to many species. Aquatic weed control costs in this country easily exceed $100 million each year.

(Source: © David P. Lewis, 2010. Used under license from Shutterstock.com)

FIGURE 14-2 Purple loosestrife.

AMPHIBIANS AND REPTILES

Some 53 species of nonindigenous reptiles and amphibians have been introduced into the United States and its territories. The vast majority of exotic reptiles and amphibians live in warm areas, primarily Florida and Hawaii, with about 30 species in Florida and 12 in Hawaii. Several of these species have had a huge negative ecological impact, particularly on island habitats. The wild fauna of any island evolves in a relative vacuum: Wildlife adapts to the predators they are faced with, and many islands develop with no natural predators. A large continental landmass like North America allows prey species more space to maneuver, but space, and therefore maneuvering room, are limited on an island. The wildlife on the islands of Guam, Hawaii, and Puerto Rico have been devastated by introduced species. The sudden introduction of a nonnative predator in such a limited area usually has a **horrendous** impact on native species, which seldom have the time to adapt to new predators.

The brown tree snake (*Boiga irregularis*), for example, was accidentally introduced into the U.S. territory of Guam after World War II (Figure 14-3). It is thought to have arrived with shipments of military equipment following **demobilization** after the war. The brown tree snake consumes eggs, reptiles, and birds and reproduces at a prodigious rate when it is faced with no natural

FIGURE 14-3 The brown tree snake is an example of a nonindigenous reptile.

predators. As a direct result of the unintentional introduction of the brown tree snake, 10 of the original 13 native forest bird species found on Guam no longer exist. Of Guam's 12 native reptile species, only three have survived. The snakes are also a major problem for small farmers on the island, consuming both eggs and chickens. The annual cost of control and research programs on brown tree snake exceeds $5 million.

BIRDS

Approximately 100 bird species have been introduced into the United States. Of these, less than 10 percent, including chickens, are considered beneficial species. Birds such as the pheasant (*Phasianus colchicus*), chukar (*Alectoris chukar*), and gray partridge (*Perdix perdix*) were introduced for sporting purposes and have proved to be valuable imports. (More information on these birds can be found in Chapter 15.) However, the vast majority of nonindigenous birds are pests. Two of the worst are the rock dove, or common pigeon (*Columba livia*), and the English or house sparrow (*Passer domesticus*). (Additional information on these birds can be found in Chapter 17.)

It is thought that the sparrow was introduced into the United States in the 1850s to control the canker worm. It proved to be a poor decision. The house sparrow competes very aggressively with a variety of native species for both food and nesting sites. Natives such as robins, yellow and black-billed cuckoos, and Baltimore orioles are often harassed. Sparrows are known to displace native wrens, bluebirds, cliff swallows, and purple martins from their nest sites. In addition to competing with native bird species, English sparrows are believed to be capable of transmitting more than 25 different human and livestock diseases. They also consume large amounts of grain, such as corn and wheat, damage fruit trees by consuming the buds, and damage plants around buildings.

FIGURE 14-4 These pigeons on a building can be both a health hazard and a nuisance.

By far the most damaging nonindigenous bird in this country, in financial terms, is the common pigeon. These pests exist in almost every city in the world and are known to harbor more than 50 human and livestock diseases. Encephalitis, parrot fever, histoplasmosis, and ornithosis are just four of the diseases pigeons spread. Pigeons are also a public health hazard and nuisance because they foul cars, statues, buildings, and sometimes people with their droppings (Figure 14-4). In rural areas they consume large amounts of grain, including corn, wheat, and grain sorghum. The estimated cost of damage by pigeons, combined with the money spent attempting to control them, exceeds $1 billion per year.

MAMMALS

Although fewer exotic mammals have been introduced into the United States and its territories than other animals, the mammals are by far the most destructive. Introduced rodents, mainly mice and rats, cost U.S. citizens $19 billion per year in destroyed property. The Norway or brown rat (*Rattus norvegicus*), the European or tree rat (*Rattus rattus*), and the common house mouse (*Mus musculus*) are three destructive nonnative rodents. Rats and mice consume or foul enormous quantities of grain annually. Their habit of chewing electrical wiring leads to numerous house and barn fires every year. There are an estimated 250 million rats (Figure 14-5) in the United States. Rats also pollute other foodstuffs and carry a number of diseases, including leptospirosis, salmonellosis, and to a smaller degree plague and some strains of typhus. They also prey on various native wildlife species, such as young birds and bird eggs. Another nonindigenous member of the rodent family is the South American nutria (*Myocastor coypus*). These large rodents weigh 8 to 18 pounds each and are very destructive to marshes, farm ponds, and other wetlands. They were first introduced

(Source: © Oshchepkov, 2010. Used under license from Shutterstock.com)

FIGURE 14-5 The rat is known to carry numerous diseases and prey on various native wildlife species.

in the 1940s as fur-bearers, to be trapped and used much like beaver was 100 years earlier. With the decline in the fur industry the population of these rodents has exploded. Nutria are essentially herbivores, consuming large quantities of marshland vegetation. As their population has grown, their damage to wetland vegetation has been dramatic. Nutria also dig extensively, harming or destroying the root systems of many plants. Their digging also often undermines the dams of farm ponds. When we think of introduced species, we might not think of cats, dogs, cattle, horses, burros, sheep, pigs, and goats. However, all these are introduced species, some of which have devastated native wildlife and their habitat. Cats, for example, are a serious threat to many of our native bird, reptile, amphibian, and small mammal species (Figure 14-6). An estimated 60 million pet cats live in the United States, along with some 30 million feral cats roaming the countryside. Cats are very efficient predators of small mammals, reptiles, amphibians, and birds. Studies have shown that a cat kills an average of five birds per year, which means potentially 450 million birds each year. Cats also kill reptiles, amphibians, and small mammals. Although cats undoubtedly kill a large number of destructive rodents, they are having a tremendous negative impact on many native bird populations. It is highly recommended that people spay or neuter their pet cats and keep them indoors as much as possible.

Horses and burros were first brought to Mexico and the southwestern United States by Spanish explorers and missionaries. Until the late 1960s and early 1970s their numbers were controlled by ranchers, the federal government, and various western states. Excess horses were gathered up or shot and sold to pet food companies for use in dog and cat food. Public outcry put an end to such control methods decades ago. Unfortunately, an effective, more

(Source: © Johanna Goodyear, 2010. Used under license from Shutterstock.com)

FIGURE 14-6 This house cat is one of the estimated 60 million pet cats in the United States.

FIGURE 14-7 The dog is seen as man's best friend, but they can become problems if left unattended in the wild.

(Source: Photo courtesy of Texas Parks & Wildlife Department.)

suitable method of control has been hard to find. The government runs a campaign promoting the adoption of wild horses and burros, but the number of adoptees far exceeds the number of people willing to adopt. Over 1 million horses and burros **languish** in government holding pens in Nevada and Utah, fed with the American tax payers' money. The destruction of native wildlife habitat and watering holes in the high desert and desert Southwest has been extensive. Native species such as the desert bighorn sheep have therefore suffered a drastic drop in numbers. Dogs, long known as "man's best friend," were first brought to this country for domestic purposes (Figure 14-7). However, many have escaped, or more likely been dumped by unthinking owners, into the wild. Wild dogs are becoming a problem in several states, including Texas and Florida. These wild dogs often run in packs and kill domestic livestock, such as sheep and goats. They also prey on deer, rabbits, and many other small mammals. Wild dogs in Texas cause approximately $5 million dollars worth of damage to domestic livestock every year. It is virtually impossible to figure out the damage these wild dogs do to wildlife, but it is surely extensive. In addition, more than 4 million people are bitten each year by feral and pet dogs. Approximately 800,000 of these cases require medical treatment, which costs an estimated $165 million each year in direct medical costs. If you count indirect costs, such as lost time at work, the total cost of dog bites in the United States each year probably exceeds $250 million. In addition, 10 to 15 people on average die of dog attacks each year. The majority of these deaths, approximately 80 percent, occurred in small children.

Individuals, state fish and wildlife departments, and landowners have introduced a number of mammals into the United States. These animals were introduced for a variety of reasons, including commercial or expected commercial usage and sport hunting. Some species, most notably the gemsbok (*Oryx gazella*), were introduced to populate unused or underused habitat. Some were imported simply because people like to look at exotic game animals. Several of these species are now more common in their adopted country than they are in their native homes. They also provide a major economic boost to some rural areas and landowners. The remainder of this chapter covers a few of the more common introduced species.

Nilgai
(Boselaphus tragocamelus)

DESCRIPTION

Nilgai are large members of the antelope family. Males (bulls) have short, smooth horns, averaging 8 to 10 inches in length. Females (cows) do not usually grow horns, but may occasionally. Nilgai stand 47 to 60 inches at the shoulder, with short hindquarters, which give them a sharply sloped appearance from shoulder to rump. Nilgai bulls are gray to brownish gray, with cows and young being brown to light brown. They may have patches of white on the face and below the chin. White below the chin may stretch into a broad white "bib" on the throat. They also have a narrow band of white that may start between the front legs, widen over the abdomen, and stop between the hind legs. Cows usually weigh 250 to 450 pounds, and mature bulls are considerably larger, averaging 300 to 600 pounds. Nilgai are extremely tough, hardy animals.

Nilgai are the largest antelope found in their native India and Pakistan. They are thought to have been imported into Texas in the 1930s as game animals and have since become one of the most abundant exotics in Texas. Large, free-ranging populations are found in South Texas, with the majority found on big ranches in Willacy and Kenedy counties.

HABITAT: Nilgai prefer fairly open areas, with scattered brush. They seem to avoid heavily wooded areas, and the south Texas brush country seems to be ideal for them.

FEEDING HABITS: Nilgai are herbivores who both graze and browse. In Texas the bulk of their diet is made up of grasses, with mesquite and oak browse important at times. Nilgai are quite opportunistic; they will eat what is available at the time.

LIFE CYCLE: Nilgai appear to breed throughout the year, but in Texas most of their mating activity occurs in December, January, February, and March. Except during the breeding season the bulls remain separate from the cow and calf groups. Cows and calves usually form herds of 10 to 12 animals, with occasional herds of 20 to 50. The gestation period is 240 to 258 days, and cows commonly have twin calves. Females do not normally breed until they are 3 years old, and bulls are sexually mature at 2.5 years. However, it is unusual for bulls to breed until they are 4 years old or so due to the competition from older, more mature bulls. Young, small, weak, or injured calves are often preyed on by coyotes, but adults are too formidable for the predators commonly found in their range. Nilgai have thin coats and do not store much winter fat, so an extended period of severe cold, although rare in south Texas, can be fatal to adults and calves. Their life span in the wild is about 8 to 10 years.

Blackbuck
(Antilope cervicapra)

DESCRIPTION

Blackbucks are a small- to medium-size antelope. The males have prominent, ringed, V-shaped, "corkscrew" horns that rise above their heads. Females do not normally have horns. The horns on males may reach 30 inches in length but are usually only 24 inches. Females and young are a light brown or tan, with males being darker along their backs. During the breeding season and as they age, mature bucks get quite dark. All blackbucks have white bellies, inner legs, chest, eye rings, and chin patch. In Texas, mature bucks average 45 to 125 pounds, with does averaging 40 to 70 pounds.

Blackbucks are another member of the antelope family, whose original range was Pakistan and India (Figure 14-8). Blackbucks were once widespread and common in the plains and open woodlands of India and Pakistan. However, expansion of human population, habitat destruction, and extensive hunting have reduced them to isolated pockets within their former range. They were introduced into Texas in 1932 in Kerr County, in the Edwards Plateau area. They are now free-ranging in at least 8 to 10 counties and confined to ranches in 86 other counties.

HABITAT: Blackbucks prefer a mixture of open grasslands and brush. This mixture of habitat affords them excellent forage as well as escape cover. They do not do well during prolonged cold spells, which restricts their range to the warm climates. The vast majority of the blackbucks in Texas are found in the Edwards Plateau area.

FEEDING HABITS: Blackbucks are both grazers and browsers, but they appear to prefer medium to short grasses. Browse species such as mesquite and live oak are also commonly eaten, with availability appearing to be the determining factor. Blackbucks seem to eat very few forbs.

LIFE CYCLE: Blackbucks breed at any time of the year, but males are most active during spring and fall. Gestation is about 5 months, with fawns born at all times of the year, with the fewest during winter. Does normally give birth to a single fawn. Does reach sexual maturity at about 8 months and physical maturity at 1 year, but they rarely breed before 2 years of age. Mature males claim a territory of 3 to 30 acres, which they defend vigorously from other males. Groups of females and fawns are allowed to graze through each male's territory, but males do not tolerate other males. Young bucks reach sexual maturity at 18 months and physical maturity at 2 to 2.5 years. Young males often form bachelor groups until they are big and strong enough to claim a territory of their own. Coyotes are the major predator of blackbucks in Texas, with the young, weak, or old most often the victims. Bobcats also take fawns. In the wild, blackbucks can live 12 to 15 years.

(Source: Photo courtesy of Texas Parks & Wildlife Department.)

FIGURE 14-8 The blackbuck antelope.

Gemsbok
(Oryx gazella)

DESCRIPTION

The gemsbok is a large, gray to dusty brown antelope, with distinctive black and white markings on face, legs, and bellies. It has a black tail that resembles a zebra or horse tail. Both males and females have impressive, straight, ringed horns that rise from the head and sweep backward. Mature adults may have horns longer than 3 feet. Mature bulls may weigh 450 pounds, and females probably average around 350 to 400 pounds. They stand 2.5 to 4.5 feet tall at the shoulder.

Gemsbok is the common name for the largest and best known of three species of African antelope (Figure 14-9). Both the scimitar oryx (*Oryx dammah*) and the white or Arabian oryx (*Oryx leucoryx*) are listed as endangered species. The gemsbok, on the other hand, is quite common in the dry savanna of southern Africa. The gemsbok was introduced to the White Sands Missile Range in southern New Mexico between 1969 and 1977. It is also found on various ranches in south Texas.

HABITAT: Gemsbok inhabit dry savannas and desert regions. In the United States the largest free-ranging population is found in the desert of New Mexico in and around the White Sands Missile Range. They are also found in the dry south Texas brush country on numerous ranches. They are especially well adapted to survive with little or no standing water.

FEEDING HABITS: Oryx both graze and browse, preferring to feed on desert grasses. In the United States they feed on tumbleweed, mesquite bean pods, yucca, buffalo gourds, and a variety of desert grasses. One of the reasons that the gemsbok has been so successful in the United States, particularly the free-ranging herd in New Mexico, is its ability to survive with little or no standing water. Gemsbok can use nearly every molecule of water stored in the plants they consume. They also tend to graze at night, when plants contain more moisture.

LIFE CYCLE: Female gemsbok reach sexual maturity at approximately 2 years of age. They have a gestation period of 9 to 10 months (260 to 300 days) and normally produce a single calf. The young mature rapidly, being weaned at 3 to 3.5 months. By 5 months the young look just like an adult, only smaller, with smaller horns. In Africa their calving is seasonal, determined by rainfall. In the United States they appear to breed year-round, producing one calf each year. Oryx horns are very formidable weapons, which they use to good effect. The gemsbok is the only hoofed mammal reported to have killed full-grown African lions. Consequently, they have few if any predators, other than people, in the United States. It is thought that mountain lions might kill the occasional young oryx, but their ranges rarely overlap.

(Source: © EcoPrint, 2010. Used under license from Shutterstock.com)

FIGURE 14-9 The gemsbok antelope.

Ibex
(Capra ibex)

DESCRIPTION

Ibex are gray to brown, with short tails and bearded chins. They stand 2.5 to 3.5 feet tall at the shoulder and are stocky and sturdily built. Males have long horns, with pronounced bumps on the outer curve, which curve up and back and may reach 40 inches in length. Females have short horns that point more straight back than those of the males. Ibex are very agile and sure-footed, and they are strong jumpers.

Several varieties of ibex are found from the Iberian Peninsula to North Africa, southern Russia, and Asia (Figure 14-10). They have been introduced into New Mexico, Texas, and other game ranches around the country. A free-ranging herd exists in the Florida Mountains, near Deming, New Mexico. They are also found on numerous game ranches and in wildlife parks.

HABITAT: Ibex prefer rough, rugged country. They are usually found in mountain ranges in southern Europe, central Asia, and northeast Africa. They have begun to move out of the Florida Mountains in New Mexico onto the plains.

FEEDING HABITS: Ibex feed on a wide variety of vegetation, both grasses and browse species.

LIFE CYCLE: Ibex typically form herds of 10 to 20 animals, usually **segregated** by sex. Females and young form social groups and are joined by the males only during the breeding season. Males are most often found in bachelor groups, which have a distinct **hierarchy** dominated by the older, larger, and stronger males. The gestation period is approximately 165 to 170 days, usually resulting in one kid, rarely twins. The kids are very active from a young age and capable of running and jumping shortly after birth. The young appear to be very gradually weaned, with no set cutoff time. Females reach sexual maturity at 1 to 1.5 years of age and males at about 2 years. In Europe, golden eagles are known to prey on the very young, but adults have few predators. In the United States mountain lions are probably the greatest threat, certainly to adults. On game farms and ranches coyotes undoubtedly take young ibex, and they may become significant predators of young ibex in New Mexico as these animals move out onto the plains.

(Source: © Vladimir Mucibabic, 2010. Used under license from Shutterstock.com)

FIGURE 14-10 The Siberian ibex.

Axis Deer
(Cervus axis)

DESCRIPTION

The axis deer, called the *chital* in its native India, was brought into Texas in the early 1930s. Axis deer have subsequently been introduced on game farms and wildlife parks in several other states. Axis deer have become the most abundant exotic deer in Texas, with free-ranging populations in 25 counties. They are confined to ranches in 60 other counties. Axis deer are medium-size, spotted deer, with males averaging about 36 inches at the shoulder and weighing 80 to 160 pounds. Females are smaller and usually do not have the male's characteristic tall antlers. Antlers consist of a main beam on each side, with three tines per side. Axis deer are gregarious animals commonly found in herds ranging from a dozen to 75 or more.

HABITAT: Axis deer prefer a mixed forest habitat, with wooded areas broken by clearings and adequate shade and water.

FEEDING HABITS: Axis deer are primarily grazers, feeding on a variety of grasses, supplemented by browse. Browse species include sumac, live oak, and hackberry.

LIFE CYCLE: Axis deer have a breeding cycle unlike any of our native deer, although bucks do emit a bugle-like call similar to our native elk. Axis does have estrous cycles throughout the year, with each cycle lasting 3 weeks. Bucks also can be in breeding condition at any time of the year and the bucks do not appear to be **synchronized.** Mature bucks can be found living with herds of young and old animals of both sexes throughout the year. Some bucks in the herd may be in breeding condition, with hard antlers, while others are in velvet or have shed their antlers. Although axis deer can breed at any time, in Texas the major breeding activity occurs in May through August, with a peak in June or July. Bucks in breeding condition wander through the herd, breeding **receptive** does, making no effort to gather or keep a harem of does. Does normally give birth to one fawn after a gestation period of 210 to 238 days. Fawns may be born at any time of year, but most are born in January through April. Twins are rare in axis deer, at least in the wild. Fawns may begin eating tender, green forage at 5 to 6 weeks of age, but they are not normally weaned until 4 to 6 months of age. Axis deer mature slowly, with most does breeding at 14 to 18 months. Males reach adult size at 4 to 5 years, but females are usually 6 before reaching full adult size. Life span in the wild is about 10 to 13 years, but animals in captivity may reach 18 to 20 years. Coyotes are probably the major predators of fawns in the United States, certainly in Texas. Adults, except in rare cases where their range overlaps that of the mountain lion, appear to have few predators.

Sika Deer
(Cervus nippon)

DESCRIPTION

Sika deer are a small- to medium-size deer, native to southeastern China, southern Siberia and the Japanese island of Hokkaido (Figure 14-11). Its coat is a drab to deep brown, almost red, mottled with white spots. The amount of spotting is highly variable, with some animals having none. In Texas their color and size is highly variable due to widespread **hybridization.** They have a distinctive white rump patch that is more noticeable when they are frightened or alarmed. Sika deer have compact bodies and legs that appear almost dainty. Males have antlers consisting of two main beams that are 12 to 24 inches long with three to four tines or points on each beam. Exceptional bucks may have horns 30 inches in length. Sika deer stand 30 to 40 inches at the shoulder and usually weigh 100 to 175 pounds.

HABITAT: Sika are a woodland deer, most often found in mixed forests in their native countries. In Texas they are found in the mixed brush and grasslands of the Edwards Plateau and south Texas. There are a dozen counties in south Texas known to have free-ranging populations of sika deer, with introductions in some 70 other counties in central and south Texas. Sika deer are also found on game farms and wildlife parks in other states.

FEEDING HABITS: Sika deer feed on a variety of grasses, leaves, and tender shoots of woody plants, depending on what is available at the time. Browse species utilized include live oak, wild plum, hackberry, and green briar. Forb use usually increases in the spring and summer.

LIFE CYCLE: Does have a gestation period of 7.5 to 8 months, with a single fawn born usually between May and August. The birth of twins is rare. The rut, or breeding season, usually peaks from early September through October but may last longer in southern climates. During the rut, males establish territories that vary in size according to the size of the buck and the habitat. Some bucks may have territories of up to 4 or 5 acres. Sika males mark their territories with a series of shallow depressions, or **scrapes**, into which they urinate. These scrapes have a strong musky odor that warns other males away. Bucks keep harems of does during the rut, and fights with rival males can be long, fierce, and occasionally fatal. The coyote is probably the main predator of the sika over most of its range in the United States, preying on the young or the weak and old. In captivity sika deer normally live 15 to 18 years, whereas in the wild their life expectancy is probably 8 to 12 years.

(Source: Photo courtesy of Texas Parks & Wildlife Department.)

FIGURE 14-11 The sika deer.

Fallow Deer
(Cervus dama)

DESCRIPTION ■

Fallow deer are a medium-size deer, native to the Mediterranean region of Asia Minor and Europe (Figure 14-12). Fallow deer are the most widespread of the world's deer, having been introduced to all inhabited continents. They are found on many game farms and wildlife parks and are represented in more than 90 counties in Texas. Adult males have large, palmate antlers, similar to those of moose, but much smaller. The buck's antlers may be 15 to 24 inches in length. Palmate areas may be 3 to 9 inches wide. Males are usually 3 to 4 years old before they develop a full set of palmate antlers; until then they grow beams with simple points. Males weigh 175 to 225 pounds, and females weigh 80 to 100 pounds. Coloration varies widely, with four major color variations. They can have rust-colored bodies, with white spotting and a white belly and rump patch. They may also be tan with white spotting and a white belly and rump patch; white with dark eyes (not a true albino); or very dark brown, almost black, with barely visible or no spotting.

HABITAT: Fallow deer spend a lot of their time feeding in grassy, open areas. In winter, they require some tree cover and brush for shelter and for food. The Edwards Plateau area of Texas, with its mixture of oak **mottes,** brush, and grasslands, seems ideally suited to fallow deer.

FEEDING HABITS: Fallow deer feed on a wide range of grasses as well as browse species, such as live oak and hackberry.

LIFE CYCLE: Rutting seems to begin in early to mid-September and continues into November, with peak breeding activity in October. The males mark out and defend a small territory from which all other rutting males are excluded. Females and their young remain in a buck's territory until the does cycle and are bred. After the rut, and for most of the year, males form bachelor groups, while does and young form their own groups. Gestation takes about 7.5 months, with most fawns born in late May and June. Single fawns are the norm, but twins also occur. Bucks reach sexual maturity at about 14 months of age, but, like most members of the deer family, they rarely compete successfully in the rut until several years later. Females are sexually mature at around 16 months and usually have their first fawn when 2 years old. Life span appears to be 10 to 15 years. Healthy adults face few predators in the United States, but the old, the weak, and the young are vulnerable to coyotes.

(Source: © Jiri Krajicek, 2010. Used under license from Shutterstock.com)

FIGURE 14-12 A fallow deer buck.

Aoudad
(Ammotragus lervia)

DESCRIPTION ◼▶

The aoudad or Barbary sheep is a relatively large sheep, native to the dry mountainous areas of North Africa (Figure 14-13). Males may weigh up to 300 pounds and stand about 36 inches at the shoulder. Females, or ewes, are considerably smaller, weighing 140 to 160 pounds. Both males and females have horns, but the males' are much larger. The horns curve outward, then backward and inward and in males may exceed 30 inches in length when measured from the base around the outside curve. Aoudad horns also have pronounced ridges or rings around them. The coat is a uniform reddish or grayish brown, with an obvious growth of long hair down the throat, chest, and the back of the upper front legs. They are extremely surefooted and agile. Aoudads live in small groups made up of young and old animals of both sexes. They were released into the wild in New Mexico in 1950 and in Texas in 1957 but were first brought to the United States around 1900 and have been reared in zoos and wildlife parks for years.

HABITAT: Dry, mountainous, rocky areas in Texas and New Mexico. A large free-ranging population exists in the Palo Duro Canyon area of the Texas panhandle.

FEEDING HABITATS: Aoudads feed on grasses, browse species, and forbs. Their diet seems to be pretty evenly split between grasses and browse, with a few forbs also being eaten. The most preferred browse species is mountain mahogany. Aoudads may use standing water if available but appear not to need it to survive.

LIFE CYCLE: The breeding season for aoudads seems to be rather long, but most breeding takes place from mid-September to mid-November. Most lambs are born from late February until late April, after a gestation period of 160 days, but some lambs may be born as late as November. First births are usually single lambs; thereafter, twins may be born if habitat conditions are good. Aoudads in the wild have an estimated life span of 10 to 14 years, but they may live longer. Some aoudads in captivity have lived for 20 years. In the wild, coyotes or eagles may take the occasional lamb, but the steep rocky terrain most aoudads inhabit makes things tough on the coyote. In areas where their ranges overlap, mountain lions are capable of taking both young and adult sheep.

(Source: © Chung-Yen, 2010. Used under license from Shutterstock.com)

FIGURE 14-13 Barbary sheep.

Feral Hog, Eurasian (or Russian) Wild Boar
(Sus scrofa)

DESCRIPTION

Wild hog coloring includes black, brownish black, red, red with black spots, and black with a white belt across the front shoulders. Generally they are dark brown or black, but color varies widely depending on the color of their escaped domestic ancestors. They can be very large, with older boars weighing 500 or more pounds. Sows are usually much smaller. Size varies greatly by region and from individual to individual and is largely a function of diet, ancestry, and age. Generally, the better the diet and the older they are, the larger they get. Feral hogs are becoming very widespread, with large numbers in several states east of the Mississippi River. Oklahoma is also experiencing a population explosion of feral swine. Feral hogs are very adaptable and prolific, which makes them very likely to continue expanding their range.

The domestic hog is a direct descendant of the Russian or Eurasian wild boar. There are three strains of wild hog in North America today: feral hogs (Figure 14-14), direct descendants of domestic swine gone wild; Eurasian wild boar; and **hybrid** crosses between these two strains. Because domestic swine descended from the Eurasian wild boar, domestic swine and Eurasian wild boar can and do readily interbreed. The resulting offspring are quite fertile, readily breeding with feral swine, domestic swine, or Russian wild boar. The first "purebred" Russian wild boars were imported into the United States for sporting purposes in the late 1800s. Despite continued importation of "pure" Russian stock in the 1980s and 1990s, it is doubtful that any pure strains exist in the wild except on fenced game ranches. Feral hogs, or domestic swine gone wild for a generation or two, are now found in 19 states. California, Florida, and Texas all have large populations. Because Eurasian wild boar and feral swine interbreed so readily, the pure Eurasian blood quickly gets diluted when these hogs are released into the wild.

HABITAT: Feral hogs seem to prefer forested areas interspersed with open areas. Forested areas they inhabit may vary from mature oak and bottomland hardwood forests to mesquite and oak scrub brush. They are often found along watercourses such as streams, creeks, and rivers. During the summer in hot, dry conditions hogs often make **wallows**, or depressions in the mud, along streams, rivers, wet ditches, marshes, lakes, and farm ponds. The main purpose of these wallows is to cool the hogs, although they probably offer some relief from external parasites. Hogs are extremely adaptable and can be found in a variety of habitats.

FEEDING HABITS: Hogs are omnivores, eating everything from snakes to grass. They consume large amounts of mast in the fall, where available. They also eat fruits, roots, mushrooms, and invertebrates. They use their powerful necks and shoulders and stout **snouts** to **root**, or dig for roots and invertebrates. Areas that have been rooted by hogs are very obvious, and hogs can be very destructive to agricultural fields and crops. Hogs are also known to eat the eggs and young of ground-nesting birds and to kill and eat fawns.

LIFE CYCLE: Feral hogs usually breed year-round. Sows reach sexual maturity at 6 to 8 months of age and normally raise two litters per year. Gestation is 115 to 118 days, and piglets are generally weaned at 2 to 3 months of age. Litters may consist of one to ten piglets, with three to five probably being about average in the wild. Piglets are preyed on by a multitude of predators, including coyotes, bobcats, black bears, and mountain lions, where their ranges overlap. Sows are generally fiercely protective of their young and are attentive mothers. Adults are not the choice prey for most predators. Average life span in the wild is probably 4 to 5 years, with some surviving 8 to 10 years or more.

FIGURE 14-14
The feral hog.

(Source: Photo courtesy of Texas Parks & Wildlife Department.)

SUMMARY

Nonindigenous species, both plant and animal, can be destructive to native plants and wildlife. We have touched on only a few of the hundreds of nonnative species now found in the United States. It is **imperative** that we be extremely cautious when introducing nonindigenous species in the future and that we do everything possible to limit the accidental introduction of species. We must continue to work hard to find control methods for introduced species and, if possible, to support **eradication** efforts for particularly damaging introduced species.

REVIEW QUESTIONS

Fill in the Blank

Fill in the blanks to complete the statements.

1. There are an estimated _____ nonindigenous species in the United States.

2. Nonindigenous species have been introduced for purposes that include _____, sport, pets, and _____ pest control.

3. Many nonindigenous species arrive as _____ aboard ships.

4. Approximately _____ native species are in danger primarily because of nonindigenous species.

5. _____ plant species can reach the point where they dominate an area.

6. _____ is changing the structure of most wetlands where it is now found.

7. It is estimated that nationwide attempts to control aquatic weeds cost more than $ _____ million each year.

8. The accidental introduction of the _____ to the U.S. territory of Guam has been devastating to the islands native wildlife.

9. There are approximately _____ species of introduced birds in the United States.

10. Two of the worst nonindigenous pest birds are the _____ and the _____.

11. It is estimated that the cost of damage and of attempts to control the common pigeon exceed $_____ per year.

12. It is estimated that introduced rodents, mainly _____ and _____, cost U.S. consumers $19 billion dollars each year.

13. The _____ was introduced in the 1940s to replace the beaver as a fur-bearer.

14. When we think of introduced species, we do not often think of _____, _____, cattle, horses, burros, sheep, pigs, or goats.

15. There are an estimated _____ million feral cats in America.

16. Feral and pet cats combined have the potential to kill _____ million birds each year.

17. Wild dogs are becoming a problem in several states, including _____ and _____.

18. More than _____ million people are bitten each year by feral and pet dogs.

19. Nilgai are native to _____ and Pakistan.

20. Nilgai are large members of the _____ family.

21. Nilgai are _____, who both graze and browse.

22. Blackbucks are a small- to medium-size _____.

23. During the breeding season, mature blackbuck males claim a territory of _____ to _____ acres.

24. _____ is the common name for the best known of the African oryx species.

25. There is a free-ranging population of gemsbok on the _____ Range in New Mexico.

26. The gemsbok is the only hoofed mammal reported to have killed full-grown _____.

27. Ibex have been introduced into the _____ Mountains in New Mexico.
28. Ibex are very _____, _____, strong jumpers.
29. The axis deer is known, as the "_____" in its native India.
30. Axis deer are free-ranging in some _____ counties in Texas.
31. Axis fawns may begin eating tender green vegetation at _____ to _____ weeks of age.
32. The _____ deer is native to southeastern China, southern Siberia, and the Japanese Island of _____.
33. Sika are a _____ deer, most commonly found in mixed forests in their native countries.
34. Sika does have a gestation period of _____ to _____ months.
35. The fallow deer is native to the Mediterranean region of _____ and Europe.
36. Fallow deer are the most widely introduced of the world's deer, now living on all _____ continents.
37. The _____ or Barbary sheep is a relatively large sheep native to the dry, mountainous areas of _____.
38. Aoudads were first introduced into the wild in New Mexico in _____.
39. Purebred Russian wild boars were introduced into the United States in the late _____.
40. Feral hogs are now found in some _____ states.
41. Three states with large populations of feral hogs are _____, Texas, and _____.
42. Hogs are _____, eating everything from snakes to grass.

Short Answer

1. What are nonindigenous species?
2. Explain how invasive pest species cost American consumers' money.
3. In what ways are some nonnative species beneficial?

Discussion

1. Explain why nonindigenous species are harming native wildlife.
2. In your opinion, what can be done to slow or stop the introduction of nonindigenous species?
3. Assuming it was possible; would you support the eradication of particularly harmful nonindigenous species, such as the burro, wild horse, or common pigeon? Why or why not?

Learning Activities

1. Select a nonindigenous pest species (preferably one found in your area). Using additional sources as necessary, write a detailed account of this species' introduction into this country and the economic impact it has had.
2. Using the same pest species you selected in the first activity, write a detailed report on possible control measures.

CHAPTER 15

Upland Game Birds

OBJECTIVES

After completing this chapter, you should be able to

- Identify the common upland game birds found in the United States.
- List the characteristics of each of the common upland game birds found in the United States.
- Identify the type of habitat each species prefers.
- List at least one major food source for each species.
- Describe some of the behavior traits of each species.

QUAIL (Phasianidae)

All of the species of quail described in this chapter have several traits in common. For example, they are all gregarious birds, enjoying the company of others of their species. This trait is perhaps stronger in some than in others, but it is present in all. They are all "covey" birds; that is, they form flocks, or coveys, that stay together during fall and winter. All species of quail we will discuss pair up in the spring. A hen selects a rooster or cock bird, and they mate and raise the young together. The family group generally stays together until late summer, at which time birds combine to form a new covey. All of these species of quail are ground dwellers, spending the majority of their time on the ground. They are all strong runners, usually preferring to run to escape danger, although this trait is much more developed in some species than others. They are also relatively strong flyers, but only for short distances. The life cycles of each of these species of quail are very similar; therefore, we present one common life cycle rather than repeat much of the same information several times.

LIFE CYCLE: Coveys break up in late winter and early spring, and the males and females pair off. The hens select a nest site, usually in a thick clump of grass or under a bush. In April, May, or June they lay one egg per day until the clutch of 12 to 14 is complete. Incubation takes 21 to 23 days and the chicks, which are not much bigger than the end of a thumb, follow the hen and rooster away from the nest site as soon as they are dry. Scaled quail and bobwhite quail can interbreed where their ranges overlap (Figure 15-1). It is unknown at this time whether these offspring are fertile, although it is doubtful. The chicks grow rapidly on a diet consisting mainly of insects and can fly short distances in a matter of days. They are essentially fully grown at 9 to 12 weeks of age, although most

FIGURE 15-1 Hybrid bobwhite and scaled quail.

(Source: Photo courtesy of Texas Parks & Wildlife Department © 2002.)

do not attain their full adult weight until 15 to 16 weeks of age. Mortality is high. The nests are destroyed and eggs eaten by raccoons, opossums, and skunks. Snakes also eat quail eggs. Should the nest be destroyed, most hens attempt to re-nest, some up to three times. However, the hens lay fewer eggs, generally with lower fertility rates, with each successive nest. Young and adults alike are prey for almost all predators. Bobcats, various birds of prey, foxes, coyotes, and house cats are major predators. Few quail live to see their first birthday. As much as 70 to 80 percent of the population may succumb to diseases, cold, starvation, and predators each year.

Bobwhite Quail
(Colinus virginianus)

DESCRIPTION

Bobwhites are small, rather plump birds that were once widespread and numerous in the eastern United States (Figure 15-2). They are the only quail native to the eastern United States. Due to a variety of factors such as changing farming practices and loss of habitat, they are not as numerous or widespread as they once were. They are a red-brown color with darker spots and scaling and lighter bellies. The males have a distinctive white throat patch and white "eyebrows," or narrow white bands from the beaks back over the top of the eyes to where the head joins the neck. The parts of the males that are white are buff colored on the females. Bobwhites are about 8 to 10 inches long, with the males having a slight crest on top of their heads. Bobwhites have a distinctive, clear, whistling call ("bob-white," or "bob-bob-white") from which they get their name.

HABITAT: They prefer field edges, particularly areas where cropland meets grassland and brush. They need considerable protective cover in order to survive.

FEEDING HABITS: Young quail feed heavily on insects; adults eat a wide variety of weed seeds, cereal grains, and insects.

(Source: Photo courtesy of Texas Parks & Wildlife Department.)

FIGURE 15-2 Bobwhite quail

Scaled Quail
(Callipela squamata)

DESCRIPTION

This quail is also known as the blue quail, and a common nickname is "cottontop," referring to the crest atop the bird's head (Figure 15-3). This crest is white on the males and buff or off-white and smaller on the females. These birds are a sandy brown to gray, with black-trimmed feathers, which gives them a distinctive "scaled" appearance. They are slightly larger than the bobwhite, being 10 to 11 inches in length and weighing 2 ounces or so more than the average bobwhite. They make a rather loud whistling call and much prefer to run than fly. Coveys are often quite large, sometimes including 40 or more birds.

HABITAT: Desert and brush areas of the southwest. Range extends from southeastern Arizona eastward through New Mexico, into southeastern Colorado and southwestern Kansas, and southward through the Oklahoma panhandle, west Texas, and into Mexico.

FEEDING HABITS: Mainly insects and seeds, particularly weed seeds.

(Source: Photo courtesy of Texas Parks & Wildlife Department.)

FIGURE 15-3 Scaled quail

California Quail
(Lophortyx californicus)

DESCRIPTION

These bluish-gray quail have a distinctive head plume that curves forward, the plume being longer and darker in males (Figure 15-4). Both males and females have a scaled look to their bellies, but the males have a dark face, with two bright white lines, and are a more slate-gray color than the females. Females tend to show less gray and more brown than the males. They are both about 10 inches in length. California quail are known for the large coveys they often form in winter, sometimes numbering 200 birds. These birds are found along the west coast from Baja

HABITAT: Desert brush areas, open rangelands, and even residential areas, most often where there is a steady supply of water.

FEEDING HABITS: Grasses, **legumes,** and seeds.

(Source: Courtesy of U.S. Fish and Wildlife Service. Photo by Lee Karney.)

FIGURE 15-4 California quail

California to Washington, and eastward into Nevada and portions of Idaho. These quail exist in a wide variety of climates.

Mountain Quail
(Oreortyz pictus)

DESCRIPTION ▪

Mountain quail are generally considered the largest quail in North America (Figure 15-5). They are 11 inches long and have a long, straight plume atop their heads. Males have a distinctive dark rust-colored throat patch outlined in white, with dark slate-gray head and breast. Their wings and back are a reddish brown, and their bellies have a scaled appearance, with black-and-white barring. The females have smaller head plumes and are duller in color. Mountain quail are found in smaller coveys than most other quail, generally 3 to 20 birds.

HABITAT: Western coastal brush thickets, dense forests, and the edges of mountain meadows. Range extends from northern Baja California up the west coast to northern Washington and into eastern Oregon, the extreme western edge of Nevada, and in several river drainages in western Idaho.

FEEDING HABITS: Mountain quail appear to eat more green vegetation in the form of grasses, legumes, and forbs than do other quail. They also eat a variety of weed, legume, and grass seeds and fruits.

(Source: © www.briansmallphoto.com)

FIGURE 15-5 Mountain quail

Gambel's Quail
(Lophortyz gambelii)

DESCRIPTION

The Gambel's quail is another of the crested or plumed quails (Figure 15-6). The males have a prominent black, forward-curving plume atop a reddish-brown cap. The black face is outlined with white, and the back and breast are gray. Females are duller colored, with a grayer tinge to their heads and faces, as well as shorter plumes. They are about 11 inches long and tend to gather in fairly large coveys, of up to 40 birds.

HABITAT: Gambel's quail are found primarily in the deserts of the southwest, ranging from extreme west Texas through southern New Mexico, Arizona, and southern California and north into southern Nevada. They also range southward well into Mexico. Isolated populations are also found in southern Colorado, Utah, and Idaho.

FEEDING HABITS: Gambel's quail feed on a wide variety of legume and weed seeds. The seeds, leaves, and flowers of many legumes are consumed. They appear to eat very little animal matter or cultivated grains.

(Source: Photo courtesy of Texas Parks & Wildlife Department © 2002.)

FIGURE 15-6 Gambel's quail

GROUSE (Tetraonidae)

This large family of birds includes prairie chickens, ptarmigans, and the sage, blue, spruce, sharptail, and ruffed grouses. The endangered Attwater's prairie chicken is a member of this family, as was the extinct heath hen. They inhabit a wide variety of North American habitats, from the arctic north to the mountains of the west, the forests of the northeast, and the Great Plains. Each species has fairly elaborate courtship displays. Prairie chickens and sage grouse are well known for their use of display grounds, or leks, to attract receptive females. As many as 100 males have been observed on a single lek, but 8 to 20 is probably the average. Often the same leks are used year after year. The ruffed grouse is well known for the males' *drumming* or rapid beating of wings to attract receptive females. This drumming usually takes place on a log or other suitable perch. The sound made by these wing beats is often likened to the muffled start of a gasoline motor. Sharptail grouse males use a dancing ground similar to the leks used by prairie chickens and sage grouse. The males congregate on these grounds and display to attract females, with the same males using the same grounds from year to year. Most female grouse, with the exception of the willow ptarmigan, get no assistance from the male in rearing the young.

LIFE CYCLE: The breeding season for the members of the grouse family ranges from late April in the southern **latitudes** to June in the more northern latitudes. After breeding takes place, the hens lay an average of one egg per day in a previously selected nest site until a clutch of 8 to 14 is complete. Incubation takes 21 to 23 days, and the chicks are led away from the nest site shortly after they dry. The chicks grow rapidly on a diet of insects and succulent green growth. As it is for most young wildlife, life is hazardous for the young chicks. Skunks, raccoons, opossums, and, where their ranges overlap, badgers destroy numerous nests and their eggs. Arctic foxes take a toll on the nests of northern-dwelling ptarmigan. If the hens dally too long on their nests, they may also be killed. Adults are killed and eaten by a wide variety of predators, such as foxes, bobcats, coyotes, and many different birds of prey. Cold spring rains, should they come before the chicks are fully feathered, can devastate entire populations. In the arid southwest and plains states, drought can be a serious limiting factor for grouse inhabiting those areas. Prolonged cold spells, particularly if accompanied by heavy snows, can also doom many birds. Individual grouse may live several years, but the average life expectancy is probably not more than 2 to 3 years. A very large number of the young live less than a year.

Ruffed Grouse
(Bonasa umbellus)

DESCRIPTION ▣

The ruffed grouse is a medium-size grouse of the forest and woodlands (Figure 15-7). They range from central Alaska south and east through the forests of Canada all the way to the east coast and then southward into the United States, as far south as eastern Tennessee and western North Carolina. They are also found along the west coast as far south as northern California and down the Rocky Mountains as far south as Utah. Minnesota, Wisconsin, and Michigan are ruffed grouse strongholds. Isolated pockets are also found in various other states. They get their name from the black feathers along their necks, which the males raise into a ruff during courtship displays. There are two fairly distinct color patterns—a red phase, in which the upper areas are a reddish brown, and a gray phase, in which those same areas are grayish. In both color patterns they are lighter and barred on

HABITAT: Preferably a wide variety of dense **deciduous** and coniferous forests, especially poplar, aspen, and birch. Greatest populations are found in climax forests, with abundant clearings or openings.

FEEDING HABITS: The buds of aspen and poplar are heavily used as food sources, as are green briar, sumac, and the buds of apple, willow, and wild cherry during winter. A wide variety of fruits and berries are eaten as they ripen. Ruffed grouse chicks, in contrast, survive mainly on insects.

(Source: Photo courtesy of Texas Parks & Wildlife Department.)

FIGURE 15-7 Ruffled grouse

the lower breast and belly. Both phases also have black feathers on the sides of the neck, as well as a band of black just before the tip of the tail feathers. Ruffed grouse are normally 16 to 17 inches in length. The females and young are quite frequently seen in the summer, and the males are often heard in the spring, when they drum for females.

Sharptail Grouse
(Pedioecetes phasianellus)

DESCRIPTION

The sharptail grouse is a large bird, closely related to the prairie chicken (Figure 15-8). They are a mottled pale brown, with some black markings, mainly spots. Their wings are white spotted and their bellies are lighter colored than the rest of their bodies. The males have a conspicuous purple air sac, which they inflate and display during courtship. They have a fairly long tail, which when spread comes to a rounded point. Other than the purple neck sacs and a slightly longer tail in the males, the two sexes look very much alike. One or more of the subspecies of sharptail can be found from central Alaska south and eastward into the Yukon territory of Canada. They are also found across much of Canada and southward into Minnesota, Wisconsin, Montana, North and South Dakota, Nebraska, and eastern Wyoming. Scattered populations are also found in Colorado, Utah, Idaho, eastern Oregon, and Washington. They have been reintroduced in western Kansas. They are about 17 inches in length.

HABITAT: Open grasslands, prairies, brush, and woodland edges are preferred habitat. They are often found in proximity to grain fields, where such fields are available.

FEEDING HABITS: They eat a wide variety of plants and some insects, such as grasshoppers. The buds of aspen, willow, birch, and ash are important winter food sources. A wide variety of fruits, seeds, and grains are eaten during the late summer and fall. Grains are particularly important in agricultural areas.

(Source: Courtesy of U.S. Fish and Wildlife Service.)

FIGURE 15-8 Sharptail grouse

Prairie Chickens

DESCRIPTION

There are two distinct species of prairie chickens, the greater (*Tympanuchus cupido*) and lesser (*Tympanuchus pallidicinctus*) (Figure 15-9). The greater prairie chicken is a large bird of 17 inches; the lesser is somewhat smaller, at 15 to 16 inches. The greater is a brownish bird with black barring on the sides and back. The belly is buff colored and lacks the dark barring. Males have a conspicuous yellow comb above the eyes and yellow air sacs that they inflate during courtship displays. Lesser prairie chickens look very similar to greaters, except that their barring is brown and their breast feathers show more brown and white barring than do those of the greater. The air sacs of the male lesser are a reddish purple. Both species have a series of pinnae feathers. These are long feathers on the neck that are raised during courtship displays. The tails of both species are rather short and rounded. Lesser prairie chickens are found primarily on the plains of eastern New

HABITAT: Prairie chickens require native prairie grasslands. Since the arrival of the steel plow, these areas have grown smaller each year. Lack of suitable habitat is a serious threat to prairie chickens.

FEEDING HABITS: Grains such as corn, sorghum, wheat, and rye are important winter food sources for today's prairie chickens. Prairie chickens do not appear to use buds as a primary food source, as many other grouse do. Succulent green growth and insects are important food sources in spring and summer.

(Source: Photo courtesy of Texas Parks & Wildlife Department © 2002.)

FIGURE 15-9 Lesser prairie chicken

Mexico, western Texas, the panhandle of Oklahoma, the southeastern tip of Colorado, and western Kansas. The greater type is found in eastern Kansas, western Missouri, central Nebraska, and South Dakota. Isolated populations are found in Illinois, Wisconsin, Minnesota, and North Dakota.

Sage Grouse
(Centrocercus urophasianus)

DESCRIPTION ▶

This is the largest North American grouse, with males weighing up to 5 pounds and being 28 inches in length (Figure 15-10). These large, dark grouse have narrow, pointed tails and feathers that reach to the base of their toes. They are dark grayish on their backs, with a dark belly, black throat, and white breast. Males have two large olive-green air sacs on the upper chest, which are inflated during courtship displays. Females are generally colored similar to the males but lack the air sacs and white breast of the males. These grouse are found wherever **sagebrush** is found in the western United States. Overgrazing and habitat destruction have greatly reduced their number, but they are still found in fairly large numbers in Wyoming, Montana, the Dakotas, southern Idaho, Nevada, eastern California, and the southeastern corner of Oregon, with smaller populations in Washington, Utah, Colorado, and northern New Mexico.

HABITAT: Sage grouse are found only where there is sagebrush habitat.

FEEDING HABITS: Adults feed almost exclusively on sagebrush. Young sage grouse eat insects, such as beetles, ants, and grasshoppers.

(Source: Courtesy of U.S. Fish and Wildlife Service. Photo by Gary Kramer.)

FIGURE 15-10 Sage grouse

Blue Grouse
(Dendragapus obscurus)

DESCRIPTION

Blue grouse live in conifer forests from southern Yukon and southeastern Alaska southward through western Canada and into Washington, Oregon, California, Idaho, western Montana, and the mountains of Wyoming, Nevada, Utah, Colorado, northern New Mexico, and northeastern Arizona (Figure 15-11). They are a large grouse, 17 to 23 inches long, with the males noticeably larger than the females. The males are a deep mottled gray, with a gray breast and belly and yellow-orange **combs** over their eyes. Both males and females have long, squared tails. During courtship the males display bare areas of their necks, which are a deep yellow to light orange. The females have smaller areas of bare skin and tend to be more brownish overall.

HABITAT: Blue grouse inhabit the western forests and forest edges. They spend a lot of time in mountain conifer forests.

FEEDING HABITS: Blue grouse depend heavily on the needles of the Douglas fir, ponderosa pine, and white fir, as well as a variety of berries and herbaceous plants.

(Source: © Muriel Lasure, 2010. Used under license from Shutterstock.com)

FIGURE 15-11 Blue grouse

SPRUCE GROUSE
(Canachites canadensis)

DESCRIPTION

Spruce grouse are widespread and are found throughout the northern spruce and pine forests (Figure 15-12). They are found in most of Alaska and Canada, and southward into the northeastern states of Maine, northern Vermont, New Hampshire, and New York. They also inhabit northern Michigan, Wisconsin, Minnesota, and the western states of Washington, far northwestern Oregon, Idaho, and western Montana and Wyoming. They are often seen on the ground or perched in trees and can be extremely tame and approachable. Adults are 15 to 17 inches long, with males being slightly larger than females. Both males and females have brown to blackish tail feathers that are unbarred, with a broad tan band at the ends. The males are a dark gray-brown on their backs, with a black throat and breast. They also have conspicuous red eye combs. Females tend to be a lighter reddish brown. They are heavily barred, except on their tails, and they lack the males' red eye combs.

HABITAT: Spruce grouse depend heavily on spruce trees, seeming to prefer young spruce forest to mature stands. They may also live in pines and prefer areas with some ground cover. The edges of bogs are also used.

FEEDING HABITS: Spruce grouse feed on the needles from a variety of trees, as well as fruits, berries, and grasshoppers.

(Source: Courtesy of U.S. Fish and Wildlife Service.)

FIGURE 15-12 Spruce grouse

Willow Ptarmigan
(Lagopus lagopus)

DESCRIPTION

The willow ptarmigan is widespread in the arctic tundra regions of North America (Figure 15-13). They are a plump, stocky grouse, about 15 inches in length. Unique among grouse, ptarmigan have feathers covering their feet during winter. Ptarmigan are also unique in that they change colors with the changing seasons, much like the snowshoe hare. Both male and female are white during the winter, but they turn a reddish brown in summer, with the female having dark barring and lighter buff bellies. The males have white wings and bellies during the summer. The males also have conspicuous red eye combs, particularly during the breeding season. They are found across most of Alaska and northern Canada.

HABITAT: Willow ptarmigan live in the arctic tundra, alpine mountains, and tundralike openings in the boreal forests of the far north. Also use low-growing scrub brush areas.

FEEDING HABITS: They feed heavily on the buds of willow, birch, and alder. They also consume insects and fruits during summer.

FIGURE 15-13 Willow ptarmigan

(Source: © Jack Conkhite, 2010. Used under license from Shutterstock.com)

Rock Ptarmigan
(Lagopus mutus)

DESCRIPTION

The rock ptarmigan is very similar to the willow ptarmigan but prefers the more barren, rocky areas of the arctic tundra (Figure 15-14). Rock ptarmigan also range farther north than do the willow ptarmigan. Both sexes are white in the winter but are mostly brown, barred with black and some white, during the summer. The red eye combs of male rock ptarmigan are somewhat less prominent than those of the willow ptarmigan. Both species are 13 to 15 inches long and have a blackish tail in all seasons. Rock ptarmigan are found in Alaska and across northern Canada. Both rock and willow ptarmigan may migrate south during the winter, although generally not very far.

HABITAT: Far northern arctic tundra habitat. Rock ptarmigan inhabit more barren, short vegetation areas than do the willow ptarmigan.

FEEDING HABITS: They feed heavily on buds and **catkins** during the winter and early spring. A wide variety of green vegetation, seeds, flowers, fruits, and berries are eaten during summer and fall. Young chicks appear to consume insects but begin eating green vegetation earlier in their lives than most grouse chicks.

(Source: Courtesy of U.S. Fish and Wildlife Service.)

FIGURE 15-14 Rock ptarmigan

Chukar
(Alectoris chukar)

DESCRIPTION ■

An introduced species originally from Europe, the chukar has done very well in the United States (Figure 15-15). Chukars have been introduced in many western states, with the largest populations in Nevada, Oregon, Washington, Idaho, Utah, Wyoming, western Colorado, and eastern California. Male and female chukars are identical in appearance, with the exception of small **spurs** on the males' legs. They have grayish backs, heads, and breasts, with a white throat and face outlined in black or dark brown. They have red bills and eyes and reddish feet. Bellies are buff and flanks are whitish, with contrasting dark, vertical barring. They are about 14 inches long and have a round, plump appearance. They prefer to run from danger, almost always going uphill. They form coveys in winter, sometimes with as many as 40 birds.

HABITAT: Bare rocky slopes, with talus slopes, bluffs, cliffs, and rocky outcroppings. Chukars flee upslope at the first hint of danger and can easily outdistance a human.

FEEDING HABITS: The seeds and leaves of cheat grass are the most important food in the chukar diet. They also eat the leaves of **alfalfa** and sweet clover when they are available.

LIFE CYCLE: The coveys usually begin to break up in February and March, and the males and females pair up. The female lays one egg every day or two until her clutch is complete. Clutch size varies considerably, from 10 to more than 20 eggs; 12 to 16 is probably about average. There is little firm evidence that the males assist the females in brood rearing, nor is there evidence to the contrary. Incubation takes about 24 days, and the hen leads the chicks away from the nest as soon as they are dry. The chicks feed on insects and succulent green vegetation and grow rapidly. Due to their speed on land and their habit of running uphill, ground-based predators are not normally a serious threat. Avian predators, however, are a different story. Because of the exposed slopes they inhabit, chukars are particularly vulnerable to birds of prey.

(Source: Photo courtesy of Texas Parks & Wildlife Department © 2002.)

FIGURE 15-15 Chukar

Gray Partridge
(Perdix perdix)

DESCRIPTION

Another introduced species that has done well in North America (Figure 15-16). A native of Europe, the gray or Hungarian partridge was first introduced to the United States before 1900. They are a medium-size bird 12 to 13 inches in length. They have a round, rather plump appearance and a complex color pattern. The sexes are similar, with a rust-colored face and throat, grayish breast, brownish back with fine white streaks, and gray flanks vertically barred in chestnut. The belly is a light gray to light brown, with males having a dark brown spot that is usually absent in females. They are a covey bird during much of the year, with coveys averaging 15 or so birds, rarely up to 30.

HABITAT: Gray partridge appear to do well in a habitat that contains a mixture of native grasses, hayfields, weed growth, and small grain fields.

FEEDING HABITS: Adults feed extensively on small grains, weed seeds, and green leafy materials. Insects make up only a small portion of the adults' diet during summer but are necessary for the chicks.

LIFE CYCLE: Coveys begin to break up in March, and the males and females are often paired before this occurs. The hen lays one egg per day until a clutch of 14 to 16 eggs is laid. When the clutch is complete, the hen incubates the eggs for 24 to 25 days. After hatching and drying, the chicks follow their parents from the nest site. The male gray partridge remains with the female and helps to rear the young. Mortality of young chicks is high, with chilling from cold spring rains a major culprit. The chicks that survive grow rapidly on a diet consisting chiefly of insects. The nests are often destroyed by skunks, raccoons, and farm equipment, but the female normally attempts to renest. Adults and chicks alike are eaten by a wide range of predators, such as foxes, bobcats, and various birds of prey.

(Source: © Karel Broz, 2010. Used under license from Shutterstock.com)

FIGURE 15-16 Gray partridge

Pheasant
(Phasianus colchicus)

DESCRIPTION

The pheasant is another introduced game bird. However, they have been here for over a century and are the descendants of many different species of pheasants (Figure 15-17). Many people therefore consider them American originals. George Washington was one of the first to import the birds, bringing several to his Mt. Vernon estate during his first presidency. However, the birds that came to stay were imported into Oregon's Willamette Valley in 1881. From an original release of 38 birds and many subsequent releases, the pheasant populated much of North America. They are large birds, with males reaching 36 inches and 2.5 to 3 pounds. Females are generally somewhat shorter and half a pound lighter. The males are unmistakable, having a dark, glossy head with bare red facial patches, an **iridescent** multicolored body that looks reddish brown from a distance, and a very long, pointed tail. The hens also have a long, pointed tail, although it is considerably shorter than a mature male's. Females are a much more drab buff, mottled with black and brown. Both sexes have short, rounded wings that propel them into an explosive flight. They are primarily a ground-dwelling bird that nests, feeds, and roosts on the ground. They are often seen feeding, sunning, or picking up grit along roadsides. Pheasants are found over a wide area of North America, with North and South Dakota, Iowa, Nebraska, and Kansas probably having the largest populations. However, they are also found in California, Oregon, Washington, Montana, Texas, the Oklahoma panhandle, eastern Colorado, and several states in the northeast. They are found in isolated populations in many other areas.

HABITAT: Pheasants prefer a mixture of croplands, grasslands, and brushlands. They also use swamps and other wetland areas as escape cover. They are most often found in **proximity** to grain crops.

FEEDING HABITS: Young chicks feed heavily on insects, but adults eat a wide variety of grains, seeds, insects, and vegetation. They prefer grains such as corn, grain sorghum, and wheat, and seeds such as sunflowers. They also feed on winter wheat when it is green and succulent.

LIFE CYCLE: Pheasants begin courtship and breeding in March and April over most of their range. A single male may breed several females that come to his territory. The hens generally lay one egg per day until a clutch of 8 to 12 is complete. Hens prefer to nest in fairly dense grassland vegetation; consequently, their nests are often destroyed by roadside mowers and hay equipment. They also nest in grain stubble, particularly wheat stubble, if it is tall enough. The hen leads the chicks from the nest site as soon as possible, and they begin to feed on a variety of insects. The chicks grow rapidly, and the family groups begin to break up in late summer or early fall. Pheasants are exceptionally hardy, capable of surviving under snow for several days. They prefer to run from danger but can fly 35 to 45 miles per hour when needed. Pheasants are killed and eaten by a wide variety of predators. Foxes, bobcats, coyotes, and various birds of prey all enjoy pheasant dinners. Skunks and raccoons destroy the nests whenever they can find them. It is estimated that less than 5 percent of all pheasants live to see their third spring. Annual turnover is 50 to 70 percent, with average life expectancy estimated at 7 to 8 months.

(Source: © Jack Scrivener, 2010. Used under license from Shutterstock.com)

FIGURE 15-17 Pheasant—rooster

Turkey
(Meleagris gallopavo)

DESCRIPTION

To the untrained eye all wild turkeys look very much the same. They are the largest game bird found in North America (Figure 15-18). Males, or toms, may reach 48 inches in length and 30 pounds or more in weight, but they probably average 42 inches and 20 to 22 pounds. Females, or hens, are considerably smaller, averaging around 30 inches and 8 to 12 pounds. Size varies considerably between species and among individuals of each species. Eastern turkeys generally produce the largest birds, with Merriam's being the smallest and the Rio Grande somewhere in between. They are an iridescent reddish bronze, with unfeathered, pale blue heads. The males have highly visible red **wattles** on their heads and necks, as well as a protruding "beard" on the breast. The beard is coarse and appears hair-like, but it is actually modified feathers. As a general rule, the older the tom, the longer and more noticeable the beard. The hens are generally smaller and duller colored, with less visible wattles; most do not have beards. These birds roost in trees and feed on the ground. They are most often seen foraging on the ground. They are very nervous and difficult to approach, possessing keen eyesight and excellent hearing.

The United States has several species of wild turkeys. The three major varieties we will deal with are the eastern turkey, the Rio Grande turkey, and the Merriam's turkey. All domestic turkeys are descendants of these wild turkeys.

HABITAT: The eastern turkey is found in mixed woodlands and along woodland edges from east Texas to the east and north throughout most of the eastern half of the United States. The Rio Grande is found from south Texas northward throughout most of the Great Plains states to Canada. They are located primarily along watercourses such as rivers, streams, and creeks. The Merriam's is a bird of the mountains and is found from far west Texas to the north and west throughout most of the mountain states of the West. All turkeys need standing water in the form of ponds, lakes, rivers, and streams to survive. Hence all turkeys, regardless of species, tend to live near water sources. Because of the available water and the big roost trees that normally line their banks, rivers, streams, and creeks are preferred habitat.

FEEDING HABITS: All turkeys feed primarily on seeds, nuts, insects, and fruits. Young birds feed heavily on insects during the early part of their lives. The types of seeds, nuts, and fruits vary between species and from location to location. However, acorns are a favorite food of adults, and grasshoppers are a favorite with young birds.

(Source: © Jeff Banke, 2010. Used under license from Shutterstock.com)

FIGURE 15-18 Wild turkeys

LIFE CYCLE: Turkeys are social birds, usually found in flocks, except during the nesting season. Spring in turkey country finds the early morning woods ringing with the "gobbles" of toms attempting to attract hens. The toms, or gobblers, as they are often called, select a spot to gobble from and to strut and display for potential mates. Toms gather a harem of hens in this manner and defend them from all rivals. Harems may range from two to as many hens as a tom can keep up with, with 8 to 10 probably being the maximum number. Around midmorning the hen leaves the tom and proceeds to her chosen nest site to lay an egg. The hens continue in this manner until egg-laying is complete. Normal clutch size varies from 6 to 20 or more eggs; average is about 8 to 12. The hen and young, known as poults, stay on the nest for about 24 hours after the last chick hatches. After this time the hen leads them from the nest, and they begin to feed on any insects they see and can catch. The poults grow rapidly and can fly short distances in about 10 days. They begin roosting in trees with the hen at about 2 weeks of age. Mortality is high. Cold spring rains claim many chicks, as do predators such as bobcats, coyotes, and foxes. Great horned owls take young and adults as they roost at night. Bobcats are a particularly damaging predator as they may climb the roost tree at night and kill several birds but feed on only one. In the wild, turkeys may live up to 5 years, but this is the exception. Three years is probably the average.

MIGRATORY UPLAND GAME BIRDS

Each of the species discussed next are migratory to some extent. They nest and raise their young in one area and move, or *migrate*, to another area to spend the fall and winter. Snipe, woodcock, and mourning doves migrate, generally southward, to spend the winter in warmer surroundings. Birds migrate for many reasons, but generally it is to take advantage of the best conditions in each area.

Mourning Dove
(Zenaida macroura)

DESCRIPTION

Mourning doves are by far our most common and widespread dove (Figure 15-19). In fact, mourning doves have the largest range of any North American game bird. Mourning doves have been found in Alaska and Greenland in summer, but they generally nest from Canada southward throughout the United States, all the way to the Bahamas. Winters are spent from Georgia to California and as far south as Panama. Mourning doves vary from 10 to 13 inches in length and 3.5 to 5 ounces in weight. They are an overall slate-gray color, with a buff to pinkish breast and belly, with occasional black spots on back and sides. They have a long, pointed tail and sharp-pointed wings that whistle at takeoff and when they are traveling at great speed. In their northern range doves begin their rather leisurely migration as early as September.

HABITAT: Mourning doves are found in a wide variety of habitats all across North America. They prefer field edges and open woodlands. They are also commonly found in urban and suburban areas.

FEEDING HABITS: Mourning doves feed on a wide variety of seeds, chiefly weed seeds. They prefer croton, sunflowers, and grains like sorghum and wheat.

LIFE CYCLE: Mourning doves may begin breeding in January or February in the southern portions of their range and as late as March and April in the north. They are **monogamous,** and the males assist the females in rearing the young, even sharing incubation chores. Mourning doves normally nest in bushes and trees, but their nests are haphazard affairs of a few twigs or sticks. Two eggs are laid, usually 2 days apart. Incubation takes 14 to 15 days, and the chicks are totally dependent on their parents until they leave the nest 30 days or so later. The young doves are fed a milky, glandular secretion from both parents' crops. After leaving the nest the young continue to feather out and mature. Generally the adult pair incubates another pair of eggs within a week of the young leaving the nest. Each adult pair may raise from two to five broods per year. However, few pairs actually succeed in raising the maximum number of young each year. Mortality is quite high. Most predation and death losses occur in the nest. The young and eggs are eaten by snakes and crows, or they are blown from the nest and eaten by ground-dwelling predators. Most adult mortality results from disease and starvation. Annual mortality ranges from 50 to 70 percent. Mourning doves may live for 6 to 9 years in the wild, although this age is rarely obtained.

(Source: Photo courtesy of Texas Parks & Wildlife Department.)

FIGURE 15-19 Mourning dove

Snipe
(Capella gallinago)

DESCRIPTION

Snipe are a rather widespread, common species, yet they are secretive and rarely seen, unless specifically sought out (Figure 15-20). They are small, averaging about 10 inches in length, with a compact build. They have short legs and a very long, straight bill. They are a mottled black-brown, with spots and bold buff-white stripes running the length of the body. They breed in Alaska, all across Canada, and as far south as northern California and New Jersey. They spend their winters far enough south to ensure unfrozen ground and some open water.

HABITAT: Snipe can be found in low-lying areas such as marshes and along stream courses, regardless of the time of year. Due to their feeding habits, snipe need soft, moist ground.

FEEDING HABITS: Snipe probe the ground with their long, straight bills, subsisting on earthworms, insects, and insect larvae for the most part.

LIFE CYCLE: Snipe begin to filter northward as early as January and February, but most eggs are laid from April to July, depending on the nesting bird's location. Four eggs are a normal clutch, and they are incubated for 18 to 20 days by the female. The young grow very rapidly and can begin flying at 14 to 18 days of age, although sustained flight usually does not occur until 20 or more days. The male may help in rearing and brooding the newly hatched chicks, although the female seems to shoulder much of the burden.

For the first week or so of their lives, the young chicks take food from their parents. By the age of 10 days or so, the young can feed themselves for the most part. Both the female and male **feign** injury in order to lure potential predators away from the nest or young. This consists of dragging a wing and flirting away, keeping just out of the predator's reach, until they feel they are at a safe distance from their young. When they are some distance off, they take flight and return to the brood, often leaving the predator wondering where its easy meal went. As with all ground-nesting birds, however, mortality is quite high. Foxes are probably the number one predator of the young and adults, but raccoons and skunks are known nest predators.

(Source: Photo courtesy of Texas Parks & Wildlife Department.)

FIGURE 15-20 Snipe

Woodcock
(Philobela minor)

DESCRIPTION

Woodcock are similar to snipe but are even more shy and reclusive (Figure 15-21). Woodcock are stockier than snipe and have a somewhat shorter bill. They are about 11 inches long, with a mottled black and buff upper body, dark brown wings with buff streaks, and a bright reddish or rust-colored belly. Their wings are shorter and more rounded than the snipe's. Their flight when flushed is usually not quite as erratic as the snipe's. Both sexes look pretty much the same, but the female is usually slightly larger. They breed from east Texas (rarely) north and east to southeastern Canada. Winters are spent in the southeast United States and southward.

HABITAT: Woodcock are found in both northern and southern wooded swamps, eastern lowland forests, and upland thickets. They prefer areas with soft, moist soils, where they can probe for food with their long bills.

FEEDING HABITS: Woodcock feed almost exclusively on earthworms, but they also eat slugs, sow bugs, and other insects during dry periods. It is estimated that 75 percent of their diet is earthworms.

LIFE CYCLE: Male woodcock do not appear to take an active role in rearing the young. The female lays a clutch of eggs, usually four, during March, April, or May and incubates them for 20 to 21 days. The young grow rapidly and can fly pretty well at 14 days of age. They begin to probe the ground for food within days of hatching and are almost full grown at 28 days of age. Young are generally on their own within 60 days. Woodcock appear to have only one brood per year, although they re-nest if their first nest is destroyed. The female woodcock fakes an injury, using a routine similar to the snipe's, to lure predators away from the young or nest. In the later stages of incubation, as the chicks are about to emerge, the females hold very tightly to the nest. This often spells their doom when a hungry fox or coyote is on the prowl. Raccoons and skunks prefer eggs, preying on the nests of all ground-nesting birds.

(Source: Courtesy of U.S. Fish and Wildlife Service.)

FIGURE 15-21 Woodcock

SUMMARY

Upland game birds are abundant, vital members of many of our ecosystems. Most have adapted well to human encroachment; however, some, most notably the bobwhite quail have not adapted well. Bobwhite numbers have declined dramatically over the past 20 to 30 years. Most birds, and particularly upland game birds, are at the top of many predators menu. It is therefore all the more vital that they have habitat of an adequate quality and quantity.

REVIEW QUESTIONS

Fill in the Blank

Fill in the blanks to complete the following statements.

1. All quail are _____ birds, enjoying the company of others of their species.
2. All species of quail are strong _____, usually preferring to run to escape danger.
3. Quail are also relatively _____ flyers, although only for short distances.
4. List two of the major predators of young and adult birds.
   (a) _____
   (b) _____
5. List two of the most common nest predators of quail.
   (a) _____
   (b) _____
6. As much as _____ to _____ of a given population may succumb to disease, cold, starvation, and predators each year.
7. _____ are the only quail native to the eastern United States.
8. A common nickname for the scaled quail is _____.
9. _____ quail have a distinctive head plume that curves forward.
10. _____ quail also have a head plume and are the largest quail in North America.
11. _____ quail are found primarily in the deserts of the Southwest.
12. _____ are a large family of birds that includes prairie chickens and ptarmigans.
13. The _____ is a medium-size grouse of the forests and woodlands.
14. The _____ grouse is a large bird, closely related to the prairie chicken.
15. _____ require native prairie grasslands.
16. The _____ is the largest North American grouse.
17. Spruce grouse are found throughout the northern _____ and _____ forests.

18. List three species of introduced (nonnative) game birds found in North America.

    (a) _____

    (b) _____

    (c) _____

19. List the three major varieties of wild turkeys found in the United States.

    (a) _____

    (b) _____

    (c) _____

20. Turkeys _____ in trees and generally feed on the ground.

21. All turkeys tend to live near permanent _____ sources.

22. Except during the nesting season, turkeys are social birds and are usually found in _____.

23. List three species of migratory upland game birds.

    (a) _____

    (b) _____

    (c) _____

24. _____ are far and away our most common dove.

25. Due to their feeding habits, _____ need soft, moist ground.

26. It is estimated that 75 percent of the woodcock's diet is _____.

Short Answer

1. Why do mourning doves, woodcock, and snipe migrate? How does this help them survive?

2. Of the introduced (nonnative) species of upland game birds, which has been most successful? Why?

Discussion

1. The quails and some other upland game birds are gregarious by nature. In your opinion, how does this help them survive?

2. Many of the species discussed in this chapter are ground-dwelling species. Raccoons and skunks are very efficient and effective predators that destroy many nests each year. Discuss ways and means to lessen the impact of these two predators on ground-nesting species.

Learning Activities

1. Draw a map of your state. Use different colors to illustrate the areas of your state that are known to hold a population of any of the species discussed in this chapter.

2. Select a species of upland game bird native to your area or state. List all the factors, such as weather, available habitat, food, and predators, which might limit the species' population in your area.

CHAPTER 16

Ducks, Geese, Swans, and Cranes

OBJECTIVES

After completing this chapter, you should be able to

- Identify the common ducks, geese, swans, and cranes found in the United States.

- Describe some of the behavior traits of each species.

- List the characteristics of each species covered in this chapter.

- Identify the type of habitat where you might find each species.

- List at least one food source for each species.

DUCKS (Anatidae)

There are two major groups of ducks found in North America—puddle, or surface-feeding ducks, and divers, or bottom-feeding ducks. This is somewhat misleading because most species of puddle ducks readily dive beneath the surface to feed. However, they do not dive as deep nor stay under as long as do the ducks in the diver group. Puddle ducks are very much at home on land, whereas divers are ungainly and awkward on land. Puddle ducks leap straight up from the surface of water when they take flight, but the diving ducks must have a "takeoff run" along the surface of the water to get airborne. The diving species feed almost entirely in the water, whereas the puddle duck species often feed on land. All species of ducks nest in fairly close proximity to water, often around lakes, marshes, and streams. Divers tend to require larger bodies of water than do puddle ducks. Puddle ducks use large lakes and saltwater bays, but they are just as likely to be found on a half-acre farm pond or a shallow **playa lake**. All of our ducks are migratory, at least to some extent, some much more so than others. Due to the similar nature of most ducks' life cycles, particularly puddle ducks, we use a common life cycle for the puddle ducks.

Life Cycle: For the most part, ducks pair up on their wintering grounds or on their migration north. When they reach their destination, usually close to where the hen was born, the hen constructs a nest of grasses, reeds, and leaves and lines it with **down**. Nests are usually very well concealed. The exception to this rule is the tree-nesting species, such as the wood duck. Wood duck hens select a nest site in a tree cavity or old woodpecker hole, often some distance from water. After hatching, young wood ducks leap from the nest to the ground, often 30 or more feet off the ground. The hen lays a clutch of 5 to 14 or more eggs, with an average of 8 to 10 eggs. When the hen begins to incubate the eggs the drake departs, leaving the hen with the sole responsibility for rearing the young. The hen leads the young ducklings to water, usually within 24 hours after they hatch. The ducklings grow rapidly, and those that survive are fully feathered by late August. The fall migration southward begins as early as September on the most northern nesting grounds. Some species, such as blue-winged teal, migrate southward very early in fall, whereas others, such as the mallard, stay in the north as long as there is open water. Mortality is quite high, especially among ducklings. The nests are destroyed and the eggs eaten by raccoons, skunks, and foxes. Foxes, bobcats, and lynxes are deadly predators of nesting hens and their broods. Numerous aerial predators also take ducks, the most adept of which is the peregrine falcon. Ducklings are also killed and eaten by turtles and large fish, such as northern pike, once they have reached the water. In the south, snapping turtles and largemouth bass take many ducklings.

PUDDLE DUCKS

Mallard
(Anas platyrhynchos)

DESCRIPTION

The mallard is our most common wild duck and the ancestor of most domestic ducks (Figure 16-1). They can breed with most domestic ducks, producing offspring in a wide variety of colors and shapes. Mallards are our largest puddle or "dabbling" duck, measuring 18 to 27 inches in length, averaging around 23 inches. The males are unmistakable, having a dark green head with a narrow white ring around the neck. Males have a chestnut breast and grayish body with several upturned feathers at the base of the tail. Both males and females have a purplish blue wing **speculum**, bordered in white. The females have a much drabber, mottled brown overall body color. Mallards range over most of North America, with a breeding range that reaches from Alaska and Greenland southward to Virginia, Texas, and northern Mexico. The vast majority nest in the prairie pothole regions of the northern United States and west-central Canada. Mallards spend the winter as far north as open water permits or as far south as Central America and the West Indies.

HABITAT: Ponds, marshes, and lakes are common mallard habitat, but mallards can be found on or around just about any body of water.

FEEDING HABITS: Mallards consume many different types of aquatic plants, insects, seeds, and small aquatic animals. They also eat acorns from flooded bottomland hardwoods and have a strong preference for grains such as corn, rice, and millet.

(Source: Photo courtesy of Texas Parks & Wildlife Department.)

FIGURE 16-1 Mallards—hen and drake.

Pintail
(Anas acuta)

DESCRIPTION

Pintails are slim, graceful ducks, with males having a long, pointed tail, from which the species gets its name (Figure 16-2). The males are very long, averaging 25 to 30 inches, while the females are 21 to 24 inches. Males have brown heads with a white line running up the neck to the head and a white breast and belly. Back and sides are grayish, with a long, pointed black tail. Females are a mottled grayish brown, similar to female mallards, but slimmer and lacking the purplish blue wing speculum of a hen mallard. Pintails breed from Alaska and Greenland south to Nebraska, northern Utah, and northern California. Some spend their winters as far north as open water allows, but many migrate great distances to Central America, the West Indies, and even the Hawaiian Islands.

HABITAT: Prairie potholes and marshes during the spring and summer, and almost any open water during the winter, including coastal saltwater marshes and bays.

FEEDING HABITS: Feeds mainly on aquatic vegetation but also eats seeds and grains when available. Pintail ducklings, like all young ducks, depend heavily on aquatic insects for their food supply.

(Source: Photo courtesy of Texas Parks & Wildlife Department © 2002.)

FIGURE 16-2 Pintail drake.

American Widgeon
(Anas americana)

DESCRIPTION ■

The widgeon is a widespread duck, but not as numerous as the mallard (Figure 16-3). They are 18 to 23 inches in length. Drakes have a conspicuous white head that has earned them the nickname "baldpate" and green bands that run from in front of the eyes to the base of the neck. The drake's sides, breast, and throat are a cinnamon-brownish color, and their bellies are white. Their wings and backs are dark brown or black, with a white shoulder patch that is visible in flight. The hens are a drab, mottled brown overall, darker over the back and lighter underneath, and with some grayish color to the head. Both sexes have white shoulder patches, which are visible in flight, as well as pale blue bills. During spring and summer the birds range from Alaska across Canada and into Minnesota, Nebraska, and California. They spend their winters on the Pacific coast, interior Southwest, the Gulf coast of Texas and Louisiana, and the Atlantic coast.

HABITAT: Commonly found on marshes, ponds, and shallow lakes.

FEEDING HABITS: Feed heavily on aquatic plants, often stealing choice tidbits from diving ducks. Also feed on seeds and grains and graze in meadows and winter wheat fields.

(Source: © David Nagy, 2010. Used under license from Shutterstock.com)

FIGURE 16-3 American widgeon.

Gadwall
(Anas strepera)

DESCRIPTION ■

The gadwall is a common but rather plain duck (Figure 16-4). It lacks the bright colors that so many of our waterfowl possess. Males are an overall grayish color, with some brown on the head. The lower tail feathers are black. Females are a mottled brown overall with a narrow gray bill that has an orange border. The best identifying mark on both sexes is the white wing speculum, which is visible in flight and unique to gadwalls. They are 18 to 22 inches in length. They can be found from North Carolina to California and northward into Canada and Alaska during the breeding season. They winter from New England south to the Gulf coast and westward across the Southwest to the Pacific coast.

HABITAT: Prefers freshwater marshes, ponds, and lakes; also found in saltwater marshes.

FEEDING HABITS: Feeds mainly on the seeds, stems, and leaves of aquatic plants. Comes ashore occasionally to feed on seeds and grain.

(Source: © Terry Alexander, 2010. Used under license from Shutterstock.com)

FIGURE 16-4 Gadwall.

Northern Shoveler
(Anas clypeata)

DESCRIPTION

The shoveler has a unique shovel-shaped bill that is longer than its head (Figure 16-5). The male has a green head, rust-colored sides and belly, and a white breast. The hen is a mottled brown overall. Both sexes show a blue-gray shoulder patch and a green speculum in flight. They are 17 to 20 inches long. The long bill gives the impression that the wings are set too far back when they are viewed in flight. These birds are rarely seen in the East, at least during the breeding season. They are found from northern California, New Mexico, and Missouri northward through western Canada to Alaska during spring and summer. In the winter they can be found along the Pacific coast, eastward to the Gulf coast, and up the East Coast as far as New Jersey on occasion.

HABITAT: Freshwater marshes and prairie potholes; occasionally found on **brackish** or saltwater marshes.

FEEDING HABITS: Uses its specialized bill to strain small aquatic organisms from ponds and marshes; also feeds on aquatic plants and seeds.

(Source: Courtesy of U.S. Fish and Wildlife Service. Photo by Donna Dewhurst.)

FIGURE 16-5 Northern shoveler—drake.

Black Duck
(Anas rubripes)

DESCRIPTION

The black duck is found mainly from central North America eastward (Figure 16-6). It is a common duck in the New England area and is closely related to the mallard. They are 19 to 23 inches long and a dark mottled brown color overall. The sexes look alike, and both sport conspicuous white wing linings and a purple speculum. Their breeding range is from North Carolina, West Virginia, Illinois, and Minnesota northward into Newfoundland, Quebec, and Manitoba. They spend the winter in New England and along the East Coast. They are rarely seen in the interior. They are often found in the company of mallards, with which they can interbreed.

HABITAT: Coastal mudflats, saltwater and freshwater marshes, lakes, streams, ponds, and tidal shallows.

FEEDING HABITS: Feeds chiefly on aquatic vegetation, such as eelgrass, wild celery, and various pondweeds. Also feeds on grasses, sedges, seeds, and grain.

(Source: Photo courtesy of USFW. Photo by Gene Nieminen.)

FIGURE 16-6 Black duck.

Wood Duck
(Aix sponsa)

DESCRIPTION ■

Sportspeople and conservation groups such as Ducks Unlimited rescued the wood duck from the brink of extinction. The draining of swamps and marshes and the cutting of timber throughout their range deprived the wood duck of nesting cavities. Wood ducks are a cavity-nesting species, often using old woodpecker holes as nest sites. Thousands of artificial nesting boxes were built and placed in suitable wood duck habitat. These efforts have allowed the wood duck to recover and become one of our more common ducks (Figure 16-7). The drakes have a crested head that is multicolored, with greens, blues, and purples. The throat and upper breast are brown, with a bright white line separating the breast from the buff-colored sides and belly. The drakes have dark brown wings and backs. The females are an overall grayish brown, with darker backs and paler breasts. Females

HABITAT: Prefer wooded ponds, rivers, streams, northern wooded swamps, and southern wooded swamps. Also found along heavily wooded lakes and marshes.

FEEDING HABITS: Wood ducks feed on a variety of aquatic plants, as well as acorns and grains when available.

(Source: Photo courtesy of U.S. Fish and Wildlife Service. Photo by David Menhe.)

FIGURE 16-7 Wood ducks–drake.

have a distinctive white eye patch and pale gray bellies. Wood ducks are one of our most beautiful ducks. They range across most of the eastern United States and north into Canada. They also nest in the western United States and north into western Canada. They are absent from much of the interior United States but are expanding their range. Wood ducks spend the winters as far north as coastal British Columbia and New England, as well as in many southern states.

Green-Winged Teal
(Anas crecca)

DESCRIPTION

The green-winged teal is the smallest of our dabbling ducks, being only about 15 inches long (Figure 16-8). Adult males are beautiful birds, with a dark chestnut head and a green band in front of and behind the eyes. They have a gray body overall, with a vertical white band on the sides of the speckled pinkish breast. Females are a dark grayish brown, and mottled overall. Both sexes have bright green wing speculums that are highly visible in flight. Green-winged teal breed from Alaska southward throughout much of Canada and down into the northern third of the United States. They spend their winters in the southern United States, Mexico, the Bahamas, and West Indies. Some remain as far north as open water allows. Green-winged teal are one of our fastest-flying ducks, reaching speeds greater than 40 miles per hour.

HABITAT: Freshwater marshes, prairie potholes, ponds, sloughs, and lakes. Prefer freshwater but does frequent saltwater marshes during the winter.

FEEDING HABITS: Green-winged teal eat a variety of aquatic plants, as well as seeds and grains. They are particularly fond of "puddling" about in very shallow water or on land for their food.

(Source: © Terry Alexander, 2010. Used under license from Shutterstock.com)

FIGURE 16-8 Green-winged teal.

Blue-Winged Teal
(Anas discors)

DESCRIPTION ■

Blue-winged teal are a small, brown duck that is quite abundant in much of central North America (Figure 16-9). Adults are 15 to 16 inches long. Mature drakes have a slategray head, with a distinct white crescent or line that runs from below the chin to above the eye. Overall body color of both sexes is a mottled brown. Both sexes exhibit a blue-gray patch on their forewings that is conspicuous in flight. The wing patches are usually brighter and more visible mature males than on females or immature birds. Blue-winged teal nest across southern Canada and south into United States, from northern New Mexico eastward to Maine. They are one of the first ducks to head south in late summer, spending their winters in the southern United States, Mexico, the West Indies, and as far south as Brazil and Chile.

HABITAT: Freshwater marshes, ponds, prairie potholes, **sloughs**, lakes, and saltwater marshes are all used by blue-winged teal.

FEEDING HABITS: Blue-winged teal feed on a variety of aquatic plants, seeds, and grains.

(Source: photo courtesy of texas Parks & Wildlife Department.)

FIGURE 16-9 Blue-winged teal.

Cinnamon Teal
(Anas cyanoptera)

DESCRIPTION

Adult drake cinnamon teal are a deep reddish cinnamon color over the majority of their bodies, with some dark brown on the wings (Figure 16-10). Females are a mottled brown overall, but both sexes have blue-gray forewing patches, similar to blue-winged teal, that are easily seen in flight. Like all teal, they are small birds, averaging about 16 inches in length, and capable of great speed in flight. They are primarily a bird of western North America, breeding from southwestern Canada south to west Texas and Mexico. The western Great Plains are generally the eastern edge of their breeding range. They spend their winters in southern California, New Mexico, Texas, and southward. They are occasionally seen in the eastern United States.

HABITAT: Lakes, ponds, saltwater and freshwater marshes, and prairie potholes.

FEEDING HABITS: Feeds on aquatic plants such as sedges and pondweed, and small organisms that it skims from the surface of the water.

(Source: Photo courtesy of Texas Parks & Wildlife Department © 2002.)

FIGURE 16-10 FIGURE 16-10 Cinnamon teal.

DIVER DUCKS

Redhead
(Aythya americana)

DESCRIPTION

Redheads are one of our more common diving ducks found on many of our larger bodies of water (Figure 16-11). The drakes have a dark cinnamon head, gray sides and back, and a black breast and rear end. The females are reddish brown overall with a whitish belly. Both sexes have a pale blue bill with a black tip. They are 18 to 22 inches long, and like all diving ducks, they must "patter," or run along the surface of the water for some distance, before becoming airborne. They breed from Alaska and western Canada to California and across to Colorado and Iowa. They rarely breed in the Great Lakes region but do use the area as winter habitat. They also winter in California and along the Atlantic coast.

HABITAT: Most commonly nest in marshes but are found on lakes, reservoirs, bays, and saltwater marshes during the winter.

FEEDING HABITS: Feed primarily at night and on aquatic vegetation. Redheads dive and feed on the stems, roots, and leaves of various forms of aquatic vegetation.

FIGURE 16-11 Redhead duck.

(Source: © Daniel Hebert, 2010. Used under license from Shutterstock.com)

Canvasback
(Aythya valisineria)

DESCRIPTION

Canvasbacks are large ducks, measuring 20 to 24 inches in length, with a long bill and a distinctive sloping forehead (Figure 16-12). Drakes have a dark cinnamon head and neck, black rear end and breast, and a whitish body. They look very similar to the drake redhead. Hens have a pale brown head, neck, and breast, with a grayish brown upper body. They breed in Alaska and southward into the Northwest Territories and Manitoba. Main breeding areas in the United States are northern California, Colorado, northwest Iowa, and southern Minnesota. They spend their winters from British Columbia southward to Arizona and on the Atlantic coast and Great Lakes region in the east.

HABITAT: Prefer to breed on large freshwater marshes; also found on lakes, reservoirs, saltwater marshes, bays, and estuaries.

FEEDING HABITS: Feed almost exclusively on aquatic plant life, chiefly wild celery.

(Source: © Daniel Hebert, 2010. Used under license from Shutterstock.com)

FIGURE 16-12 Canvasback duck.

Ring-Necked Duck
(Aythya collaris)

DESCRIPTION

A small, stocky diving duck, the ring-necked duck measures 14 to 18 inches in length (Figure 16-13). The male has a black breast, head, and back, with pale gray flanks and a vertical white mark on the side of the breast. The female is a pale grayish brown with a light-colored ring around the eye. Both sexes have a gray bill with a white ring at the base and a black tip. Ring-necked ducks are more common in the interior of the country than on either coast. They breed from northern New England, westward to Washington and north across much of Canada to Alaska. They spend their winters virtually throughout the United States wherever they can find open water.

HABITAT: Prefers lakes, reservoirs, and inland waters. Rarely seen near salt water.

FEEDING HABITS: Aquatic plants and their seeds, snails, and insects.

(Source: © 7877074640, 2010. Used under license from Shutterstock.com)

FIGURE 16-13 Ring-necked duck.

Common Goldeneye
(Bucephala clangula)

DESCRIPTION

Common goldeneyes are stocky, compact diving ducks, measuring 17 to 20 inches in length (Figure 16-14). The drakes have a dark green, glossy head, with a roundish white spot near the base of the bill. Their backs are black, while their breasts and sides are white. The males also have white areas on the backs and wings that are most visible during flight. The hens have a light brown head and gray back and sides. Both sexes have rounded heads, short bills, and white wing patches that are visible during flight. Their breeding range extends from Alaska, southward across much of Canada, and into the northern portion of the United States. They are one of several species of diving duck that nests in tree cavities. They spend their winters along both coasts and across the interior of the United States.

HABITAT: Freshwater lakes, reservoirs, and rivers.

FEEDING HABITS: Aquatic plants, insects, and **mollusks**.

(Source: Photo courtesy of Texas Parks & Wildlife Department © 2002.)

FIGURE 16-14 Common goldeneyes.

Bufflehead
(Bucephala albeola)

DESCRIPTION

The bufflehead is the smallest duck in North America, averaging 13 to 14 inches in length (Figure 16-15). They are very stocky and are often called by their nickname, "Butterball." The drakes have a dark greenish purple head, black back, and white breast and sides. Males also have a large white patch that reaches from behind the eye to the back of the head. The white wing patches of the males are visible in flight, and the females have a small white area at the base of the wings that is visible during flight. Hens are an overall grayish brown, with an elongated white spot on the face. Buffleheads breed from Alaska southward across Canada into the northern portion of the United States. Buffleheads are another diving duck that uses cavities in trees as nest sites. Their winters are spent along both coasts as well as in the interior United States.

HABITAT: Inland waterways, lakes, and reservoirs, as well as bays and **inlets** along both coasts during the winter.

FEEDING HABITS: Many forms of small marine life, as well as minnows, insects, and snails.

(Source: © 7877074640, 2010. Used under license from Shutterstock.com)

FIGURE 16-15 Bufflehead.

Greater Scaup
(Aythya marila)

DESCRIPTION

The greater scaup is 17 to 21 inches long and is most often seen in large groups or "rafts" on large inland lakes and saltwater bays (Figure 16-16). The drake has a very dark greenish black head and a black breast and rump, with speckled gray back and white sides. The female has a broad white band encircling the face at the base of the bill but is dark brown overall. Both species have a pale blue bill and a white wing stripe that is visible during flight. Breeding range appears to be confined to Alaska and arctic Canada. Scaup spend the winters along both the Atlantic and Pacific coasts and on the Great Lakes.

HABITAT: Freshwater marshes, lakes, reservoirs, and ponds during spring and summer. Bays, large inland lakes, and sheltered coastal areas, as well as open ocean during the winter.

FEEDING HABITS: Mollusks, insects, and aquatic plants such as eelgrass and wild celery.

(Source: © Marci Paravia, 2010. Used under license from Shutterstock.com)

FIGURE 16-16 Greater scaup.

Lesser Scaup
(Aythya affinis)

DESCRIPTION

The lesser scaup is closely related to the greater and closely resembles it, except for a higher crown to the head (Figure 16-17). The adult male has a black rump, tail, and breast, with a dark purplish black glossy head. The back is speckled with gray and white, and the lower sides are pale gray or white. The hen has a dark brown head, with a white patch at the base of the bill. Overall body color is brownish gray. Lesser scaup are 15 to 18 inches in length. They breed from Alaska southward across Canada to the Great Lakes region, northern California, Nevada, and Colorado. They spend the winters along both coasts as well as on interior lakes.

HABITAT: Lakes, reservoirs, inland waterways, and freshwater marshes. Found on rivers, ponds, and protected coastal waters. The lesser scaup has more of a preference for freshwater than does the greater.

FEEDING HABITS: Very similar to that of the greater scaup.

(Source: © Ian Davies, 2010. Used under license from Shutterstock.com)

FIGURE 16-17 Lesser scaup.

GEESE (ANATIDAE)

Each of the species of geese discussed here has several similar characteristics. Geese, with the exception of the smaller species of Canada geese, are large birds. They have longer necks than ducks do and their nesting grounds are farther north than most species of ducks. Therefore, goose populations are less affected by droughts than are duck populations. Most geese feed heavily in fields and retire to water to drink and roost. Both male and female geese are identical in appearance, unlike ducks. Most male geese assist the females in rearing the young. Both sexes can be very aggressive in defense of their nests and young.

Life Cycle: Most geese nest above the tree line on the arctic tundra of Alaska and Canada. Geese pair up on their wintering grounds and begin the long flight north in late winter or early spring. Some, especially Canada geese, are known to mate for life. However, should its mate die or be killed, a goose will select another. The female constructs a nest of grasses, lined with down, usually not too distant from a body of water. The female may lay 4 to 12 eggs, but 6 to 10 are a normal clutch. The female incubates the eggs while the **gander** stands guard over her and the nest. Both parents lead the newly hatched **goslings** to water and watch over them as they grow. By late summer or early fall the young geese are fully **fledged** and are ready for the long migration south. Geese stay in the north as long as food and open water are available. Because most geese nest beyond the northern limit of most predators' ranges, the arctic fox and various birds, such as gulls, are the main nest predators. However, geese are extremely alert and attentive to their nests and do not hesitate to attack any potential predator. Adults are killed and eaten by foxes, lynxes, eagles, and occasionally wolves. Geese tend to group in large flocks during the fall and winter and are very difficult to approach. They possess excellent hearing and eyesight. In the wild, geese may live 10 years or more.

Canada Goose
(Bronta canadensis)

DESCRIPTION

The Canada goose is easily the most recognized waterfowl in North America (Figure 16-18). Until recently they were the most numerous. Canada geese come in a wide variety of sizes because there are well over a dozen separate subspecies in North America. Some, such as the cackling goose, are no larger than a mallard duck. Others, such as the giant Canada, may stand 45 inches tall and weigh 15 pounds. They are all, however, the same basic color. Head and neck are black, with a bright white cheek patch that extends under the chin. Overall body color is a grayish brown, with white lower tail feathers. The white cheek patch makes the Canada goose easy to distinguish from all other geese. Most people have seen the large Vs of geese winging southward and heard their rich, musical honking. They breed from the Aleutian Islands off Alaska across most of the Alaskan and Canadian arctic tundra. Local populations can be found nesting throughout the United States. Canada geese return to the site where they were raised, to raise their own young; therefore, they have been introduced across much of

HABITAT: Arctic tundra, freshwater marshes, lakes, reservoirs, rivers, and even city parks.

FEEDING HABITS: Young goslings feed heavily on insects early in their lives. Adults eat a wide variety of succulent green grasses, winter wheat, and grains, being especially fond of corn.

(Source: Photo courtesy of Texas Parks & Wildlife Department.)

FIGURE 16-18 Canada goose.

(continued)

Canada Goose
(Bronta canadensis)

America. Canada geese spend the winter across much of the lower 48 states, with large concentrations in the Chesapeake Bay area, the rice fields of south Texas, and Louisiana, and the grain fields of west Texas.

Snow Goose
(Chen caerulescens)

DESCRIPTION

The snow goose comes in two distinct color phases, with the dark one known as "blue" goose (Figure 16-19). For a number of years it was actually thought that these two geese, because of the differences in their colors, were different species. We now know this to be untrue, and in fact the two color phases interbreed. The blue color phase snow goose has a white head and neck and a slate-gray body. The light color phase is pure white overall, except for black wing tips. Both color phases have black wing tips. It is a fairly large goose, averaging 28 to 30 inches in length. They often migrate in extremely large flocks and are seen in flight in long wavy lines. Only in the past few years have snow goose populations exceeded those of the Canada goose, making the snow goose the most numerous goose in North America. They breed on the arctic tundra of Alaska and Canada. They winter mainly along the Gulf coast of Texas and Louisiana, the interior Southwest, California, and along the Atlantic coast from New Jersey to the Carolinas.

HABITAT: Arctic tundra, freshwater marshes, lakes, and saltwater marshes.

FEEDING HABITS: Young feed primarily on insects early in their lives. Adults feed on a variety of **sedges** and grasses, as well as seeds and grain.

(Source: Courtesy of U.S. Fish and Wildlife Service. Photo by Dave Menke.)

FIGURE 16-19 Snow geese.

Greater White-Front Goose
(Anser albifrons)

DESCRIPTION

These are medium-size geese, averaging 27 to 30 inches in length (Figure 16-20). They have a largely grayish brown body with a white patch on the face, just back of the pinkish orange bill. The lower rump just under the tail is white, and the belly has a variable amount of black spotting or barring, which accounts for their other name of "speckled-belly geese." They breed in arctic Alaska and Canada and spend their winters in the western United States. Occasionally seen on the Atlantic coast, they winter mainly in Pacific coast states and the Gulf coasts of Texas and Louisiana. They are often seen in large flocks and will intermingle with other geese.

HABITAT: Arctic tundra and marshes in spring and summer; lakes, bays, open country, and saltwater marshes in winter.

FEEDING HABITS: Young feed on insects, and adults graze on a variety of grasses and sedges. They are often seen in fields and pastures, grazing like cattle and are also known to feed in stubble fields on waste grain.

(Source: Photo courtesy of Texas Parks & Wildlife Department © 2002.)

FIGURE 16-20 Greater white-front geese.

Brant
(Bronta bernicla)

DESCRIPTION ■

Brant are more compactly built and have shorter necks than most geese (Figure 16-21). They also inhabit salt water almost exclusively. They are 23 to 30 inches in length. Brant have a dark blue-black head and neck, with a white collar or spot on their necks. The birds found in the eastern part of North America tend to have a smaller spot of white on the neck and whitish bellies. The birds along the West Coast have a larger white collar and blackish bellies. Birds on both coasts have a largely grayish brown back and white under-tail **coverts**. They breed on the tundra of Alaska and Canada. Some eastern birds nest on the coast of Greenland. Winters are spent along the east and west coasts of the United States.

HABITAT: Tundra and coastal islands during the breeding season; coastal marshes, estuaries, and sheltered coastal bays and inlets during the winter.

FEEDING HABITS: Grazes on various sedges and grasses during spring and summer; feeds almost exclusively on eelgrass in winter.

(Source: Courtesy of U.S. Fish and Wildlife Service. Photo by Don Becker, USGS EROS.)

FIGURE 16-21 Brant.

CRANES (GRUIDAE)

Sandhill Crane
(Grus canadensis)

DESCRIPTION

An extremely large bird, which may stand 4 feet tall, sandhill cranes are virtually unmistakable (Figure 16-22). Adults are gray overall, with a bare red patch on their foreheads, and occasional rust-colored spots on their sides. Immature birds lack the bare red patch and are more reddish brown. Males and females are identical in appearance, except that the females may be slightly smaller. They have very long legs and a wing span of 60 or more inches. They nest in northern Alaska and across Canada. They winter in California, throughout the Southwest to Texas. They may also be seen in Florida. Sandhill cranes travel in very large, usually noisy flocks.

Sandhill cranes are the only cranes that can legally be hunted at this time. They are numerous and can often be seen feeding in the same fields with geese, although the cranes and geese do not usually intermingle.

HABITAT: Sandhill cranes breed in marshes and wetland prairies and winter in open country, often roosting in shallow water at night.

FEEDING HABITS: They feed on a variety of vegetation, as well as grains and seeds. They also feed on mice and insects when available.

LIFE CYCLE: These birds migrate northward in large numbers, using the Platte River in Nebraska as a staging area. They nest in wetland prairies and marshes across much of Canada. The female selects a nest site and constructs a nest of grasses and reeds. The female lays and incubates two eggs, and after they hatch both male and female rear them. The young grow rapidly on a diet of insects, frogs, and any other small creatures they can catch. They have few natural enemies because their long, pointed bills are formidable weapons, capable of inflicting considerable damage on any potential predator. Adults are fearless in the defense of their young, and any wounded bird is quite dangerous. Humans are their greatest enemies, hunting them for the excellent sport and food they provide. They are extremely alert, however, possessing excellent eyesight and hearing, and thus nearly impossible to approach. They also tend to stay in large flocks, which give them many eyes and ears for their defense.

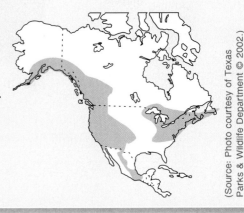

FIGURE 16-22 Sandhill cranes.

(Source: Photo courtesy of Texas Parks & Wildlife Department © 2002.)

SWANS (ANATIDAE)

Whistling or Tundra Swan
(Olor columbianus)

DESCRIPTION

The tundra swan is the most widespread of the swans in North America (Figure 16-23). It is a large bird, being 48 to 55 inches in length. Adults are solid white, with a black bill and a yellow spot in front of each eye. Immature birds are a pale gray, with a black-tipped, pinkish bill. Adults and juveniles alike have the characteristic long swan neck. They are often seen migrating in large flocks and give a melodious "hoo-hoo, hoo-hoo-hoo" call. Tundra swans nest where the name suggests, on the arctic tundra of Alaska and Canada. They spend their winters on the Pacific and Atlantic coasts.

HABITAT: Freshwater marshes, lakes, reservoirs, and other inland waters. They winter on saltwater marshes and coastal bays.

FEEDING HABITS: Feeds mainly on various forms of aquatic vegetation. Tundra swans use their long necks to reach below the water surface for roots and **tubers** of aquatic plants.

LIFE CYCLE: After spending the winter in the South, the tundra swan migrates north to nesting grounds in the Arctic in early spring. The mated pair constructs a huge nest of mud and grasses into which the female lays 4 to 6 eggs. The male guards the female and the nest during incubation. Both parents assist in rearing the young and very aggressively defend the young and the nest. Because of their formidable defense, natural predators are few, although arctic foxes may snatch an unwary young or sick bird. The young grow rapidly and are ready to migrate southward with the adults in early fall.

(Source: Photo courtesy of Texas Parks & Wildlife Department.)

FIGURE 16-23 Whistling swan.

Trumpeter Swan
(Olor buccinator)

DESCRIPTION

The trumpeter swan is the largest of North America's waterfowl (Figure 16-24). It is 60 inches or more in length, with an 8-foot wing span. The adults are solid white with black bills. Immature birds are a dusky gray-brown. There are two distinct populations. One is resident on interior lakes of Idaho, Montana, and Wyoming, and the other nests in arctic Alaska and Canada and winters along the coast from Alaska to Washington. These are immense birds, with gracefully arched necks.

HABITAT: Inland lakes, reservoirs, coastal bays and inlets, and tundra.

FEEDING HABITS: Feed on a variety of aquatic plants, as well as grasses and other succulent green growth.

LIFE CYCLE: Adults build a huge nest of mud, grass, and feathers. A female normally lays a clutch of 4 to 6 eggs, which she and the male fearlessly guard. The young swans feed on any insect they can catch, and they grow rapidly. They have few predators. The adults are so large and aggressive that they will attack a human who approaches the young or nest too closely. Swans generally live relatively long lives.

(Source: Photo courtesy of U.S. Fish and Wildlife Service.)

FIGURE 16-24 Trumpeter swan and young.

Mute Swan
(Cygnus olor)

DESCRIPTION

This large swan was imported from Europe and is fairly common in parks and inland lakes in the Northeast (Figure 16-25). A very large bird, 60 inches or so in length, it has a very long curved neck. Adults are pure white, with an orange bill that has a black knob at its base. The knob is larger and more prominent in males. Immature birds are a gray-brown color. Mute swans can be very aggressive, particularly in defense of their nests.

HABITAT: Inland lakes, ponds, and parks.

FEEDING HABITS AND LIFE CYCLE: Similar to other swans, but they are non-migratory.

(Source: Photo courtesy of Texas Parks & Wildlife Department © 2002.)

FIGURE 16-25 Mute swans.

SUMMARY

The majority of our ducks, geese, swans, and cranes are doing extremely well. Farming practices have helped to increase waterfowl numbers, especially geese in recent years. Waste grain is an excellent source of protein and energy for waterfowl. Consequently, most are returning to their nesting grounds each spring in excellent physical condition which allows larger clutches of eggs to be laid and healthier hatchlings. Snow geese numbers have increased to the point that they are doing very serious damage to their arctic nesting grounds. However, drought and continued destruction of the prairie potholes on which ducks depend for nesting habitat will continue to be a serious issue.

REVIEW QUESTIONS

Fill in the Blank

Fill in the blanks to complete the following statements.

1. The two major groups of ducks found in North America are puddle ducks and _____.

2. _____ ducks leap straight up from the surface of water when taking flight.

3. Diving species of ducks feed almost exclusively in the water, whereas puddle duck species often feed on _____.

4. Duck _____ are generally very well concealed.

5. The _____ are a cavity-nesting species, with the hen using an artificial nesting box or a natural tree cavity.

6. List three common nest predators of waterfowl.

   (a) _____

   (b) _____

   (c) _____

7. List two aquatic predators of ducklings.

   (a) _____

   (b) _____

8. The _____ is our most common wild duck.

9. Pintail drakes have a long, pointed _____, from which the species got its name.

10. List five common species of puddle ducks.

   (a) _____

   (b) _____

   (c) _____

   (d) _____

   (e) _____

11. List five common species of diver ducks.

   (a) _____

   (b) _____

   (c) _____

   (d) _____

   (e) _____

12. Both male and female _____ look alike, unlike ducks.

13. Most geese nest above the tree line on the _____ of Alaska and Canada.

14. The _____ is the main four-legged predator of most goose nests.

15. The young of both ducks and geese grow rapidly on a diet consisting primarily of _____.

16. Our most common saltwater or sea goose is the _____.

17. The _____ crane is the only crane that can legally be hunted at this time.

18. The _____ is the most widespread of the swans found in North America.

19. The _____ is the largest of North American waterfowl.
20. Most geese and swans can be very aggressive, especially in defense of their _____.
21. The _____ is a large swan imported from Europe.

Short Answer

1. What are some of the more common foods eaten by adult puddle ducks?
2. Why do diver ducks generally prefer larger, deeper bodies of water than do puddle ducks?
3. What are some of the more common foods eaten by adult diver ducks?

Discussion

1. Discuss the advantages and disadvantages of waterfowls' migratory nature.
2. In your opinion, how does the migratory nature of waterfowl make them more difficult to manage? Why?

Learning Activities

1. Compile a list of the species of puddle ducks that nest, migrate, or winter in your area or state, with their approximate numbers. What types of habitat do they utilize?
2. Compile a list of predators native to your area or state that might enjoy a duck dinner. Does this predation affect waterfowl numbers in your area? If so, how?
3. Select a species of waterfowl not discussed in this chapter but native to your area, if possible. Using your local library and state fish and wildlife agency for information, write a detailed account of the species. Include a description, preferred habitat, life cycle, and feeding habits. Present your report to your class.

Useful Web Site

<http://www.mbr-pwrc.usgs.gov>

CHAPTER 17

Songbirds and Other Common Birds

OBJECTIVES

After completing this chapter, you should be able to

- Identify some of the birds and songbirds commonly observed in the United States.
- Describe some of the behavior traits of each species.
- Identify the type of habitat where each species might be seen.
- List the characteristics of each species.

INTRODUCTION

Several hundred species of birds are found in the United States. Many of these are common only in certain areas or regions. As we do not have the space necessary to discuss all of them, 20 birds have been selected. Selection was based mainly on how widespread and common the birds are. Admittedly, some birds may be more common in your state or area than the ones listed here. If you want additional information on a bird, a trip to the library may be in order. Many excellent references are available at most public libraries that cover all birds native to the United States. An excellent online source for bird identification is the Web site of the Patuxent Bird Identification Information Center at www.mbr-pwrc.usgs.gov. Avian predators and shorebirds, egrets, and herons are covered in Chapters 18 and 19. Most, but not all, of the birds in this chapter can be classified as **songbirds**. The remainder are birds that are fairly widespread and commonly observed.

Winter losses of songbirds to disease, starvation, and extreme cold are quite high. To help our avian friends through the hard times, we can maintain feeders with a good mix of seeds. Once the birds become accustomed to a feeder, however, it must be stocked regularly. Allowing feeders to become empty can cause birds that have come to rely on the feeders severe hardship and even death.

Northern Mockingbird
(Mimus polyglottos)

DESCRIPTION

Common from Maine to northern California, the mockingbird is an active bird, usually easily observed (Figure 17-1). It is about 10 inches in length, with a fairly long tail. Mockingbirds have a largely gray body with a whitish gray breast and belly. The wings and tail are darker, with white patches that are visible during flight. Mockingbirds often perch on fences, power lines, and the outer branches of bushes and trees, where they are very visible. They are very vocal, with a multitude of calls, and they mimic other bird calls. They have also been known to mimic squeaky car doors, barking dogs, and other common neighborhood noises. They are generally non-migratory, normally found in the same area year-round.

HABITAT: They inhabit cities and suburbs, as well as areas of the countryside where adequate brush and trees are available. Mockingbirds use a tremendous variety of habitat, ranging from desert to forest thickets to residential areas.

FEEDING HABITS: They eat a wide variety of seeds, berries, and insects.

LIFE CYCLE: Courtship, breeding, and nest building occur in spring. Females lay 4 to 6 eggs in a loosely built nest of twigs, leaves, or even rags. As with most songbirds, both parents cooperate to rear the young. Parents are very protective of the nest and young, fearlessly attacking potential predators. Many a house cat has been set upon while simply strolling around the neighborhood. Snakes, house cats, raccoons, and various birds of prey are the most common predators of mockingbird eggs, young, and adults.

(Source: Courtesy of U.S. Fish and Wildlife Service. Photo by Ryan Hagerty.)

FIGURE 17-1 Mockingbird.

Meadowlark

DESCRIPTION

There are two distinct species of meadowlarks, the eastern (*Sturnella magna*) and western (*Sturnella negleeta*) (Figure 17-2). However, the ranges of the two species frequently overlap, and the two are very difficult to tell apart. To further complicate matters, the western meadowlark has been expanding its range eastward for a number of years. The best way to distinguish the two species is by their voice or song. The western meadowlark's song has a flutelike quality, with a distinct twang, while the eastern's call is more bubbling in nature. Both birds are about 10 inches in length and have a distinctive yellow breast with a black crescent shape in its center. Their upper parts are a mixture of buff, brown and white streaks and blotches. Their heads have several black and white stripes running lengthwise. One of these two meadowlarks is found throughout most of the United States.

HABITAT: They prefer open fields, meadows, and plains. Meadowlarks in northern areas may migrate southward during the winter.

FEEDING HABITS: Feeds on a variety of seeds, insects, and berries. Young are mainly fed insects early in their lives.

LIFE CYCLE: As with almost all birds, breeding and brood rearing occur during the spring and summer. The nest is built on the ground, of grass, and normally well hidden in a clump of tall grass. Four to six eggs are laid, with up to three clutches per year being raised. Meadowlarks are prey for a variety of predators, with skunks and various snakes being the main egg eaters. Foxes, various birds of prey, and bobcats probably all enjoy adult meadowlarks from time to time.

(Source: Photo courtesy of Texas Parks & Wildlife Department.)

FIGURE 17-2 Meadowlark—western.

American Robin
(Turdus migratorius)

DESCRIPTION

Most people are familiar with this member of the **thrush** family (Figure 17-3). The bright rusty red breast and belly of the adult males is very distinctive. The upper parts of the male are a brownish gray, with a tendency to become black on the head. The throat is white, with dark streaks, and the lower belly is white. Females are similar in color but duller than the males. Robins are about 10 inches in length and are often seen in large flocks during the winter. These are common birds, breeding from Alaska to Newfoundland and southward across much of the United States. The northern-most birds migrate south in the winter, with birds wintering in most of the United States.

HABITAT: They are commonly found in parks, suburbs, and urban areas, as well as forests and open country. Robins are often seen hopping along the ground as they search for food.

FEEDING HABITS: They feed heavily on worms but also eat insects, berries, and seeds.

LIFE CYCLE: Nests are sloppy affairs, constructed in the spring, in which the female lays an average of four eggs. Both parents take part in rearing the young and may raise two broods in a year. As with most birds that nest in trees and shrubs, snakes and egg-eating birds, such as crows, are the main predators.

(Source: Photo courtesy of Texas Parks & Wildlife Department.)

FIGURE 17-3 American robin.

Yellow-Headed Blackbird
(Xanthocephalus xanthocephalus)

DESCRIPTION

A conspicuous, common blackbird, the yellow-headed is often found in marshes and surrounding farmland (Figure 17-4). Males are jet black, except for a white wing patch and a bright yellow head, neck, and upper breast. The females have a dull yellow breast and throat with a brownish body. These birds are loud, having a raspy, squeaky call. They are also very gregarious, often seen in large flocks, especially in winter. They are about 9.5 inches long and often perch on reeds in their marsh habitat to sing. They breed from western Illinois westward to southern California and north to central Canada. They are migratory, and winters are typically spent from southern California to west Texas.

HABITAT: Commonly seen around marshes, grasslands, and waterways.

FEEDING HABITS: Insects, seeds, and berries. Grains such as corn, rice, and wheat may be eaten when available.

LIFE CYCLE: Nests are constructed in low-growing bushes or in marsh reeds. Four to six eggs are laid, with both parents working to rear the young. The young and the eggs are eaten by a variety of snakes, as well as raccoons and nest-raiding birds.

(Source: © Wolfgang Zintl, 2010. Used under license from Shutterstock.com)

FIGURE 17-4 Yellow-headed blackbird.

Red-Winged Blackbird
(Agelaius phoeniceus)

DESCRIPTION ■

These are very common birds, breeding from Alaska south and east across much of Canada and throughout the United States (Figure 17-5). Males are coal black, with a bright red shoulder patch edged in yellow. Females are streaked and speckled with brown, buff, and black, for an overall mottled brown appearance. They are 8 to 9 inches in length and very gregarious in the winter, often congregating in huge, noisy flocks. They are migratory, spending the winter from the northern United States southward.

HABITAT: Red-winged blackbirds inhabit marshes and wet meadows during most of the year but may be found in a wide range of areas during the winter.

FEEDING HABITS: They feed on grubs and other insects, seeds, and grain.

LIFE CYCLE: These blackbirds build their nests in the reeds and bulrushes commonly found in their marshland habitat. They also use bushes and brush as nest sites. Three to five eggs are laid, and rearing is shared by the parents. Both young and adults are food for a variety of predators, with snakes, skunks, and raccoons enjoying their eggs and birds of prey taking adults.

(Source: Photo courtesy of Texas Parks & Wildlife Department © 2002.)

FIGURE 17-5 Red-winged blackbirds.

European Starling
(Sturnus vulgaris)

DESCRIPTION

The European starling is an introduced bird, originally native to Europe (Figure 17-6). They are very aggressive and have spread throughout the United States. Males and females are similar in appearance, being black, with a long yellow bill. Plumage varies some from spring to winter, with the birds having an iridescent quality in the spring and a tendency to have light speckles over very dark, glossy plumage in the winter. They are 8 to 9 inches in length, with a short, square tail. They are very gregarious and often gather in huge flocks during the winter.

HABITAT: Starlings are very adaptable and can be found in almost any habitat, but seem to prefer those with large human populations. Parks and urban and residential areas are favorite spots. They are commonly seen on lawns and other short-grass areas.

FEEDING HABITS: Starlings eat a wide variety of insects, seeds, fruits, and berries. They also feed on grain and worms.

LIFE CYCLE: Starlings nest in any cavity they can find, building a nest of grass and leaves lined with feathers. Five to eight eggs are generally laid in the spring, and both parents rear the young. Despite being preyed on by a wide variety of predators, from egg-eating snakes to house cats, starlings are very common and are considered serious pests in most areas.

(Source: Courtesy of U.S. Fish and Wildlife Service. Photo by Luther C. Goldman.)

FIGURE 17-6 Starling.

Common Grackle
(Quiscalus quiscula)

DESCRIPTION

Grackles are common, highly gregarious birds often seen in parks, cities, and residential areas (Figure 17-7). They are 12 to 13 inches long, with a long, wedge-shaped tail. They are commonly seen walking along the ground, with stiff, rather deliberate strides. Adults are a dark glossy black with bright yellow eyes. They have short, strong bills. Grackles are loud, noisy birds that are easily seen. They breed from central Canada south and eastward across much of the eastern two-thirds of the United States. They are migratory to some extent and spend their winters from Minnesota to New England and west to Texas. The boat-tailed and great-tailed grackles are also found in various portions of the United States.

HABITAT: Often found in residential areas and parks, they also inhabit forested areas and open countryside.

FEEDING HABITS: Grackles are omnivores; they eat seeds, insects, fruits, and berries. They also rob and eat eggs and young from other birds' nests. They even catch and kill the young of other birds after they are fledged and have left the nest, but cannot fly very well yet.

LIFE CYCLE: In the spring grackles build their nests of grasses and weeds, typically in coniferous trees. Four to seven eggs are laid, and the young are tended to by both parents. Nests are preyed on by snakes, but grackles are very aggressive in defense of their nests. Adults are killed by a variety of birds of prey, house cats, and foxes.

(Source: Photo courtesy of Texas Parks & Wildlife Department © 2002.)

FIGURE 17-7 Common grackle.

Common Crow
(Corvus brachyrhynchos)

DESCRIPTION ▶

This is a very familiar species, found virtually throughout the United States (Figure 17-8). Crows are jet black and 17 to 20 inches in length. They are very intelligent, gregarious birds, often found in large numbers around farmlands. Crows are often seen foraging on the ground, where they move along with stiff hops or deliberate strides. They have strong claws and a short, thick bill. They are somewhat migratory in nature but may be found anywhere in the United States that has a food source in winter. Crows typically feed and roost together in large numbers during the winter.

HABITAT: Crows can be found virtually anywhere, in a wide variety of habitats. They tend to prefer agricultural areas and areas with large trees for roosting and nesting. They are also commonly seen along watercourses and the edges of lakes. Crows can inhabit a great variety of areas, from deserts to mountains to forests to plains. They are very adaptable.

FEEDING HABITS: Crows are omnivores, eating virtually anything they can find, plant or animal, alive or dead. They feed on insects, fruit, nuts, berries, and farm grain. They also kill and eat many young birds and small mammals. They are particularly destructive to the nests of smaller, less aggressive songbirds.

LIFE CYCLE: Crows construct a rather large nest of sticks and twigs, lined with grasses and feathers. Nests are usually constructed well up in a tall tree. Three to nine eggs are laid and the young are cared for by both parents. The young are **voracious** eaters, and the parents are kept very busy. Both young and adults are killed by birds of prey. Some owls are especially adept at picking crows off their roosts at night.

(Source: Photo courtesy of Texas Parks & Wildlife Department.)

FIGURE 17-8 Common crow.

Black-Billed Magpie
(Pica pica)

DESCRIPTION

A striking black and white bird, the magpie is unmistakable (Figure 17-9). It has a long, tapered tail, with white on the belly and back. The head, upper breast, wings, and tail are black. Magpies have glossy plumage and wing patches that flash during flight. They are about 19 inches in length. These birds are noisy and gregarious. They are found from northern New Mexico and California north to central Alaska and across much of the Great Plains.

HABITAT: They inhabit open country and the wooded areas along watercourses. Magpies are frequently spotted along roadsides.

FEEDING HABITS: Magpies are opportunists and eat what they can find, including insects, seeds, berries, and grain. They also feed on **carrion**, seeming to be particularly fond of road kill.

LIFE CYCLE: Nests are built of sticks, mud, and grass during early spring, generally in a tree. Five to seven eggs are laid. The male and female cooperate in rearing the young. Magpies are quite protective of their nests, although an occasional snake or egg-eating avian predator may succeed in destroying a nest. Adult birds are preyed on primarily by hawks and owls.

(Source: Courtesy of U.S. Fish and Wildlife Service. Photo by James C. Leupold.)

FIGURE 17-9 Black-billed magpie.

Blue Jay
(Cyanocitta cristata)

DESCRIPTION

Blue jays are strikingly handsome birds found throughout much of the United States (Figure 17-10). The upper body is bright blue, barred with black and white on the tail and wings. The breast, sides, and belly are grayish white, with a conspicuous black collar around the upper breast and neck. Blue jays are 11 to 12 inches long and have a conspicuous crest atop their heads. They are very noisy and gregarious in the fall, but generally quiet and secretive during the spring nesting season. They are found virtually throughout the eastern two-thirds of the United States but are only rarely seen in the Northwest. Birds that nest in the northernmost areas of the range may migrate south in the winter but many remain all year.

HABITAT: Residential areas, parks, suburbs, and forested areas.

FEEDING HABITS: Blue jays eat a wide variety of seeds, nuts, berries, and insects. They are also known to prey on the eggs and young of other birds.

LIFE CYCLE: Nests are well constructed of twigs and leaves and lined with feathers. The male and female vigorously defend their three to six eggs from predators. The young are reared by both parents. Snakes and house cats raid blue jay nests in search of eggs and young birds. Adults may be preyed on by house cats, bobcats, and a variety of avian predators.

(Source: Photo courtesy of Texas Parks & Wildlife Department.)

FIGURE 17-10 Blue jay.

Northern Cardinal
(Cardinalis cardinalis)

DESCRIPTION ■

Widespread across the eastern two-thirds of the United States, cardinals are one of our most familiar songbirds (Figure 17-11). They are one of the few songbirds that sing in any month of the year. The males are unmistakably bright red, with a black mask around the face and throat. Females are an overall gray-green on the majority of the body, with a tendency for variable amounts of red mixed in the tail, wings, and crest. Both sexes have a conspicuous crest on the head, although the male's is more pronounced. At about 9 inches in length and possessing a short, thick bill, cardinals are easy to identify. Cardinals are non-migratory, generally spending the year in a relatively small area.

HABITAT: Cardinals can be found in a wide variety of habitats, from thick forested areas to parks and residential areas.

FEEDING HABITS: Cardinals feed on a variety of insects, as well as fruits, seeds, grains, and berries. They feed on the ground and in bushes and trees, also visiting bird feeders regularly during the winter.

LIFE CYCLE: Nests are constructed by the female, for the most part, and consist of twigs, leaves, string, hair, and feathers. Cardinal nests are often located in thickets or dense brush. Four eggs seem to be an average clutch. As with most songbirds, the female incubates the eggs while the male tends to, and protects her and the eggs. Cardinals may raise two or occasionally three broods in a year. They are beset by the usual predators—snakes, house cats, foxes, and bobcats.

(Source: Photo courtesy of Texas Parks & Wildlife Department.)

FIGURE 17-11 Northern cardinal—male.

Eastern Bluebird
(Sialia sialis)

DESCRIPTION

As the only bright blue bird commonly found in the eastern half of the United States, the bluebird is unmistakable (Figure 17-12). The male is bright blue over the head, neck, back, and tail, with a white belly and rich chestnut breast. Females are very similar, but duller overall, with more brown on the back. Bluebirds are small, averaging only about 7 inches in length.

HABITAT: They prefer open country, not heavily forested, with scattered trees. They are particularly fond of perching on fence posts and telephone poles. Bluebirds are migratory, as a general rule, spending the winters far south of their nesting site.

FEEDING HABITS: Robin-like, in that they feed heavily on insects and grubs. They also eat berries and fruits.

LIFE CYCLE: Bluebirds are a cavity-nesting species and prefer old woodpecker holes or hollow limbs. They use nesting boxes if the larger and more aggressive European starling does not steal its chosen site. Nests are lined with fine grass; and three to seven (usually five or six) eggs are laid. The male tends to the female while she incubates the eggs. Bluebirds may raise more than one brood per year. Largest losses to predators are probably from the many egg-eating snakes.

(Source: Photo courtesy of Texas Parks & Wildlife Department © 2002.)

FIGURE 17-12 Eastern bluebird.

House Sparrow
(Passer domesticus)

DESCRIPTION

The house sparrow was introduced from Europe in the mid-1800s and has since spread throughout the United States and much of Canada (Figure 17-13). It is our most familiar, widespread sparrow, common in cities, towns, and rural areas. The male has a gray cap, a gray-white belly, and a black "bib" on the breast. The back is a dark brown and it has a black bill. Females are brownish overall, darker and streaked over the back and lighter buff on the breast and belly. They are small birds, averaging about 6 inches in length. These birds are very noisy and gregarious, often seen in large flocks, especially in winter. For the most part, they are non-migratory.

HABITAT: Residential areas, parks, suburbs, and agricultural areas. Very widespread, although they are not generally found in dense forested areas.

FEEDING HABITS: Sparrows feed on nearly anything that is edible. Seeds, insects, grain, fruit, and berries are all eaten. Sparrows also feed on garbage and food scraps when possible.

LIFE CYCLE: Sparrows are prolific breeders, often raising several broods per year. Their nests are messy affairs of grass and feathers. Both sexes construct the nest and raise the young. Four eggs are about average, with the female handling incubation chores. Snakes undoubtedly consume some eggs, but house cats probably account for most adult predation.

(Source: Photo courtesy of Texas Parks & Wildlife Department.)

FIGURE 17-13 House sparrow—male.

Common Yellowthroat
(Geothlypis trichas)

DESCRIPTION ◼

Although it is a common bird, found virtually throughout the United States, it is secretive and is heard more often than seen (Figure 17-14). Because it is only about 5 inches long, the yellowthroat is difficult to observe. The male has a bright yellow throat and upper breast with a black "mask" over both eyes. Overall body color is gray-green, almost olive. Adult females are drabber, lacking the black mask of the male and having a smaller yellow area on the throat. Yellowthroats stay low to the ground, often darting about in marsh reeds and low-growing bushes. They are migratory, spending their winters in the southern United States and Mexico.

HABITAT: Commonly found in freshwater marshes, thickets, and along watercourses, as well as open country with scattered trees and brush.

FEEDING HABITS: Feeds heavily on insects, particularly caterpillars, grasshoppers, flies, moths, and spiders.

LIFE CYCLE: Both parents participate in the rearing of the young and are very attentive. The female assumes sole responsibility for nest building and incubation, but the male is very helpful in rearing the young. Nests are usually built in low-growing vegetation, reeds, and bushes, or on the ground in a clump of grass, normally in late spring. Nest raiders such as snakes and skunks, if they can reach the nest, undoubtedly take many eggs.

(Source: Photo courtesy of Texas Parks & Wildlife Department © 2002.)

FIGURE 17-14 Common yellowthroat.

Purple Martin
(Progne subis)

DESCRIPTION

This is the largest **swallow** found in the United States, measuring 8 inches or so in length (Figure 17-15). They are commonly seen across most of the United States, although they are not so abundant in the Rocky Mountain region or Alaska. They nest virtually throughout the United States, often in **colonies**, particularly in the eastern portion of their range. The males are a dark black, which is so black as to look purplish at times. Females are lighter in color, with whitish bellies. Purple martins have a shallow fork in the tail and triangular wings that are long in relation to their body size. They are migratory, moving north in the spring and south during the fall. Winters are spent in South America for the most part.

HABITAT: Prefer open areas, in rural or residential regions.

FEEDING HABITS: Purple martins consume great quantities of insects. All types of flying insects are taken, with the martin doing all its feeding on the wing. Purple martins are fast fliers and very agile in the air. They snatch flies, dragonflies, mosquitoes, and nearly any other flying insect they can catch.

LIFE CYCLE: Martins are cavity nesters and prefer holes in cliffs or human-erected "martin houses." In the western portion of their range they tend to use old woodpecker holes in the trees found along watercourses. The nest is lined with grasses, leaves, and feathers and is built by the male and female together. Three to seven eggs is the average number laid. Although the female seems to handle most of the incubating chores, the males are attentive and help to rear the young. Both parents are needed to keep the hungry chicks fed.

(Source: Photo courtesy of Texas Parks & Wildlife Department © 2002.)

FIGURE 17-15 Purple martins often use houses made by humans.

Common Flicker
(Colaptes auratus)

DESCRIPTION

The common flicker or north-ern flicker is our most common and widespread **woodpecker** (Figure 17-16). It is also rather large, measuring in the 12- to 13-inch range. Common flickers are widespread, breeding from Alaska east and south across most of Canada and throughout the United States. These birds are rather conspicuous, with a slow, direct flight, often high above the ground, above tree-top level. They are also often seen hopping along the ground. Males have a gray head, with a small red "mustache" stripe below the eyes. The upper body is barred in gray, brown, and black. There is a conspicuous black crescent shape on the breast. Belly and underwings are a white, buff color. Females look very similar but lack the red mustache stripe. The bill is long, powerful, and chisel shaped.

HABITAT: Open areas, as well as forested habitats.

FEEDING HABITS: Flickers feed on ants, insects, berries, and fruits, with the fruit of the wild cherry being a favorite as are the berries of poison ivy.

LIFE CYCLE: Nests are built from holes excavated in trees, usually 3 inches or so in diameter and up to 24 inches deep. The female lays anywhere from three to fifteen eggs, with five to nine being the average. Both par-ents assume responsibility for rearing the young, and both take part in the incubation of the eggs. Nest predation is not normally a serious problem, but adults can become victims of house cats, foxes, and birds of prey.

(Source: Photo courtesy of Texas Parks & Wildlife Department © 2002.)

FIGURE 17-16 Common flicker.

Pileated Woodpecker
(Dryocopus pileatus)

DESCRIPTION

A large, crested woodpecker, it is 16 to 18 inches long, with a conspicuous red crest on its head. Adults are largely black, with black-and-white stripes on the face and the red crest on their heads. Males also have a small red stripe, or "mustache," on their faces. Pileated woodpeckers are more often heard than seen. They are shy and secretive by nature, but their loud, rolling call and the rat-a-tat-tat of their drilling on trees carries quite a distance. They are equipped with a large, pointed, and powerful bill with which to accomplish their drilling. They

HABITAT: Found almost exclusively in forested habitats, they may occasionally be observed in open areas. They are absent from the desert Southwest and other areas without adequate forest habitat.

FEEDING HABITS: Pileated woodpeckers feed heavily on boring insects and their larva, which they extract from the limbs and trunks of dead trees with their powerful bills.

LIFE CYCLE: These birds are, of course, a cavity-nesting species. Nest holes are usually 3 to 4 inches in diameter and 12 to 30 inches deep. The female lays three to six eggs on a bed of wood shavings left over from the boring operation. Both the male and female feed and rear the young. Nest predators are not usually a serious threat to these cavity-nesting birds, but the young and adults may be killed by birds of prey as well as foxes, house cats, and bobcats.

(*continued*)

are widespread, from central Canada south to northern California and eastward to the Gulf Coast states.

FIGURE 17-17 A male pileated woodpecker at his nest hole.

(Source: Shutterstock)

Black-Capped Chickadee
(Parus atricapillas)

DESCRIPTION

The **diminutive** black-capped chickadee is a very active and common bird that frequents bird feeders during the winter (Figure 17-17). Only 4.5 to 5.5 inches in length, they are very industrious and busy little birds. They are found from Alaska throughout most of Canada and the northern United States, as far south as northern New Jersey, northern Oklahoma, and northern California. They have a distinctive black "cap" atop their heads, as well as a black throat. Large white cheek patches are evident, with the upper body parts gray and whitish-to-buff belly and breast.

HABITAT: These birds can be quite tame and are found in forested areas, residential areas, and parks on a regular basis. They do not normally migrate any great distance, instead spending the icy winter in their northern habitat. They can, however, always use the help of a well-stocked feeder.

FEEDING HABITS: Black-capped chickadees feed on a tremendous variety of insects, as well as seeds such as sunflower. Seeds and suet are particularly important during the winter, when insects are not available.

(Source: © Bruce MacQueen, 2010. Used under license from Shutterstock.com)

FIGURE 17-18 Black-capped chickadee.

(*continued*)

Black-Capped Chickadee
(Parus atricapillas)

LIFE CYCLE: Chickadees are another cavity-nesting species. Cavities are usually excavated in a rotten stub of a limb, but chickadees occasionally use an old woodpecker hole. The female lays 5 to 10 eggs and both sexes share incubation duties. The chicks demand constant work on the part of the male and female to satisfy their demands for food.

Whippoorwill
(Caprimulgus vociferus)

DESCRIPTION

These birds are largely nocturnal, and although fairly widespread and common, are not often seen (Figure 17-18). They get their name from their distinctive call, commonly heard at dusk and during the night. Whippoorwills are a member of the **nightjar** family of birds and have short, rounded wings. They are 9 to 10 inches in length, with a wingspan of 16 to 19 inches. Adult males are an overall mottled gray-brown, with white outer tail feathers, black chin, and buff under parts. Females are similar to males but have buff outer tail feathers instead of white.

HABITAT: These birds prefer open forest habitat and forest edges. They are almost entirely nocturnal, flying low to the ground, with a darting, twisting flight.

FEEDING HABITS: The whippoorwill feeds entirely on insects it catches on the wing, during its nocturnal flights.

LIFE CYCLE: Nests are simple affairs, usually just a depression among some leaves, often under a bush. Two eggs are a normal clutch, and both parents cooperate in rearing the young. The young are apparently not fed during the day due to the whippoorwill's nocturnal habits. The normal nest predators of ground-nesting birds, such as skunks and snakes, destroy any nests they can find.

(Source: Photo courtesy of Texas Parks & Wildlife Department © 2002.)

FIGURE 17-19 Whippoorwill.

Rock Dove
(Columba livia)

DESCRIPTION

The rock dove is the pigeon so common in our cities and agricultural areas (Figure 17-19). It is an introduced species, originally from Europe. Rock doves, or feral pigeons, as they are also called, are 12 to 13 inches in length. They come in a great variety of colors, with dark gray head and gray body being common, but all-white, speckled, red, and black are also seen. These birds are stocky and thickly made. They commonly use human-made structures such as building ledges and bridge supports as nesting and roosting structures. Rock doves are found from South America northward across most of Mexico, the United States, and Canada, to Alaska.

HABITAT: Often found in cities, parks, residential areas, as well as rural areas.

FEEDING HABITS: Pigeons feed on a wide variety of items, from seeds and grains to garbage and human food scraps. In rural areas grains are favorite food items; in more urban areas, bread, hamburger, and scraps are eaten.

LIFE CYCLE: Nests are usually haphazard affairs of sticks, twigs, and a few feathers. Rock doves commonly nest under bridges and highway overpasses. Two eggs are a normal clutch, but several broods may be raised each year. The male and female participate in the rearing of the young. The young are fed a milky secretion by the female for the first few weeks after hatching. House cats and peregrine falcons are common predators.

(Source: Courtesy of U.S. Fish and Wildlife Service. Photo by Lee Karney.)

FIGURE 17-20 Rock dove.

SUMMARY

Each bird in this chapter is preyed upon by a variety of avian predators, house cats, bobcats, and foxes. The young and eggs are eaten by snakes; by other birds such as crows and jays; and by raccoons and skunks if they nest on or near the ground. Disease, cold, and starvation all take a toll on these birds. Although they may bring great joy to our lives through song or actions, they themselves seldom live joyous lives. Life in the wild is a daily struggle for survival, and there are no retirement homes for the old or sick. Songbird numbers have been in decline for a number of years, with habitat loss, lawn chemicals, and domestic and feral cats all probably being contributing factors.

REVIEW QUESTIONS

Fill in the Blank

Fill in the blanks to complete the following statements.

1. _____ is a common bird from Maine to northern California.

2. There are two distinct species of _____, the eastern and the western.

3. Most people are familiar with this member of the thrush family, known as the _____.

4. A conspicuous, common blackbird, the _____ is often found in marshes and surrounding farmlands.

5. Male red-winged blackbirds are coal black, with a bright _____ shoulder patch, edged in yellow.

6. An introduced bird, 8 to 9 inches in length, the _____ was originally native to Europe.

7. Grackles are common, highly _____ birds, often seen in parks, cities, and residential areas.

8. _____ are jet black and 17 to 20 inches in length.

9. A striking black and white bird, the _____ is unmistakable.

10. _____ are strikingly handsome birds found throughout much of the United States.

11. The male _____ is an unmistakable bright red, with a black mask around the face and throat.

12. The _____ is the only bright blue bird commonly found in the eastern half of the United States.

13. The _____ was introduced from Europe in the 1800s and has spread throughout the United States.

14. A secretive bird, the _____ is heard more often than seen, even though it is found virtually throughout the United States.

15. The _____ is the largest swallow found in the United States.

16. The _____ is our most common and widespread woodpecker.
17. A large crested woodpecker, the _____ is 16 to 18 inches in length.
18. The diminutive _____ is a very active and common bird that regularly visits feeders, particularly during the winter.
19. A largely nocturnal bird, the _____ is fairly widespread and common.
20. The _____ is the pigeon so common in our cities and agricultural areas.

Short Answer

1. Find and list five cavity-nesting species discussed in this chapter.
2. Find and list five tree-nesting species discussed in this chapter.
3. What advantages, if any, do the cavity-nesting species enjoy? Why?
4. What are some disadvantages that the ground-nesting species have to deal with?

Discussion

1. Given the feeding habits of most of the birds covered in this chapter, what can we do to ensure an adequate diet for them during the winter? Why is this important?
2. Discuss and compare the advantages and disadvantages of each of the nesting systems used by the birds covered in this chapter (ground nests, cavity nests, and nests built in trees or brush).

Learning Activities

1. Choose three species of songbirds not covered in this chapter and write a brief account of their breeding, feeding, and nesting habits. Also include a description of the species and their normal range.
2. Construct or purchase a bird feeder. Fill your feeder with feed and observe it for a period of at least 2 weeks. Record each of the species that visit your feeder during this period. Note: This exercise will work best if done during the fall or winter. However, it is very important to continue to fill your feeder after your observation period is over. This is important because the birds will have began to rely on it as a secure source of food.

Useful Web Site

<http://www.mbr-pwrc.usgs.gov>

CHAPTER 18

Avian Predators

TERMS TO KNOW

accipiter

avian

clutch

coniferous

harrier

juvenile

regurgitated

OBJECTIVES

After completing this chapter, you should be able to

- Identify the avian predators commonly found in the United States.
- Describe some of the behavior traits of each species.
- Identify the type of habitat where each species might be found.
- List some characteristics of each species.
- List some of the prey species that each predator species might utilize.

INTRODUCTION

Eagles, owls, hawks, falcons, and vultures are **avian** predators. These birds subsist on meat, fish, other birds, mammals, or reptiles. Some other bird species, such as the crow, could also be considered avian predators due to their feeding habits. However, crows are omnivores: they eat grains as well as eggs, young birds, and small mammals. The birds covered in this chapter are true carnivores. They survive entirely, or almost so, on various types of flesh. These birds are at the top of the food chain. They have few predators themselves other than larger avian predators or humans.

Vultures are included in this chapter even though they are carrion eaters for the most part. As far as is known, the turkey vulture is strictly a carrion eater, but there have been numerous reports of black vultures attacking sick, injured, or young animals. In some cases, the birds descend on their victim in a fairly large group and attack the animal's vulnerable areas. The eyes, belly, and genitals seem to be preferred sites. Black vultures appear to be much more aggressive than turkey vultures, and their range is continuing to expand. Each of the birds in this chapter has exceptional eyesight, and with the possible exception of owls, this is the primary sense used while hunting. Each bird species also possesses good hearing, especially the owls. Because they hunt primarily at night, owls use their hearing much more so than do other birds of prey. All birds of prey appear to have a poorly developed sense of smell.

American Bald Eagle
(Haliaeetus leucocephalus)

DESCRIPTION

Adults are unmistakable, being second in size among North American raptors only to the California condor (Figure 18-1). Adults average around 36 inches in length and typically have a wing span in excess of 6 feet. Females are noticeably larger than males, as with most birds of prey. The distinctive white head and tail of the adult is easily recognized, as is the large yellow bill. Overall body color is a brown-black. **Juvenile** birds are an overall brown, with various amounts of white on the tail, belly, and under their wings. The young bird lacks the white head and tail of the adult and generally does not obtain adult plumage until 4 or 5 years of age. Consequently, young bald eagles can be mistakenly identified as golden eagles. Bald eagles range from Alaska southward across Canada and in suitable habitat throughout the United States. They spend their winters far enough south to ensure open water.

Our national bird, the bald eagle, is a magnificent bird. It was once common throughout North America. The pesticide DDT brought the bird perilously close to extinction in the lower 48 states in the late 1960s. Although it was never in any serious danger in Canada or Alaska, its numbers plunged in the southern portions of its range. Protection as an endangered species and the banning of the pesticide DDT has allowed bald eagle numbers to rebound in recent years. From an estimated low of 2,400 nesting eagles in the contiguous states in 1980, bald eagles have increased to more than 6,000, and they continue to increase in number today. However, this is a far cry from the estimated 25,000 to 75,000 bald eagles thought to have inhabited the lower 48 states when the first Europeans arrived.

HABITAT: Bald eagles are usually found in the areas around lakes, rivers, and sea coasts. They prefer very large trees for roosting and nesting sites, but will also use coastal cliffs. They can often be seen perched high in trees, particularly those that offer an unobstructed view of a hunting area.

(continued)

FEEDING HABITS: Bald eagles feed heavily on fish, but they also take waterfowl and small mammals on occasion. They are also carrion eaters, and they are known to steal the catches of other fish-eating birds, such as the osprey.

LIFE CYCLE: Bald eagles typically mate for life but will take another mate should their first die. They build massive nests of sticks and branches that may be reused year after year. These nests may reach 10 feet in diameter and weigh up to 2,000 pounds. One to four eggs are laid, typically two, with the female handling incubation chores. The males are attentive to their mates and assist in rearing the young. The adults are very protective of their young and nest. The young grow rapidly and by late summer or early fall are capable of caring for themselves. Mortality among young eagles is quite high, with flying accidents and starvation taking a toll on the birds. Bald eagles are thought to live for 30 or more years in the wild.

(Source: Photo courtesy of Texas Parks & Wildlife Department.)

FIGURE 18-1 American bald eagles.

Golden Eagle
(Aquila chrysaetos)

DESCRIPTION ■

Golden eagles are large, soaring birds that are fairly common in the western half of the United States (Figure 18-2). Unlike the bald eagle, which is found only in the United States, golden eagles are also found in Europe. They are 30 to 36 inches in length, with a wing span of 6 feet or more. Adults are an overall brown, with a golden tinge to the feathers of the head and back of the neck that is visible only at close range. Juveniles have various amounts of white on the underside of their wings and white at the base of the tail. Adult plumage is not usually obtained until the fifth year. Golden eagles have a darker bill than do bald eagles. Birds in northern areas migrate southward during the winter.

HABITAT: Golden eagles inhabit open areas, plains, mountainous regions, and canyons and prefer remote areas, removed from human disturbance. They nest on cliffs, ledges, and in large trees.

FEEDING HABITS: Golden eagles feed mainly on large rodents and rabbits, although they are also known to take bighorn sheep lambs, fawns, and mountain goat kids.

LIFE CYCLE: Golden eagles also appear to mate for life and may use the same nest year after year. There are records of nesting sites in Europe that have been used continually for hundreds of years, obviously not by the same pair, but rather by successive pairs of birds over the years. A **clutch** normally consists of two eggs, with the female handling the majority of the incubation while the male tends to her and protects the nest. Both parents are extremely protective of the nest and young. Young eagles grow rapidly and by late summer or early fall are independent of their parents. Like most young birds of prey, the young may fly from the nest to a nearby perch, where they are cared for by the adults for another couple of weeks or so. Mortality among young birds is fairly high. Like all birds of prey, the most dangerous period occurs shortly after the young have left the nest and before they have fully developed their hunting skills.

(Source: Photo courtesy of U.S. Fish and Wildlife Service. Photo by Tom Smylie.)

FIGURE 18-2   Golden eagle.

Osprey
(Pandion haliaetus)

DESCRIPTION

The osprey is a large, fish-eating hawk that is widespread in the United States and throughout the world (Figure 18-3). Ospreys average around 23 inches in length and are easy to identify in flight due to their white under parts. Adults have a largely white head, with a brown eye stripe and white throat and belly. Juveniles may be a buff color where the adults are white. Ospreys nest from Alaska all across Canada and the United States. They spend the winter in the southern United States to as far south as Argentina.

HABITAT: Ospreys are found on or around most any inland body of water, from lakes and reservoirs to marshes and rivers. They are also found along the coast and in saltwater marshes, particularly during the winter months.

FEEDING HABITS: They feed almost exclusively on fish, which they catch by hovering above the water, then diving, talons first, in pursuit of their prey.

LIFE CYCLE: Ospreys construct rather large nests and are prone to reuse a nest from year to year, with the nest growing larger each successive year. Preferred nest sites are large trees, but ospreys also nest on cliffs and other sites. Two to four eggs are an average clutch, and the parents are very attentive to the nest and their young. The young birds mature rapidly and are soon fishing for themselves, although success is usually slow in coming. The young may dive and get a drenching many times before they succeed in catching their first fish.

(Source: Photo courtesy of U.S. Fish and Wildlife Service. Photo by Bart Foster.)

FIGURE 18-3 Osprey.

Red-Tailed Hawk
(Buteo jamaicensis)

DESCRIPTION

The red-tailed hawk is the most common large hawk in North America (Figure 18-4). They breed from southern Alaska and Canada throughout the United States and spend the winter from southern Canada to Central America. Red-tailed hawks average around 22 inches in length with a 48- to 52-inch wing span. Coloration is highly variable, depending on the area. In general the adults are dark to pale brown on upper body and head, with white to buff below, and a distinct reddish tail. Their call, a harsh, screaming "keeey-err" or "kee-aherrr," is commonly heard. They often are seen soaring, particularly during migration, or sitting on a high perch, such as a telephone pole.

HABITAT: Red-tailed hawks inhabit a wide variety of habitats, from plains to open woodlands. They are rarely found in dense forests.

FEEDING HABITS: They feed primarily on rodents, rabbits, snakes, and other ground-dwelling small creatures. They occasionally take a bird; particularly one caught feeding on the ground.

LIFE CYCLE: Red-tailed hawks begin to prepare for the nesting season at an early date, usually having laid their eggs by late March or early April. The nest is usually large and made of twigs, sticks, and branches; it may be reused from year to year. Two to four eggs are a normal clutch. Both parents take part in rearing the young, which are fairly slow to mature. The young are fed a diet of rodents, rabbits, snakes, and ground squirrels. They spend a great deal of time on high perches, scanning the ground for a potential meal rather than soaring over the ground in search of one.

(Source: Photo courtesy of U.S. Fish and Wildlife Service.)

FIGURE 18-4 Red-tailed hawk.

Northern Goshawk
(Accipiter gentilis)

DESCRIPTION

The goshawk is the largest North American **accipiter** (Figure 18-5). They are widespread but uncommon and rarely seen except in winter, when they must travel greater distances in search of prey. Adults average 23 inches in length, with a wing span of about 42 inches. Mature birds have pale under parts, barred with fine lines of gray and distinct whitish eyebrows, and slate gray-blue upper bodies. Juvenile birds are brownish above, with pale underbellies streaked with brown. Goshawks nest primarily in Alaska and Canada, with some birds found in New England, the northern Appalachians, and Great Lakes area. They spend the winter in essentially the same areas, although they sometimes venture farther south.

HABITAT: Goshawks prefer dense forests but are also found along woodland edges and even in open country during the winter.

FEEDING HABITS: Goshawks feed on squirrels, rabbits, and birds ranging in size from small songbirds to pheasants.

LIFE CYCLE: Beginning in early spring, usually in a tall tree, a rather large nest is constructed of twigs and sticks. Two to five eggs are laid, with both parents taking part in rearing the young. The adults are very protective of their nest and young, ferociously attacking any intruders. By late summer the young goshawks are ready to venture out into the world, and the parents spend less and less time tending to their needs. When attempting to take a bird on the fly, goshawks do not normally dive on their prey—they overtake it in level flight. Most other birds of prey that feed on birds dive on their intended victim from above, thus allowing them to be swiftly overtaken. However, the goshawk possesses great speed and determination and can overtake most other birds.

(Source: Courtesy of U.S. Fish and Wildlife Service. Photo by Phil Detrich.)

FIGURE 18-5 Northern goshawk.

Swainson's Hawk

(Buteo swainsoni)

DESCRIPTION

This crow-sized hawk is commonly found in the western half of the United States (Figure 18-6). They average about 21 inches in length, with a wing span of around 48 inches. This makes them somewhat smaller than the red-tailed hawk, with which they commonly share habitat. These birds have highly variable plumage, usually brown on upper parts and buff to white underneath. However, dark-phase adults may have little difference in coloration between the upper and under portions of their bodies. Juveniles, at least in the light-color phase, generally show more buff or white markings. These birds nest from Alaska southward across western Canada and the western United States. They are highly migratory, generally spending their winters in South America. Swainson's hawks are rarely seen in the east.

HABITAT: Grasslands, prairies, plains, and meadows, usually with few trees.

FEEDING HABITS: They feed on a variety of rodents, as well as grasshoppers and locusts.

LIFE CYCLE: Swainson's hawks begin nesting in early spring, often using their nest from the previous year. The nest may be in a tree, low-growing bush, or on the ground. Two to four eggs make up the average clutch. The parents share the responsibility of rearing the young, although like most birds of prey the female appears to handle incubation duty. The young hawks grow quickly and are ready to follow the adults south when migration time arrives in late fall. Swainson's hawks often form large flocks during migrations and are most visible during these periods.

(Source: Courtesy of U.S. Fish and Wildlife Service. Photo by Gary R. Zahn.)

FIGURE 18-6 Swainson's hawk.

Northern Harrier
(Circus cyaneus)

DESCRIPTION

Easily identified in flight by its long wings and long tail, the northern harrier is the only member of the **harrier** group of hawks found in North America (Figure 18-7). They are about the size of a large crow, averaging 22 inches in length, with a 42-inch wingspan. Adult females have brown upper bodies and are buff below, with brown streaks. Adult males generally have a pale gray head, neck, back, and breast, with a white belly. Both males and females have the characteristic white rumps. This is a widespread and common hawk of open areas, particularly in the West. Harrier hawks are often called marsh hawks and were formerly known by that name. They breed from Canada southward across the United States and into Mexico. They spend their winters from South America to as far north as southern Canada.

HABITAT: Prefers marshes, open grasslands, and prairies.

FEEDING HABITS: Northern harriers feed primarily on mice and other rodents but also eat snakes and other small animals. Their flight when hunting is very distinctive. They glide low over the ground, rarely flapping their wings, until they spot prey, at which time they pounce on the victim.

LIFE CYCLE: Harriers nest on the ground, usually in dense grass or marsh vegetation. The males and females often arrive at the nesting site already mated and begin raising their young shortly after their arrival. Four or five eggs are an average clutch, but as few as two or as many as seven may be laid. Incubation is primarily the female's responsibility, although the male may assist and often takes part in nest building. The male's primary function is to provide his mate and offspring with food. As the young grow and mature, the male is hard-pressed to meet the demand, and the female begins to hunt as well. The young birds grow rapidly and are usually ready for the fall migration by late summer.

(Source: © Jemini Joseph, 2010. Used under license from Shutterstock.com)

FIGURE 18-7 Northern harrier.

Prairie Falcon
(Falco mexicanus)

DESCRIPTION

A medium-size falcon averaging 19 inches in length, the prairie falcon is widespread and quite common in the western United States (Figure 18-8). They are unmistakable in flight, having long wings, a long tail, and conspicuous black areas on the underside of their wings. No other falcon has these black "wing pits." Adults are brown overall, with whitish or buff markings on lower portions of the body, and a dark eye patch. Brown streaks are evident on under parts. They breed throughout much of the western United States, from northern Texas to southern Canada. Winters are spent in much the same areas, but they may migrate eastward to the Mississippi River and occasionally farther east.

HABITAT: Open country, prairies, and grasslands.

FEEDING HABITS: Feeds primarily on small mammals, such as mice, rats, and ground squirrels and birds, and from small songbirds to pigeons.

LIFE CYCLE: Prairie falcons are particularly fond of cliffs and mountainous areas, using them for nest sites and roosts. The nest is constructed of twigs, leaves, and grass on a cliff, in a tree, or in another suitable site. Two to five eggs make up the average clutch. Parents cooperate in rearing the young, but the female assumes the majority, if not all, of the incubation. The male hunts and provides food for his mate and their brood. The young are almost full-grown by late summer and are beginning to test their flight abilities. By early fall they are pretty much on their own. Prairie falcons tend to hunt by soaring at considerable height and then diving on prey they spot with their keen eyesight.

(Source: © Bufo, 2010. Used under license from Shutterstock.com)

FIGURE 18-8 Prairie falcon.

Cooper's Hawk
(Accipiter cooperii)

DESCRIPTION ▶

The Cooper's hawk is a medium-size, widespread hawk, about the size of a crow (Figure 18-9). They average about 16 inches in length, with relatively short wings of 28 inches or so. The adult is a slate gray over the shoulders and back, often with a darker cap on its head. It has rusty white barring on breast and belly. The tail is long and rounded, with a wide white band at the tip. It closely resembles the smaller sharp-shinned hawk, which has a squared tail. These hawks breed throughout the United States and up into southern Canada. Winters are spent from the central United States southward to Central America.

HABITAT: Prefers mixed clearings and deciduous forests. Also found in open country and occasionally in coniferous forests.

FEEDING HABITS: The Cooper's hawk is extremely powerful for its size and feeds on a wide variety of game birds as well as rabbits. If larger prey is scarce, rodents and insects such as grasshoppers are taken. The Cooper's hawk is often called a chicken hawk, and not without cause. They are not above plucking a farmer's chicken from the barnyard.

LIFE CYCLE: Cooper's hawks tend to place their nests quite high, constructing them of sticks and twigs, often making them quite bulky. Four eggs seem to make up the average clutch. The female appears to handle incubation while the male protects her and the nest and provides food. As with many other birds of prey the female may begin incubation when the first egg is laid, making for a wide variation in the sizes of chicks. The young progress rapidly and are ready to join their parents on hunting forays within 45 days or so. The parents and young may migrate southward together, hunting as they go.

(Source: © Al Mueller, 2010. Used under license from Shutterstock.com)

FIGURE 18-9 Cooper's hawk.

Sharp-Shinned Hawk
(Accipiter striatus)

DESCRIPTION ■

The smallest member of the Accipiter clan found in North America, it is also the most common (Figure 18-10). Averaging only 14 inches in length, with a wing span of only 21 inches, the sharp-shinned hawk is indeed small. The adults are slate-gray on their upper parts, with rusty barring on breast and belly. The tail is long, squared, and has a narrow white band at the tip. The females are noticeably larger than the males. They breed from Alaska across Canada and southward to the southern United States. Winters are spent from Central America to the northern United States.

HABITAT: Prefers dense, coniferous forests but also inhabits deciduous forests and open areas.

FEEDING HABITS: Feeds primarily on small birds such as sparrows, warblers, and starlings, but will also feed on rodents and insects.

LIFE CYCLE: Sharp-shinned hawk nests are usually built high up in a tree, typically a **coniferous** one. The nest is constructed of twigs, leaves, and strips of bark, and four to eight eggs are laid. The female handles incubation while the male hunts and protects the nest, which he does fearlessly. The young birds grow rapidly, soon requiring three or four small birds or rodents each per day. This keeps both parents very busy. The young join their parents, and hundreds of other sharp-shinned hawks, for their annual migration southward during the late summer.

(Source: Photo courtesy of Texas Parks & Wildlife Department © 2002.)

FIGURE 18-10 Sharp-shinned hawk.

American Kestrel
(Falco sparverius)

DESCRIPTION ◼

This little falcon, often called a sparrow hawk (Figure 18-11), is widespread and fairly common. Averaging only about 11 inches in length, with a 23-inch wing span, the kestrel is the smallest falcon found in North America. This bird is easily recognized by the bright rusty back and tail and blue-gray wings. It has a whitish belly and dark barring down the back and tail. It is also easily identified by its distinctive hunting style: hovering above the ground until prey is sighted, then diving swiftly to take the prey. Kestrels are found from Alaska south across Canada, throughout the United States, and into Mexico. They may spend the winter in the same locations, but northern birds are migratory.

HABITAT: One of the few predatory birds that live in close proximity to humans. Although they prefer open rangeland, farmland, and forest edges, kestrels also reside in urban areas and city parks.

FEEDING HABITS: Subsists largely on insects, such as grasshoppers and crickets, but also takes mice and small rodents. They are also known to feed on small birds on rare occasions.

LIFE CYCLE: The kestrel is fairly unique among birds of prey, with the exception of owls, because they lay their eggs in holes in trees. When a hole is used, little nest building actually occurs in the hole—the eggs are laid on the floor of the hole. They also nest in the cracks and crevices of buildings, and occasionally in another bird's old nest. Four to seven eggs may be laid, with four or five being about average. The female takes care of the majority of the incubation, but the male may help, which is rare among birds of prey. The young grow rapidly on a diet of rodents and insects and have voracious appetites, like most young birds of prey. The parents spend countless hours in search of food for their brood. By late summer, the young kestrels are generally ready to venture off on their own.

(Source: Photo courtesy of U.S. Fish and Wildlife Service. Photo by Tom Smylie.)

FIGURE 18-11 American kestrel.

OWLS

Owls are a large group of predatory birds that hunt largely at night or at dusk and dawn. They are very specialized, having large round eyes with excellent light-gathering properties and large ears that enhance their excellent hearing. There are many species of owls in North America, but due to their nocturnal habits, they are rarely seen by most people. A few of the more widespread and common owls are covered in this chapter.

Great Horned Owl
(Bubo virginianus)

DESCRIPTION

The great horned owl is one of the largest owls in North America (Figure 18-12). Only the great gray owl is larger. They average 22 inches in length and have a wing span of around 55 inches. Great horned owls are also widely distributed, ranging from South America northward throughout the United States and Canada. Their large ear tufts give the appearance of horns. Adults are a mottled gray-brown on the upper body and buff colored below, with dark brown to black barring overall. Juveniles are very similar to adults but tend to show more reddish buff than adults. Prominent ear tufts, large yellow eyes, white throat patch, and large overall size are distinctive features. These birds are largely nocturnal and are rarely seen during the day, unless disturbed from their roosts.

HABITAT: Widely distributed, residing in deserts, forests, open country, and even in residential areas and parks.

FEEDING HABITS: This extremely powerful bird routinely kills animals as large as a porcupine. It feeds primarily on rodents, rabbits, skunks, and other small- to medium-size mammals. However, it also eats birds, frogs, snakes, and lizards. There is at least one documented account of a great horned owl attacking, killing, and making off with a large barnyard tomcat. It is also known to feed on turkeys and crows, which are plucked from roosts and nests during the night.

LIFE CYCLE: Great horned owls are one of the earliest nesting birds. They may begin nesting as early as late January. Two to five eggs may be laid, with two to four being most common. The nest is usually on a cliff, ledge, or the abandoned nest of other birds, such as crows or herons. They rarely build their own nests. The females are very devoted to their nests, having been known to sit tight through late spring snows. The young are voracious eaters and grow rapidly. Both the male and female work to rear the young. Great horned owls are not known to be migratory to any great extent.

(Source: Photo courtesy of U.S. Fish and Wildlife Service. Photo by Robert Drieslein.)

FIGURE 18-12 Great horned owl.

Barred Owl
(Strix varia)

DESCRIPTION

These large owls, averaging around 20 inches in length, have a wing span of 44 inches or so (Figure 18-13). Barred owls are vocal compared to other owls, and their "hoot" is often heard from late afternoon to early morning. They lack the ear tufts of the great horned owl and are an overall gray-brown color. They are barred with white on the back and their buff underparts are streaked with black. They are common in the eastern United States, and their range in the West is expanding. They range from the Gulf Coast northward across the eastern United States to southern Canada, as far west as eastern British Columbia. Their call is a distinctive, "hoo-hoo-hoo".

HABITAT: They seem to prefer northern and southern wooded swamps but are also found in other wooded areas.

FEEDING HABITS: They feed on a variety of rodents, rabbits, and birds.

LIFE CYCLE: Barred owls are like great horned owls in that they nest in the abandoned nests of other birds. However, they seem to prefer hollows in trees and rarely add any nesting material to these hollows. Two to four eggs generally make up a clutch, with the female apparently taking care of incubation. Within a few weeks of hatching, however, it takes both parents' considerable hunting abilities to keep the youngsters fed. The young grow fairly quickly and by late summer or early fall are beginning to hunt for their own meals. These birds are largely nocturnal and are difficult to observe during daylight hours.

(Source: © Jill Lang, 2010. Used under license from Shutterstock.com)

FIGURE 18-13 Barred owl.

Barn Owl
(Tyto alba)

DESCRIPTION ■

Barn owls have a distinctive heart-shaped, whitish face encircled with brown (Figure 18-14). It is long and slim compared to other owls, averaging 16 to 20 inches in length, with wings longer than those of a crow. The upper body is golden brown to buff, with obvious black and white spotting. The lower body is usually whitish but may exhibit some buff on the breast. The barn owl also lacks ear tufts. It is widespread, breeding throughout the United States, but is not often seen, as it is largely nocturnal. Barn owls are not particularly migratory, but northern birds do drift southward during the winter.

HABITAT: Open and semi-open habitat, forests, and urban and residential areas. Often found in close proximity to human populations.

FEEDING HABITS: Feeds mainly on rodents, such as mice, rats, gophers, and other small mammals.

LIFE CYCLE: The barn owl acquired its name from its tendency to build nests in old barns and attics. It also uses natural cavities such as hollow trees, caves, and shallow ledges. The barn owl tends to lay rather large clutches of 5 to 11 eggs over a period of several days. Like most birds of prey, the female begins to incubate the eggs as soon as one is laid. This results in a staggered hatching, and young birds in the same nest are a variety of sizes. Due to the large number of young in the average barn owl nest, and to their voracious appetites, the adults catch large numbers of mice, rats, and other pests to feed their growing brood. This makes the barn owl a very beneficial species.

(Source: Photo courtesy of Texas Parks & Wildlife Department.)

FIGURE 18-14 Barn owls.

Burrowing Owl
(Athene cunicularia)

DESCRIPTION ▸

This small owl is difficult to mistake (Figure 18-15). It measures only about 10 inches in length and is typically seen on the ground, standing upright on rather long legs. It has a rounded head and lacks ear tufts. Mature birds are an overall sandy brown, with white spots. Lower breast and belly are often whitish with brown barring. It is primarily a western bird, ranging from Manitoba south to Texas and west to California. Burrowing owls are not as numerous as they once were, due primarily to the persecution of prairie dogs, whose burrows they often use.

HABITAT: Lives in open country such as prairies, grasslands, and deserts. In urban settings, frequents airports and golf courses.

FEEDING HABITS: Feeds mainly on small rodents and insects.

LIFE CYCLE: Although known as the burrowing owl, this owl more often uses abandoned prairie dog or ground squirrel burrows. Five to seven eggs make up the average clutch, but as many as 12 may be laid, with the female handling incubation duty. The bird is largely nocturnal, but the male often hunts day and night when trying to keep a brood of young fed. As the young birds reach maturity they can often be seen around the entrance to the burrow, apparently sunning themselves. Daylight appears to have little adverse effect on these owls.

(Source: © Bob Blanchard, 2010. Used under license from Shutterstock.com)

FIGURE 18-15 Burrowing owl.

VULTURES

Two species of vultures are found in the United States: the turkey vulture and the black vulture. The turkey vulture is found throughout the United States and is somewhat larger than the black vulture. The black vulture is primarily a southeastern bird, being resident from Pennsylvania to Texas. Black vultures are currently expanding their range.

Both types of vulture are carrion eaters and are considered Mother Nature's garbage disposals. Although their consumption of dead and decaying animals may seem revolting, they serve a very useful purpose. It is almost certain that without these two birds, our roadsides would smell considerably worse than they do. They are two of the few species that appear to have suffered little adverse effect from the presence of humans. In fact, they may benefit from the excellent supply of food we provide for them. In many states the number of deer killed by motorists rivals, or exceeds, the number harvested by hunters each year. This does not take into account the thousands of raccoons, opossums, rabbits, dogs, cats, and other small animals that lose their lives on our highways each year. It would be hard to find someone who has not seen one or both of these vultures enjoying such a meal beside (or in the middle of) one of our highways. These birds are also commonly seen soaring, often at great altitude, in search of carrion. Both vultures use their excellent eyesight to spot potential meals and are often found in groups as they follow one another to carrion. Where their ranges overlap, both species may be observed feeding together.

Turkey Vulture
(Cathartes aura)

DESCRIPTION ■

These are large birds, averaging 28 inches or so in length and possessing a wing span of up to 72 inches (Figure 18-16). Only the California condor and our two eagles are larger. Both the juveniles and adults are dark brown to black, with unfeathered heads. The adult has a red head; the immature bird has a dark head. Both young and adult have a white-tipped bill. These birds range from southern Canada, throughout the United States, Mexico, and into Central and South America. Northern birds migrate southward during the winter.

HABITAT: Virtually anywhere carrion can be found, except within cities. Prefers open country and is often observed along roadsides.

FEEDING HABITS: Feeds exclusively on the flesh of dead animals.

LIFE CYCLE: These birds are not noted nest builders. They lay their eggs in a hollow log, cave, cliff, or ledge, usually with no attempt to build a nest. One to three eggs are usually laid, with two being most common. The young are fed a nourishing diet of **regurgitated** carrion and mature rapidly. They are usually ready to take flight and search for their own meals by late summer.

(Source: Photo courtesy of Texas Parks & Wildlife Department.)

FIGURE 18-16 Turkey vulture.

Black Vulture
(Coragyps atratus)

DESCRIPTION ▮

Only slightly smaller than the turkey vulture, at about 25 inches, with a wing span of 60 inches or so, the black vulture is another large, soaring bird (Figure 18-17). They are solid black, with a gray, unfeathered head. Black vultures also have shorter tails than turkey vultures. They range throughout the southeastern United States and southward to South America. Their range appears to be expanding westward and frequently overlaps that of the turkey vulture.

HABITAT: Very similar to that of the turkey vultures—almost anywhere carrion can be found, except urban areas.

FEEDING HABITS: Feeds primarily on carrion but also known to prey on the young of colonial-nesting herons and egrets, as well as any young or weak small mammal or seriously injured larger prey. They are quite aggressive, and will not always wait for their prey to die before beginning to feed.

LIFE CYCLE: Essentially the same as the turkey vultures.

(Source: Photo courtesy of Texas Parks & Wildlife Department © 2002.)

FIGURE 18-17 Black vulture.

SUMMARY

Avian predators are vital to a healthy wildlife community. To maintain a healthy population of avian predators we must be careful and considerate in our use of pesticides and other pollutants. They also require some areas relatively free from human disturbance and encroachment. These birds are almost universally sensitive to changes in their environments; consequently, their populations must be carefully monitored to ensure their survival. For years, hawks, owls, and other birds of prey were poisoned and shot indiscriminately as vermin. They were thought to prey heavily on barnyard livestock, such as chickens and lambs, as well as game birds. Although it is undoubtedly true that they do on occasion prey on these species, particularly game birds, avian predators are extremely unlikely to have any serious effect on a species' population. If we continue to educate ourselves and others we may finally lay to rest the old adage that "the only good hawk is a dead hawk."

REVIEW QUESTIONS

Fill in the Blank

Fill in the blanks to complete the following statements.

1. Eagles, owls, hawks, falcons, and vultures are _____ predators.

2. Each of the birds in this chapter has exceptional _____.

3. Each of these birds also has good hearing, but the _____ have excellent hearing.

4. Bald eagles feed heavily on _____ but also take _____ and small mammals on occasion.

5. Golden eagles are large, soaring birds, fairly common in the _____ half of the United States.

6. The _____ is a large, fish-eating hawk that is widespread throughout the United States and the world.

7. Ospreys feed almost exclusively on _____.

8. The _____ is the most common large hawk in North America.

9. The goshawk is the largest North American _____.

10. The _____ hawk is a crow-size hawk commonly found in the western half of the United States.

11. The _____ is the only member of the harrier group of hawks found in North America.

12. A medium-size falcon, averaging 19 inches in length, the _____ is widespread and quite common in the western United States.

13. The _____ is a medium-size, widespread hawk, about the size of a crow.

14. The _____ is the smallest member of the accipiter clan in North America.

15. The _____ is often called a sparrow hawk.

16. _____ are a large group of predatory birds that hunt largely at night or at dusk and dawn.

17. The _____ is one of the largest owls in North America.

18. _____ owls are vocal compared to other owls, and their "hoot" is often heard from late afternoon to early morning.

19. _____ are distinctive-looking owls, with a heart-shaped, whitish face, encircled by brown.

20. Barn owls acquired their name from their tendency to build their nests in _____ and _____.

21. Although known as the _____ owl, these owls more often use abandoned prairie dog or ground squirrel burrows.

22. Two species of vultures are found in the United States, the _____ and the _____.

23. Both of these birds are _____ eaters.

Short Answer

1. Find and list two species discussed in this chapter that feed on fish. What types of habitat might these species utilize?

2. What can we do to help ensure the survival of avian predators?

3. Our national symbol, the bald eagle, almost became extinct in the lower 48 states. How did this happen? Why? What have wildlife managers done to help restore the species?

Discussion

1. In your opinion, what sorts of benefits do people receive from avian predators?

2. Discuss the effects of pesticides and pollution on avian predators. What have we done to counter these negative factors? What else might we do?

Learning Activities

1. Select a species of avian predator found in your area. List the types of habitat where it is found in your area. What are some likely prey species? What are some likely nesting areas?

2. Select an avian predator discussed in this chapter. Using this book, the local libraries, and additional classroom texts, write a detailed account of your chosen species. Present your report to your class.

Useful Web Site

<http://www.mbr-pwrc.usgs.gov>

CHAPTER 19

Common Shorebirds, Herons, and Egrets

TERMS TO KNOW

amphibians

crayfish

crustaceans

plume

secretive

OBJECTIVES

After completing this chapter, you should be able to

- Identify some of the common shorebirds, herons, and egrets.
- Describe some characteristics of each of these species.
- Identify the type of habitat where each species might be found.
- Describe some of the individual habits of each bird.

INTRODUCTION

Shorebirds constitute a very large group of birds, with a tremendous variety of species. It is easy to see why there are entire books devoted only to shorebirds. We will cover just a few of the more common and widespread species. These species may not necessarily be found in your area year-round, but most can be observed during migration periods. Several of these species may never venture near a "shore" and in fact are more likely to be found in upland pastures. Many species live in both habitats at one time of the year or another. The bitterns and most of the herons and egrets are more likely to be found on or near water. Many of the species discussed in this chapter are migratory, at least to the extent of moving far enough south during the winter to find open water. It should be noted that detailed research on the breeding and life cycles of many of the birds in this chapter is not yet available. However, it is safe to assume that common nest predators, such as skunks, raccoons and snakes will take the eggs and young of herons and egrets. However, particularly among the ground nesting species, floods probably account for most losses. Many of the birds in this chapter, especially the herons, are large enough to defend their nests from most would be predators.

American Bittern
(Botaurus lentiginosus)

DESCRIPTION

The American bittern is a shy, **secretive** bird that is rarely seen (Figure 19-1). They prefer freshwater and saltwater marshes, making them doubly difficult to observe. They are, however, widespread, breeding locally from the southern United States to Canada and spending their winters along both coasts of the United States, as well as in the interiors of many states. They average about 28 inches in length, with rather short legs and a heavy, plump body. Adults are streaked, with reddish brown and yellow buff from their throats to their bellies. They tend to have a black stripe along the side of the neck and a dark brown back with fine spotting. In flight, the feathers are dark along the back edge and outer half of the wings.

HABITAT: Freshwater and saltwater marshes.

FEEDING HABITS: Feeds on fish, frogs, **crayfish**, snakes, mice, and other small creatures.

LIFE CYCLE: Builds a nest of marsh vegetation and grasses, usually among marsh reeds, but rarely in low-growing bushes. Four to seven eggs are an average clutch.

(Source: Photo courtesy of U.S. Fish and Wildlife Service. Photo by Gary Kramer.)

FIGURE 19-1 American bittern.

Least Bittern
(Ixobrychus exilus)

DESCRIPTION

The least bittern is perhaps even more difficult to observe than the American bittern (Figure 19-2). It is much smaller than the American bittern, is equally secretive, and prefers the same marsh habitat. Averaging only 13 inches are so in length, the least bittern is our smallest heron. The adult has black on the back and head, with the back of the

HABITAT: Freshwater and saltwater marshes, with a preference for fresh water.

FEEDING HABITS: Feeds on small fish, crayfish, and **amphibians**.

LIFE CYCLE: Builds a nest of twigs and reeds, usually among marsh vegetation. Three to six eggs make up the average clutch.

neck being chestnut buff. It is yellow buff below, with white streaks running from throat to abdomen. The bill and slim legs are yellow. If one is fortunate enough to flush a least bittern, it seldom flies far, preferring to fly short distances with quick wing beats. Least bitterns often run along the tops of marsh reeds.

(Source: Photo courtesy of Texas Parks & Wildlife Department.)

FIGURE 19-2 Least bittern.

Green-Backed Heron
(Butorides striatus)

DESCRIPTION

Green-backed herons are very common in almost all freshwater and many saltwater habitats from southern Canada throughout most of the United States (Figure 19-3). Because of its small size, around 18 inches in length, and generally solitary nature, the green-backed heron is often overlooked. It has rather short, yellow legs; a compact, solid build; and long neck. It is most often startled by streamside fishermen, at which time it flushes into quick flight, with a surprised squawk. The top of the adult head and its upper wings are a dark shiny green, while the feathers of the back are gray-green. The chin is usually white and the breast streaked with white. The remainder of the body is an overall gray-brown.

HABITAT: Prefers freshwater marshes, streams, ponds, and lakes but is also found in saltwater habitats such as marshes, mudflats, and tidal shallows.

FEEDING HABITS: Feeds on a wide variety of fish, crayfish, frogs, and other amphibians.

LIFE CYCLE: It is generally a poor nest builder, constructing a twig-and-stick platform in a bush or low-growing tree. Three to six eggs appear to be the average clutch.

(Source: © EcoPrint, 2010. Used under license from Shutterstock.com)

FIGURE 19-3 Green-backed heron.

Tricolored Heron
(Egretta tricolor)

DESCRIPTION

The tricolored heron is a slender, tall heron, generally standing 26 to 28 inches tall and having a wingspan of about 36 inches (Figure 19-4). They are quite distinctive, with slate gray necks and backs and a white rump and belly. They tend to show some pinkish maroon coloring around their throats, and in breeding adults along their backs. Tricolored herons have long legs and a long pointed bill. They range throughout much of the eastern United States, particularly along the Atlantic and Gulf coasts.

HABITAT: Prefers saltwater and freshwater marshes, tidal flats, and mudflats and likes to wade in water as deep as possible while foraging.

FEEDING HABITS: Feeds primarily on fish, but also takes frogs, and other amphibians. Like most true herons, tricolored herons feed by stalking their prey in shallow water or by standing absolutely still and allowing their prey to come to them.

LIFE CYCLE: Nests in colonies in marshes and wooded swamps. The nest is usually a flat structure of twigs and sticks. Three to five eggs make up the average clutch.

(Source: Photo courtesy of Texas Parks & Wildlife Department.)

FIGURE 19-4 Tricolored heron.

Little Blue Heron
(Egretta caerulea)

DESCRIPTION

This heron exhibits a rather odd color scheme. The immature birds are snow white, but the adults are a dark blue, almost black color (Figure 19-5). One would think that it would be in the best interests of the young birds to be less conspicuous! Both adults and immature birds are about 24 inches in length, with long legs and a bluish bill. These birds can be found from the Gulf Coast states northward to New England. They generally spend their winters along the coast from New Jersey south to Mexico.

HABITAT: Utilizes both freshwater and saltwater marshes, wooded swamps, mudflats, and tidal shallows.

FEEDING HABITS: Feeds on small fish, mollusks, frogs, crayfish, and insects that it catches while stalking the shallows or lying in wait for its prey.

LIFE CYCLE: Like many herons and egrets, little blue herons may nest in colonies. The nest is generally a simple affair of twigs and sticks. Two to five eggs appear to be about average.

(Source: Photo courtesy of Texas Parks & Wildlife Department © 2002.)

FIGURE 19-5 Little blue heron.

Great Blue Heron
(Ardea herodias)

DESCRIPTION ▶

This is the largest heron in North America, standing 46 inches or more. The great blue heron is large, heavy, and quite unmistakable (Figure 19-6). It is very common and widespread in almost all freshwater habitats and can often be observed wading slowly or standing absolutely still in shallow water as it forages for food. If it is standing still, the great blue heron can easily be overlooked. There appear to be two color phases: overall medium gray and white. Adults may be a medium gray overall, with white face and white atop the head. A black band or "eyebrow" extends from above the eye to a narrow black **plume** at the back of the head. The white color–phase birds are solid white. Both color phases have long pale yellow legs and long, sharp bills. Great blue herons are found virtually throughout the United States and in southern Canada. Birds in more northern latitudes move far enough southward to find open water during the winter.

HABITAT: It can be found in all freshwater habitats, from lakes to marshes to ponds and streams.

FEEDING HABITS: Feeds on a wide variety of fish, amphibians, and insects, as well as snakes, mice, and other rodents. During heavy spring rains these birds are particularly fond of "mousing" in low-lying pastures, which become soggy or flooded, forcing the mice to seek higher ground.

LIFE CYCLE: These birds are most often solitary, but nesting is a colonial affair, with several nests in a relatively small area. Nests are usually in tall trees, near water, often in wooded swamps. Three to six eggs are an average clutch.

(Source: Photo courtesy of Texas Parks & WildlifeDepartment.)

FIGURE 19-6 Great blue herons.

Cattle Egret
(Bubulcus ibis)

DESCRIPTION

The cattle egret is a European species that derives its name from its habit of feeding in, around, and on the backs of cattle (Figure 19-7). Cattle egrets were first observed in South America around 1880 and in Florida about 1952. Since they first arrived they have greatly expanded their range. They are a common sight throughout most of their present range, and they continue to expand their range each year. Cattle egrets can be found from Texas to Michigan, as far north as Maine, and all along the Gulf coast. Cattle egrets are gregarious birds, most often seen in flocks flying to and from roosting areas. They are shorter and more compact than other egrets, with shorter, more muscular legs and shorter, thicker bills. In the winter they are solid white with a yellow orange bill, feet, and legs. During the breeding season adults are equipped with light orange plumes atop their heads and along their breasts. Some longer plumes down the lower back give the appearance of an orange-buff blanket across the back. These birds average 20 inches in length. Cattle egrets utilize upland grasslands much more than do other herons and egrets. They range from central Texas eastward to Maine and as far north as Michigan. Birds in the more northern limits of their range spend their winters in the south.

HABITAT: Freshwater marshes and open grasslands.

FEEDING HABITS: Cattle egrets feed on a wide variety of insects, from ticks to grasshoppers. They often feed among herds of cattle, picking off the insects the cattle flush as they graze. They are also known to sit atop cattle and pick parasites, such as ticks, off the host, hence their rather common nickname "tick bird." Cattle egrets have also been observed killing mice and other rodents if the opportunity presents itself. They are also thought to feed on the young and eggs of ground-nesting birds, but overall they are beneficial birds.

LIFE CYCLE: These birds normally nest in colonies, usually around water. Nests are constructed from sticks and twigs. Three to five eggs are about average.

(Source: Photo courtesy of Texas Parks & Wildlife Department © 2002.)

FIGURE 19-7 Cattle egret.

Great Egret
(Casmerodius albus)

DESCRIPTION

These large white birds are almost as big as the great blue heron, standing around 39 inches tall, but they are much slimmer (Figure 19-8). Great egrets are snow-white, with dark legs and feet and a pale yellow bill. During the breeding season adults have long white plumes down their backs, which extend past their tails. These plumes are much shorter during the winter. Egrets nest from the Long Island area west to southern Minnesota and Oregon and from there southward to Florida and Texas. They spend their winters along the Pacific, Atlantic, and Gulf coasts.

HABITAT: Freshwater and saltwater marshes, wooded swamps, mudflats, and tidal shallows.

FEEDING HABITS: Feeds on fish, crayfish, snakes, frogs, and insects.

LIFE CYCLE: These birds generally nest in colonies in marshes or wooded swamps. Three to five eggs are laid on a platform of sticks and twigs.

(Source: Photo courtesy of Texas Parks & Wildlife Department © 2002.)

FIGURE 19-8   Great egret.

Snowy Egret
(Egretta thula)

DESCRIPTION

These gregarious, medium-size herons can be found in many freshwater and saltwater habits (Figure 19-9). They are snow-white and stand about 24 inches tall. During the breeding season the adults have plumes along the back of their necks and down their backs. The plumes along the back extend past the tail and curl upwards. The bill is slim and black as are the legs, while the feet are yellow. They breed from Florida north to Long Island and in the Mississippi River Valley. In the West they are found from Colorado westward to central California. Winters are spent along the Atlantic and Gulf coasts in the east and in southern California and along the Colorado River in the west.

HABITAT: Freshwater and saltwater marshes, mudflats, tidal shallows, and wooded swamps.

FEEDING HABITS: Feeds on a wide variety of aquatic prey, such as fish and crayfish, as well as a variety of insects.

LIFE CYCLE: Nests in colonies in marshes and wooded swamps. Three to five eggs are laid in a nest consisting of sticks and twigs, located in a fairly large tree.

(Source: Photo courtesy of Texas Parks & Wildlife Department.)

FIGURE 19-9 Snowy egret.

Killdeer
(Charadrius voriferus)

DESCRIPTION ■

These extremely common members of the shorebird family are seldom seen near a shore except during their winter migrations (Figure 19-10). Killdeer spend a great deal of their time in upland pastures. These birds are easily recognizable both from their appearance and their loud vocalizations. They seem to call "kill-dee, kill-dee" constantly and bob their heads up and down while standing on the ground. Killdeer average 10 to 11 inches in length. They have white throats, breasts, and bellies, with two conspicuous black bands, one around the neck and one across the breast. The forehead is white, and they have an obvious white "eyebrow" above the eyes. The top of the head and the remainder of the upper back are gray-black. A bright white stripe goes across each upper wing, from the body to the wing tip. The rump and lower tail are a buff reddish color. They have pale gray legs and a relatively short dark bill. Killdeer nest from southern Canada to central Mexico; they can be found virtually throughout the United States. They spend their winters from the southern two-thirds of the United States to as far south as Venezuela.

HABITAT: Most commonly found in open pastures and prairies; also utilizes mudflats and lake shores during the winter.

FEEDING HABITS: Feeds on a wide variety of insects, grubs, and worms.

LIFE CYCLE: Nest is often on bare, rocky soil in a shallow depression. Their eggs are olive with dark spots and often closely resemble the pebbles around them. Four eggs are almost always what the female lays. Female killdeer defend their nests by faking an injury, luring the potential predator a safe distance from the nest or young. When this has been accomplished, they return to the young. While the female is engaged in acting, the male is most often diving and screaming at the intruder.

(Source: Photo courtesy of U.S. Fish and Wildlife Service. Photo by Gary Kramer.)

FIGURE 19-10 Killdeer.

Sanderling
(Calidris alba)

DESCRIPTION

This common shorebird is sometimes known as the "wave chaser" for its habit of dashing about in the surf, following receding waves in pursuit of prey (Figure 19-11). Sanderlings are 8 to 9 inches long, with a straight, fairly short, black bill and gray-black feet and legs. The lower breast, belly, and sides are white, while the back, head, throat, and upper breast is reddish brown, with abundant black speckling intermingled with white. In flight, a bright white stripe is evident along the length of the wings. There are considerable variations in plumage color between summer and winter, as well as between individual birds. Sanderlings are primarily spring and winter visitors to the lower 48 states. They nest in Alaska and arctic Canada and winter on both coasts of the United States as well as on interior lakes and rivers.

HABITAT: Primarily utilizes beaches and mudflats in the lower 48 states.

FEEDING HABITS: Feeds on insects, **crustaceans**, and shoreline insects, such as sand fleas.

LIFE CYCLE: Normally nests on bare ground, near water. Four eggs are a normal clutch and they are typically laid in a shallow depression lined with grass.

(Source: Photo courtesy of Texas Parks & Wildlife Department © 2002.)

FIGURE 19-11 Sanderling.

FEEDING HABITS: They also utilize much the same food sources—insects, worms, and crustaceans—and feed in much the same way, by probing with their bills.

HABITAT: They all utilize much the same habitat: freshwater marshes, mudflats, and beaches. Even the upland sandpiper, which spends most of its time in upland pastures and prairies, utilizes these areas during the winter.

LIFE CYCLE: Each of these species are ground-nesting birds, and most lay their eggs in shallow depressions lined with leaves, grass, or seaweed. Three to five eggs are about average, with four being most common. We discuss here four species of sandpiper that are widespread and common.

SANDPIPERS

One or more species of this large family of shorebirds is present throughout North America. Most of the members of this family nest to the north of the lower 48 states, but some, such as the upland sandpiper, nest over large portions of the United States. More than a dozen members of the sandpiper family migrate through or winter in the United States. With a few noted exceptions, these birds have very similar characteristics.

Upland Sandpiper
(Bartramia longicauda)

DESCRIPTION ■

Upland sandpipers have a unique shape, which coupled with their preference for prairies and upland pastures, make them easy to identify (Figure 19-12). They have a near pigeon-size body, averaging 12 inches or so in length, but a long neck and small head. They have comparatively long, pale yellow legs and a short, thin bill. Upland sandpipers also have the habit of sitting atop fence posts and poles. They often hold their wings out for a brief period after they land. Adults are an overall dark brown, mottled with black above, with whitish belly and buff breast streaked with brown. The feathers of the wings and back have a golden tan fringe that gives the birds a scaled appearance. Upland sandpipers nest across most of the lower 48 states, Canada, and Alaska. They migrate virtually throughout the United States and spend their winters in South America. With their habit of nesting in pastures throughout much of the United States, upland sandpipers are probably the most commonly observed member of the sandpiper family during the summer.

(Source: Photo courtesy of Texas Parks & Wildlife Department © 2002.)

FIGURE 19-12 Upland sandpiper.

Spotted Sandpiper
(Actitis macularia)

DESCRIPTION ■

Another very common member of the sandpiper clan (Figure 19-13), this small bird measures about 7.5 inches. They breed throughout most of Canada and the northern United States, and winters along the Atlantic, Pacific, and Gulf coasts, as well as inland across much of the southern United States. It has the killdeer habit of bobbing its body up and down while holding its head still. They are found in almost all freshwater habitats, including small streams. These small sandpipers are most often seen alone. The adults are whitish on the lower breast and belly, with roundish dark spots. The white tends to turn to buff on the upper breast, and the upper parts are dark greenish brown, with fine black barring and spots. Spotted sandpipers have a distinctive pinkish orange bill, with a dark tip and a whitish buff stripe running the length of both wings.

(Source: Photo courtesy of U.S. Fish and Wildlife Service. Photo by Gary Kramer.)

FIGURE 19-13 Spotted sandpiper.

Western Sandpiper
(Calidris mauri)

DESCRIPTION ▶

The western sandpiper, although more abundant in the western United States, is nonetheless common along the East Coast, particularly during the fall migration (Figure 19-14). These small birds are only around 6.5 inches in length. The western sandpiper likes mudflats but is also fond of fresh-water marshes and beaches. Adult birds have white bellies and a mixture of grays, browns, and blacks on their upper parts. The fall color pattern is lighter overall than the spring color pattern and lacks the rust tinges that breeding adults exhibit during the spring. Western sandpipers nest in northwest Alaska and migrate and winter along both the Atlantic and Pacific coasts.

(Source: Photo courtesy of U.S. Fish and Wildlife Service. Photo by Lee Karney.)

FIGURE 19-14 Western sandpiper.

Least Sandpiper
(Calidris minutilla)

DESCRIPTION ▶

This is our smallest sandpiper, measuring just 6 inches in length (Figure 19-15). It is also one of our more widespread. Least sandpipers breed from Alaska, across northern Canada. They spend their winters along the southern portions of both coasts as well as inland. They are particularly widespread during spring and fall migrations. Least sandpipers are found in a variety of freshwater and saltwater habitats but are usually found near cover of some kind. They have a very fine black bill about the length of their head or slightly shorter. Their upper parts are mottled brown black with grayish fringes. Their bellies are white, while their breasts and necks are buff, streaked with fine brown lines. Least sandpipers forage quickly, probing and picking with their bills.

(Source: Photo courtesy of Texas Parks & Wildlife Department © 2002.)

FIGURE 19-15 Least sandpiper.

Greater (Tringa melanoleuca) and Lesser (Tringa flavipes) Yellowlegs

DESCRIPTION

As one might guess from their names, the primary difference between the greater and lesser yellowlegs is one of size (Figure 19-16). The greater yellowlegs is about 14 inches in length, and the lesser is only around 10.5 inches in length. The greater is also the more darkly marked of the two birds. Both birds have long, dark bills and long, bright yellow legs. Adults have a white belly and a blackish brown upper body with an abundance of white dots. Breast, head, and neck are finely lined with brown on a white background. Both of these birds breed in Alaska and central Canada and migrate and winter along both coasts. The greater yellowlegs is much more likely to be seen inland and is a quite common migrant across much of the United States. It often feeds by sweeping its bill back and forth in shallow water. Lesser yellowlegs are very active wading birds and seem to prefer freshwater ponds and marshes.

HABITAT: The greater yellowlegs inhabits virtually all freshwater habitats, as well as mudflats and saltwater marshes. The lesser yellowlegs may be found in freshwater and saltwater marshes, ponds, mudflats, and even northern wooded swamps.

FEEDING HABITS: A variety of insects, mollusks, crustaceans, and tiny fish.

LIFE CYCLE: Normally lays four eggs in a shallow depression on the ground, not too far from water.

(Source: Photo courtesy of Texas Parks & Wildlife Department © 2002.)

FIGURE 19-16 Lesser yellowlegs.

Long-Billed
(Limnodromus scolopaceus) and
Short-billed (Limnodromus griseus)
Dowitchers

DESCRIPTION

These two birds are nearly identical in size and very similar in appearance (Figure 19-17). As one might suspect, the long-billed dowitcher has a slightly longer bill than does the short-billed dowitcher, although it is nearly impossible to tell the difference in the field. Long-billed dowitchers are also slightly larger, averaging 11.5 inches to the short-billed dowitcher's 11 inches. Both birds have medium-length grayish legs and long, gray-black bills. Adults of both species have an overall mottled appearance, with various amounts of white on the belly. The back and sides of both birds are heavily spotted and barred with varying amounts of black. Breeding adults of both species tend to have reddish under-parts, with the short-billed dowitcher generally exhibiting more reddish color. The ranges of both birds are very similar: They nest in Alaska and Canada and spend their winters along the Pacific, Atlantic, and Gulf coasts of the lower 48 states. The long-billed dowitcher is more prone to using freshwater habitats than is the short-billed and it winters in inland areas. The short-billed is almost exclusively a coastal creature during the winter. Dowitchers are gregarious by nature and so are often found in large numbers. During migration periods, these two shorebirds are very widespread and common in North America.

HABITAT: Both of these birds can be found in much the same habitat. Mudflats, freshwater and saltwater marshes, and ponds are all utilized.

FEEDING HABITS: A variety of creatures, from worms and crustaceans to insects.

LIFE CYCLE: Four eggs are a normal clutch, and they are typically laid in a shallow depression in moss or short grasses.

(Source: Photo courtesy of Texas Parks & Wildlife Department © 2002.)

FIGURE 19-17 Short-billed dowitcher.

Long-Billed Curlew
(Numenius americanus)

DESCRIPTION

This is the largest of our shorebirds, measuring 22 to 24 inches in length (Figure 19-18). Due to its size and very long, downcurved bill, the long-billed curlew is fairly easy to identify. It has long, grayish legs and is an overall mottled brown buff, with light cream underparts. Feathers are heavily streaked and dotted with light brown and buff on a darker brown background. The undersides of the long-billed curlew's wings are a cinnamon brown, visible when the bird is on the wing. It is often seen on mudflats during the winter, probing the soft ground for worms and other prey. It is capable of twisting and turning its bill to follow worm burrows. Although the long-billed curlew's range is somewhat restricted in the east, usually along the coast from the Carolinas south during the winter, its range to the west is quite extensive. These birds breed from across much of the southwest to California and northward into southern Canada. Their winters are spent in California and eastward into northern Louisiana.

HABITAT: Grasslands, prairies, saltwater marshes, beaches, and mudflats.

FEEDING HABITS: Feeds heavily on worms and insects it probes from the earth with its long, flexible bill.

LIFE CYCLE: Long-billed curlews nest on the ground in grasslands and low-lying pastures. Four eggs seem to be a normal clutch and little effort appears to be made at nest construction. The eggs are laid in a shallow depression, as is common among shorebirds, relying on the hen's camouflage and low grasses for protection from predators.

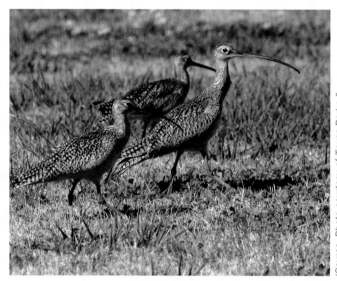

(Source: Photo courtesy of Texas Parks & Wildlife Department.)

FIGURE 19-18 Long-billed curlew.

Ruddy Turnstone
(Arenaria interpres)

DESCRIPTION ■

This gregarious shorebird is a rather common winter visitor to our coasts (Figure 19-19). Although the ruddy turnstone nests in Alaska and northern Canada, it winters along the Pacific, Atlantic, and Gulf coasts, where it is quite numerous. Ruddy turnstones are striking in color, combining a bright mixture of white, rust, and black. The white abdomen, belly, and lower breast and black upper breast turns into a complex pattern of black and white on the neck and head. The back and the base of the wings are a mixture of bright rust and black. They have bright orange legs and a short, slightly upturned black bill. Their legs are also relatively short. Ruddy turnstones are often observed in flocks along the coast during the winter months.

HABITAT: They prefer rocky, pebble beaches to sandy beaches, which they also inhabit. They also frequent mudflats and are occasionally observed in upland habitats.

FEEDING HABITS: Feed on crustaceans, small mussels, washed-up seaweed, and other small marine creatures that they secure by overturning small stones and pebbles.

LIFE CYCLE: Nests are usually a small hollow lined with leaves. Four eggs are a normal clutch.

(Source: Photo courtesy of Texas Parks & Wildlife Department © 2002.)

FIGURE 19-19 Ruddy turnstone.

SUMMARY

Each bird in this chapter is subject to the normal array of predators. Numerous species of snakes prey on their eggs and occasionally on their young. Skunks, raccoons, and foxes all destroy nests, eating both the eggs and the nest occupant, if possible. Foxes, bobcats, and house cats all prey on the adults, as do some birds of prey; but disease, cold, starvation, accidents, and other environmental factors probably result in more deaths than all the predators combined. We have abundant populations of most of these species. Unless we destroy too many inland and coastal marshes and other wetland habitats, we should continue to have viable populations for the future.

REVIEW QUESTIONS

Fill in the Blank

Fill in the blanks to complete the following statements.

1. The _____ is a shy, secretive bird that is rarely seen.
2. The _____ is perhaps more difficult to observe than the American bittern.
3. The green-backed heron is very common and is found in almost all _____.
4. The _____ is quite distinctive, with a slate-gray neck and back and a white rump and belly.
5. The little blue heron exhibits a rather odd color scheme. The immature birds are _____, whereas the _____ are a dark blue, almost black color.
6. The _____ is the largest heron in North America.
7. Great blue herons are very common and widespread, being found in almost all _____ habitats.
8. The _____ is a European species that derives its name from its habit of feeding in, around, and on the backs of cattle.
9. The _____ is a gregarious, medium-size heron that can be found in many freshwater and saltwater habitats.
10. Killdeer spend a great deal of their time in _____.
11. The _____ is a common shorebird, often known as the "wave chaser."
12. One or more species of _____ is present throughout North America.
13. The greater yellowlegs is about _____ inches in length, and the lesser is only around _____ inches in length.
14. The long-billed dowitcher is more prone to using _____ habitats than is the _____.
15. The _____ is the largest of our shorebirds, measuring 22 to 24 inches in length.
16. The _____ is a gregarious shorebird and is a rather common visitor to our coasts.

Short Answer Questions

1. List the common foods eaten by the great blue heron. What type of habitat might you expect to find these birds feeding in?
2. List the common foods eaten by many shorebirds. What type of habitat might you expect to find these birds feeding in?

Discussion

1. Several of the species discussed in this chapter nest in colonies. In your opinion, what are some of the advantages on colonial nesting? What are some of the disadvantages?
2. What factors do you believe most affect the populations of the birds discussed in this chapter?

Learning Activities

1. Select two species of herons, shorebirds, or egrets not discussed in this chapter. Using your local library, school library, or additional classroom texts, write a short biography of each bird. Include such information as habitat requirements, feeding habits, breeding or mating habits, and a physical description of each species.

2. Identify at least three potential areas of heron, egret, or shorebird habitat in your area. Write a detailed description of each habitat and identify species using these areas if possible.

Useful Web Site

<http://www.mbr-pwrc.usgs.gov>

CHAPTER 20

Reptiles

OBJECTIVES

After completing this chapter, you should be able to

- Describe some of the common characteristics of reptiles.
- Identify the type of habitat where reptiles might be found.
- Describe some common habits of reptiles.
- List some of the common reptiles found in the United States.
- Describe some differences between the various types of reptiles.

INTRODUCTION

One major characteristic of reptiles is their inability to regulate their internal body temperature. Reptiles are "cold-blooded," that is, their body temperature is directly related to the outside air temperature. During periods of cold weather, reptiles cannot maintain their body temperature, as mammals and birds do. Of course, their blood is not actually cold—**ectothermic** is a more appropriate term.

Because they are unable to maintain their body temperature, reptiles have a somewhat more restricted range than do most mammals or birds, particularly in the United States. Our most abundant populations of snakes, lizards, and turtles are found in the southeastern and southwestern United States, which have warm temperatures and relatively mild winters. During cold weather most reptiles are forced to **hibernate** and seek shelter

316

from the cold. As the outside temperature drops, reptiles become sluggish and **lethargic.** Because they are ectothermic, reptiles are absent from the Arctic and other areas that are permanently cold. Perhaps because of their inability to adapt to colder climates, North America has fewer species of reptiles than of mammals and birds. There are fewer than 300 species of reptiles found north of Mexico.

The reptile family contains tremendous variety. Turtles, lizards, snakes, and alligators are all reptiles, yet each of these related animals differs greatly from the others. Despite their vastly different shapes, most reptiles have a few things in common. For example, most reptile species feed on other animals, with the vegetarian land tortoises being noted exceptions. Reptiles have lungs instead of gills for breathing air. Most reptiles reproduce by laying eggs, although all poisonous snakes in North America, with the exception of the coral snake, produce live young. The eggs of reptiles have shells that are either hard, like a hen's egg, or leathery, like a pair of soft leather gloves. Reptiles have no free larval stage, as most amphibians do, and the young have a dry skin, consisting of scales or bony plates, as do the adults. Reptiles do not need to return to water to breed, as amphibians do, and fertilization is internal. Reptiles have a somewhat larger range than do the amphibians, being found over large areas of North America. Perhaps the main reason for their greater range is that they are not nearly as dependent on fresh water as are amphibians. Lizards are particularly abundant in the Southwest, whereas turtles and snakes are most abundant in the East and Southeast.

American Alligator
(Alligator mississippiensis)

DESCRIPTION

The American alligator is widely distributed throughout the South, being found in coastal and inland waterways from southern North Carolina south and west to Texas (Figure 20-1). They are large, lizard-like creatures with four short, strong legs. They have well-developed webbing between their toes, which aids them in swimming. Alligators tend to be an overall brown-green. Because of their distinctive appearance, they are virtually unmistakable. However, in portions of Florida the pet trade has resulted in the introduction of the spectacled caiman, and the American crocodile

Perhaps the most readily identifiable reptile in North America is the alligator. Alligators are members of the *Crocodilian* family, which includes gavials, crocodiles, and alligators. The gavials are common in the rivers of India, where they feed primarily on fish. Crocodiles are widespread in Africa, Australia, Asia, throughout the Pacific area, and in the Americas. Alligators have a somewhat more restricted range, being most commonly found in North America. However, their close cousins, the caimans, are widespread in South and Central America.

HABITAT: Alligators are nearly always found near water. They are aquatic by nature and are most often found in and around rivers, swamps, and ponds. They are also found in coastal marshes.

FEEDING HABITS: Alligators are carnivores. They survive on a diet that consists largely of meat, with the occasional fish being taken. Alligators eat whatever they can catch, from small mammals to deer, feral hogs, dogs, and (rarely) humans. They most often hunt by lying in wait just below the

can be found in limited numbers in the Florida Keys. The caiman is much smaller than both the American alligator and crocodile and has a bony ridge between its eyes. The American crocodile has a more tapered snout than does the alligator, and the fourth tooth in its lower jaw is much larger than the others and is visible from the side when its jaws are closed. The alligator has a more blunt snout, and no teeth are visible when its jaws are closed. Alligators can grow to enormous proportions, with adults occasionally reaching 19 or more feet in length. This great size is rare, but 6 to 10 feet is common.

surface and a short distance from shore for potential prey to come to the water's edge to drink. With a short, amazingly quick rush, they seize their prey and drag it into the water, to be drowned and eaten.

LIFE CYCLE: In contrast to most reptiles, the American alligator takes a fairly active role in the care of its young. The female constructs a nest of plant materials and soil on land, but not too far from water. These nests may be 3 feet or more in height. The eggs are laid in a hole or depression in the top of the nest and covered over with additional vegetation. Clutches average 30 to 40 eggs, with as few as 2 or as many as 80 being laid. Female alligators normally remain close to the nest site, protecting it from predators until the young hatch. When the female hears the young hatching, she removes some of the top covering, allowing the young to escape to the water. Young alligators are only about 9 inches long and lead very dangerous lives until they are large enough to fend for themselves. The young and eggs are preyed on by raccoons, skunks, and black bears. Adult alligators also feed on the young. Adult alligators, certainly those that are large enough to fight off other adult alligators, have few enemies other than humans.

(Source: Courtesy U.S. Fish and Wildlife. Photo by Dick Bailey.)

FIGURE 20-1 Alligators.

TURTLES

Turtles are readily identifiable reptiles, which nearly everyone has encountered. It has been stated that a turtle is simply "a reptile with a shell." The turtle's shell may be hard and rigid or it may be more pliable and leathery to the touch. The rigid variety is made up of a bony layer with hard scales or scutes overlaying it. Most inland turtles have hard, bony shells. A turtle's shell is typically divided into two parts, the carapace and the plastron. The carapace is the upper half or "roof" of the shell; the plastron is the lower half or "bottom" of the shell. These two parts are usually fused, giving the turtle its solid shell. In some turtles there are a pair of hinges on the plastron that allow the turtle to completely close its shell once its legs are pulled inside.

Each turtle has four well-developed limbs, unlike some other reptiles. These limbs vary according to the habitat of the turtle. Land tortoises have short, strong limbs that are not webbed but carry claws. These allow the tortoise to traverse its habitat, although not at great speed! The more aquatic freshwater turtles normally have webbed feet and longer claws. Turtles tend to have very strong jaws, which can inflict damage on the unwary. All turtles breathe air with lungs, like mammals and birds. Species found in aquatic habitats can also absorb oxygen from water. This is accomplished with their mouths and **cloaca,** which have a surface similar to a gill. As water flows over these surfaces, oxygen is absorbed. This feature, coupled with an ability to slow their metabolisms, allows them to remain under water for long periods. It is believed that turtles do not hear particularly well but have good eyesight and a good sense of smell. Many scientists believe that turtles are our most intelligent reptiles.

Identification of individual species of turtles can be quite difficult, particularly among the aquatic species. It may be necessary to examine the turtle's plastron in order to identify it. Although turtles do not have teeth, they have powerful jaws and are best left alone. To pursue these reptiles in their native habitat, it is best to secure a field guide or other suitable reference of the reptiles in your area and carry it with you in the field. Although there are 40 to 50 species of turtles in the United States and Canada, few species are widespread. We will concentrate on the species that are more common and likely to be seen. Some turtles have many subspecies, which may be virtually indistinguishable, particularly to the untrained eye. Because their habitats, feeding habits, and life cycles are all very much alike, we give one description of each. Some common aquatic species are the alligator and common snapping turtles, several species of painted turtles, eastern and western box turtles, 10 or more subspecies of map turtles, the slider, the stinkpot, and the river cooter. The gopher tortoise is perhaps our most common land turtle, and the spiny soft-shell and smooth soft-shell are common soft-shell aquatic turtles.

Habitat: All North American aquatic species prefer similar habitats. The edges of ponds, streams, rivers, lakes, and swamps are favorite homes. Most aquatic turtles prefer slow-moving or still waters with muddy bottoms. Land or terrestrial turtles, on the other hand, prefer the edges of fields, pastures, gardens, and open woodlands. The best chance of seeing turtles in a natural setting probably occurs during spring. At this time of year they are most active, searching for mates and nesting sites or simply foraging after coming out of their winter hibernation period. Turtles are often observed sunbathing on partially submerged logs and along the banks of rivers, streams, and ponds.

Feeding Habits: With the exception of the land tortoises, the majority of turtles are meat eaters. They feed mainly on fish, but larger individuals can pull ducklings under. Such prey is drowned and then consumed. Turtles also feed on crustaceans, mollusks, and insects. Some aquatic species may also feed on aquatic vegetation. Land tortoises are grazers, feeding on grasses, forbs, and tender young shoots of plants.

Life Cycle: There is not a great deal of research available on many species of turtles. Some species are virtually unstudied, particularly in regard to reproductive behavior. Because much turtle behavior occurs underwater or at night, detailed studies of their habits are difficult. However, there is enough data available to be able to draw some broad conclusions about turtle behavior.

Turtles generally breed from April to November over much of their range. Peak mating and nesting activity usually occurs in June. Female turtles lay eggs, either hard or soft shelled, in a nest that they dig in moist soil. Nests are usually 4 to 8 inches deep and flask- or bottle-shaped. The number of eggs laid varies greatly among different species and even more so among individuals within a species. Generally 3 to 30 eggs are the extremes in clutch size, with 12 to 20 probably being a good average.

Incubation generally takes from 9 to 18 weeks. The eggs and young are on their own once the eggs are laid. After hatching, the young burrow out of the nest and, with the exception of the land tortoises, make their way to the nearest available water. Raccoons, opossums, and skunks prey heavily on the eggs and young. Once the young reach the water they are food for the larger fish, as well as adult turtles. Most turtles spend the colder months in hibernation, usually in mud and aquatic vegetation and often under an overhanging mud bank or ledge. Little is known about the life expectancy of turtles. However, it is almost certain that mortality is high among the young. Conversely, adults have few enemies other than humans and probably live for many years. The land turtles or tortoises are thought to live particularly long lives, sometimes 30 or more years. All adult turtles have powerful jaws and are quite capable of defending themselves. Snapping turtles in particular should be handled with great care, or not at all. Larger individuals have enough power in their jaws to sever fingers.

Common Snapping Turtle
(Chelydra serpentina)

DESCRIPTION

This large turtle is widespread and common throughout much of the South and Southeast (Figure 20-2). Adults in the wild have shell sizes from 8 to 18 inches and may weigh 40 to 50 pounds. Individuals kept in captivity and fed well may weigh considerably more. The upper shell or carapace color varies from almost black to a light brown. However, older adults may be so covered with algae that their true color is impossible to discern. These turtles are fond of basking in the sun. The common snapping turtle has a large head and powerful jaws.

FIGURE 20-2 Snapping turtle.

(Source: © Dennis Donohue, 2010. Used under license from Shutterstock.com)

Alligator Snapping Turtle
(Macroclemys temminckii)

DESCRIPTION

This is the largest freshwater turtle in North America (Figure 20-3). Shell length varies from 13 to 25 inches or more, and individuals weighing nearly 200 pounds have been recorded. When not covered with algae, these turtles are gray to dark brown. Unlike most other turtles, the alligator snapping turtle seldom if ever basks in the sun. It has a long, rounded tail; a pronounced hooked beak; and very powerful jaws. These giants are found in most southeastern states and northward into Missouri and are common in the Mississippi River.

(Source: Photo courtesy of Texas Parks & Wildlife Department.)

FIGURE 20-3 Alligator snapping turtle.

Painted Turtles
(Chrysemys picta)

DESCRIPTION

Painted turtles, as one might guess from the name, are well marked and colorful (Figure 20-4). There are numerous species of painted turtles in the United States. On most painted turtles each scute is outlined in yellow or red, and the head, neck, and legs are striped in red, yellow, or greenish yellow. Like most turtles, older adults may be so covered with algae that their brilliant colors are not visible. Size varies considerably among painted turtles, but they are generally 4 to 10 inches in length. Painted turtles are well represented in southern and eastern portions of the United States, and subspecies are also found in some western states.

(Source: Photo courtesy of U.S. Fish and Wildlife Service. Photo by Gary M. Stolz.)

FIGURE 20-4 Western painted turtle.

Map Turtles
(Graptemys)

DESCRIPTION

There are numerous species of map turtles in the southeastern United States, with some species ranging northward into the Great Lakes states and southern Canada (Figure 20-5). They range from 4 to 12 inches in length, and females are generally twice the size of males. Another feature common among the various subspecies is that the females often lay more than one clutch of eggs in a single season. Most of the map turtles enjoy basking and are often seen in groups doing just that. Most species have

(Source: © Ryan M. Bolton, 2010. Used under license from Shutterstock.com)

FIGURE 20-5 Map turtle.

brown to olive carapaces, yellow strips along the neck, and fairly large heads. Map turtles are generally found in areas with numerous sunbathing sites, such as partially submerged logs and rocks.

Sliders
(Trachemys)

DESCRIPTION

The red-eared slider may be the most familiar turtle in America (Figure 20-6). Although there are several subspecies of sliders, the red-eared were once commonly sold as pets. These turtles are most often dark green, with a variety of light and dark markings. The red-eared's distinctive characteristic is a red stripe behind each eye. Sliders prefer quiet backwaters with ample basking sites. They are gregarious in their basking, often stacking three high to

(Source: Photo courtesy of Texas Parks & Wildlife Department © 2002.)

FIGURE 20-6 Red-eared sliders.

catch some sun. Sliders also prefer habitats with soft, muddy bottoms and plenty of aquatic vegetation. One or more of the slider species is present throughout much of the southeastern United States as well as most of Texas.

River Cooter
(Pseudemys)

DESCRIPTION ▸

A variety of cooters are found in the United States, but the river cooter is probably the most widespread (Figure 20-7). As the name suggests, these are river turtles, preferring slow- to moderate-flowing rivers and streams. They are found throughout most of the drainage systems in the Southeast. The carapace is generally an olive color, with pale yellow or orange dividing each shield. The plastron is usually yellow, with feet, legs, and head being dark brown to black. There are usually numerous yellow or orange yellow stripes on the head.

(Source: Photo courtesy of John D. Willson.)

FIGURE 20-7 River cooter.

Common Musk Turtle
(Sternothaerus odoratus)

DESCRIPTION ▸

This fairly common but rather small turtle averages less than 5 inches in length (Figure 20-8). They are generally found from Florida north to southern Canada and westward to south Texas and Oklahoma. These turtles can be found in almost any body of water, including slow-moving streams. They are generally brown-black, but their true color is often obscured by algae or caked-on mud. Two pale yellow stripes run from the snout to the base of the neck, one above the eye and one below.

(Source: Courtesy of Isabelle Francais.)

FIGURE 20-8 Musk turtle.

Common Mud Turtle
(Kinosternon)

DESCRIPTION ◗

This turtle is found from New York to eastern Illinois and south to the Gulf of Mexico (Figure 20-9). However, it is not thought to inhabit Florida or to be common west of a line from eastern Illinois to the Gulf of Mexico. They are an overall olive brown, with a brown head, neck, and legs. These are small turtles, rarely longer than 4 inches. They prefer pond bottoms and are almost entirely aquatic, leaving the water only to nest.

(Source: Courtesy of Isabelle Francais.)

FIGURE 20-9 Mud turtle.

Soft-shell Turtles
(Trionychidae)

DESCRIPTION ◗

There are several species of spiny and smooth soft-shell turtles (Figure 20-10). These turtles have a soft, leathery shell that does not afford them the protection hard-shell turtles enjoy. Soft-shells are highly aquatic, leaving their watery homes only to nest and to bask in the sun. They are quite fond of basking, using logs or tree limbs that have fallen into the water. They are also very good swimmers, capable of outswimming such prey as fish and frogs. They prefer muddy bottoms with adequate aquatic vegetation and are found in streams, rivers, and ponds. Soft-shells can grow quite large, with shell lengths of 18 inches and weights of up

(Source: Photo courtesy of Texas Parks & Wildlife Department © 2002.)

FIGURE 20-10 Spiny soft-shell.

Soft-shell Turtles
(Trionychidae)

DESCRIPTION ▶

to 30 pounds. They tend to be olive green or olive brown, with older specimens being largely dark brown. The head, neck, and legs are often marked with a variety of black dots and lines, while the shell is often outlined in pale yellow or black. Most soft-shells have a characteristic pointed nose or snout, quite different from the average turtle's blunt snout. Soft-shell turtles are most abundant in the southeastern United States, but some species are found in more western locales.

Box Turtles
(Terrapene)

DESCRIPTION ▶

Although there are quite a few species of box turtles, we will deal with two of the more common and widespread, the eastern and western box turtles. The western box turtle is quite widespread, ranging from Indiana and Illinois westward to the Rocky Mountains and southward to New Mexico, Arizona, and northern Mexico. It is particularly common on the Great Plains and in areas with sandy soils. The eastern box turtle is also quite widely distributed, ranging from Maine to Michigan and southeastward to Tennessee and Georgia. These terrestrial turtles eat a variety of insects, berries, and other vegetation.

(Source: Photo courtesy of Texas Parks & Wildlife Department © 2002.)

FIGURE 20-11 Ornate box turtle.

All the box turtles (Figure 20-11) have high, domed carapaces, which are often flattened on top. Box turtles also tend to be quite colorful, with a base color of olive brown and a variety of yellow lines and markings on the carapace. The western box turtle often has a well-marked plastron, too, showing many different patterns of yellow stripes and lines. These turtles are often observed in early morning or late evening and are most active during the spring and after rain showers. These turtles are not especially large, averaging about 5 inches in length, but they have the short, powerful legs and clawed feet of a land-dwelling turtle.

Gopher Tortoise
(Gopherus polyphemus)

DESCRIPTION

The gopher tortoise is a true terrestrial, land-dwelling tortoise (Figure 20-12). Gopher tortoises are slow herbivores and exhibit greater intelligence than their aquatic relatives. They are also rather large—8 to 12 inches long and 9 pounds or more. They are an overall dull olive brown, though younger individuals may exhibit yellow blotches or spots in each shield. These tortoises range from Arkansas to South Carolina and south into Florida. They are most abundant in areas with sandy soils and are noted for digging long burrows in which to take refuge. Gopher tortoises are well equipped for digging, with very strong legs and clawed

(Source: © Rose Thompson, 2010. Used under license from Shutterstock.com)

FIGURE 20-12   Gopher tortoise.

feet. Though the young and eggs of the land-dwelling turtles and tortoises are vulnerable to certain predators, most notably raccoons, opossums, skunks, and feral hogs, adults have few predators other than humans. Most predators may scar the shells with their teeth but rapidly lose interest in attempting to kill and eat a tortoise or box turtle. Their shells are simply too hard.

LIZARDS

Lizards are the largest single group among the reptiles. In the United States lizards are largely confined to warmer climates. Good populations of lizards can be found throughout the southern and eastern states as well as in the Southwest. Because they are ectothermic they do not survive well in colder climates.

Lizards are scaled reptiles that normally exhibit visible eardrums and have moveable eyelids, unlike snakes. Lizards also have four legs, with claws on their toes and a nonexpandable jaw. Most have visible scales, although some have scales so small as to give them a smooth-skinned appearance and touch. These are general characteristics and may not hold true in all species. The glass lizards, for example, do not have legs and may at first glance resemble a snake. Snakes, however, do not have movable eyelids, ears, or legs, and they have greatly expandable jaws. Lizards resemble salamanders, but salamanders are amphibians, lack scales, and do not have claws on their toes, as do the lizards.

Many lizards have a rather unique defense strategy. If they are caught by a predator, their tails break off. The cast-off part wriggles about for several minutes, providing a distraction that is often sufficient to allow the lizard to escape. Should a lizard lose its tail in this manner, it usually grows another one. However, the new tail is normally shorter and the color and scale patterns may be noticeably different from the original parts. The spinal column in the new tail consists of cartilage rather than bone.

Most lizards have a good sense of smell and excellent eyesight. However, in spite of their external ears, they appear to have poor hearing, and they have little or no voice. Unlike most turtles, lizards generally have a good set of teeth. Their teeth are usually all alike and situated along the edges of their jaws, but occasionally on the **palate.** These teeth are seldom used for chewing. Instead, they are used to grasp and hold prey. There are only two known poisonous species of lizards, one of which is the Gila monster, a native of the American Southwest. There are no poisonous lizards in the eastern United States.

Habitat: Because many species of lizards lead conspicuous lives, almost everyone who lives in their range has seen one. Conversely, some species of lizards, most notably the skinks and glass lizards, are fairly secretive and are seldom observed unless specifically searched for. Some species that are more active, and thus more likely to be observed, are the collared and fence lizards. Horned lizards and racerunners are also more commonly observed. Common spots to see lizards are along fences, among old ruins, on stumps and logs, in abandoned buildings, and in piles of dead leaves and brush. With the exception of our coldest climates, various species of lizards can be found in nearly any habitat. Some species are adapted to warm, moist, wooded, and humid climates; most of these live in the Southeast. Other species are adapted to hot, dry, and sparsely vegetated habitats. These can be found in Great Plains states and the desert Southwest. Many of the eastern species of lizards are **arboreal,** that is, they spend the bulk of their time off the ground in trees and bushes. Most species of lizards like the sun and can often be observed while they bask in its rays. Because they are ectothermic, almost all lizards hibernate during the winter months. They are generally most active shortly after emerging from hibernation, which is a good time to see them.

Feeding Habits: Some lizards are herbivores, eating buds, flowers, and other vegetation, but most are carnivorous. They eat a large number of insects, from grasshoppers to mealworms and crickets, as well as spiders, worms, and other lizards.

Life Cycle: The habits of many lizards are not well known, but like most reptiles, lizards are egg-layers. The size of the clutch depends largely on the species involved. Numbers from 1 to 21 are reported by various authorities. The eggs are typically laid in sandy soil and left to hatch, but this is not always the case. Some species lay their eggs in leaf litter and other debris found on the ground. The females of some of the glass lizards, more commonly known as glass snakes, are believed to incubate their eggs, as are some other species of lizards. Some species lay multiple clutches during the course of a spring and summer. The mating season for most lizards is generally in the spring or early summer. As a general rule the young are on their own immediately after the eggs are laid. Most lizards, particularly the smaller species, are thought to live fairly short lives. They are preyed on by a great many predators, including other lizards. The eggs and young are eaten by such predators as skunks, raccoons, and opossums. Young and adults are taken by a wide variety of birds, as well as by bobcats, house cats, and virtually every other small- to medium-size carnivore. Many species of lizards are incredibly fast, while others are equipped with excellent camouflage. These skills help them to survive.

We will discuss only the species that are the most widespread and most likely to be seen. If you develop a particular interest in lizards, you can choose from many excellent texts that are devoted solely to lizards. A good Web site to gain additional information on reptiles is www.livescience.com.

Collared Lizard
(Crotaphytus collaris)

DESCRIPTION

These lizards are well distributed throughout the western states and are found as far east as Missouri and as far north as southern Idaho (Figure 20-13). Collared lizards prefer semiarid, rocky habitat. They have rather short, thick bodies, with tails that are usually more than twice the body length. Their heads are large and they are quite **pugnacious**, often attacking and eating other lizards not much smaller than themselves. Their hind legs are much longer and more powerful than their fore legs, making them proficient jumpers.

(Source: Photo courtesy of Texas Parks & Wildlife Department © 2002.)

FIGURE 20-13 Texas collared lizard.

Collared lizards come in several colors, but all have two distinct black "collars" around their necks. Males are usually some shade of yellow or green, having a bluish throat patch in areas west of the Rockies and an orange one in areas east of the Rockies. Females are often slate gray and their neck collars are usually less conspicuous than those of the male. During the mating season, the females generally exhibit red dots and the males have their brightest colors. They range from 4 to 14 inches in length. As stated earlier, they are quite pugnacious and may attack rather large prey. Although they consume other lizards and are thought to be largely carnivorous, they also eat the blossoms of some plants, as well as tender shoots and leaves.

Glass Lizard
(Ophisaurus)

DESCRIPTION ■

Glass lizards are widely distributed in North America, ranging from Nebraska eastward through much of the eastern two-thirds of America (Figure 20-14). They are found as far south as northern Mexico and as far north as Wisconsin. Glass lizards are more commonly called glass "snakes." This misnomer arises from the fact that they have no visible limbs, unlike other lizards. At first glance they do indeed resemble snakes, but they have external ear openings and movable eyelids. These two features, along with other less-noticeable ones, make them lizards.

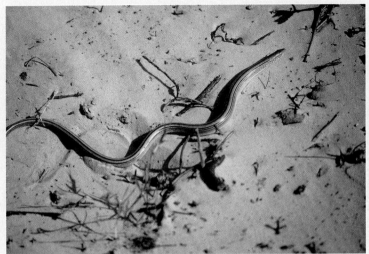

(Source: Photo courtesy of Texas Parks & Wildlife Department © 2002.)

FIGURE 20-14 Western glass lizard.

Glass lizards range in size from 20 to 35 inches, with occasional specimens being longer. They have a slender, serpentine body, which may be pale yellow to olive or brown. They often have dark spots or stripes along their scales. These lizards quite commonly shed their tails when pursued and caught by a predator. The shed portion thrashes about rapidly and generally holds the attention of the predator until the lizard can escape. Glass lizards eat insects, birds' eggs, worms, and grubs. Their preferred habitat is grassland, open woodland, or prairie.

Green Anole
(Anolis carolinensis)

DESCRIPTION ■

Although this lizard is commonly known as a **chameleon**, it is not a true chameleon (Figure 20-15). However, it does have the ability to change its color to match its surroundings. The green anole may, in fact, be any shade from dark green to dark brown. In new surroundings it can change colors quickly. Males have a prominent throat patch or **dewlap** that is pinkish and can be raised during the mating season. Females may have a pinkish area on the throat. Green anoles are common throughout the southeastern United States. They range from about 2 inches at birth to 8 inches or so as adults.

(Source: Photo courtesy of Texas Parks & Wildlife Department © 2002.)

FIGURE 20-15 Green anole.

These lizards are arboreal, typically residing in trees, shrubs, and bushes. They also are commonly seen on fences and the sides of buildings. Their food consists mostly of spiders and insects. They are very speedy and often stalk within a short dash of their prey. Their breeding season may last for up to 7 months, with the females laying single eggs in leaf litter and other debris. Green anoles require shade and prefer a warm, moist environment.

Fence Lizards
(Sceloporus)

DESCRIPTION

There are several species of fence lizards, one or more of which can be found across much of America (Figure 20-16.) These lizards are most often brown to gray, with older specimens appearing black. Their most notable feature is a bright blue throat on the males and a prominent blue spot on each side of the belly. The blue belly patches may also be bordered in black, and most species have various dark markings, spots, blotches, or stripes down the back. Fence lizards are found in a wide variety of habitats, from wooded rocky areas to fences, fallen logs, and stumps.

(Source: © Michael Ledray, 2010. Used under license from Shutterstock.com)

FIGURE 20-16 Fence lizard.

Fence lizards are very good climbers and generally make a dash up the nearest tree if caught on the ground. They are primarily insect eaters, often foraging for food on the ground. They range from 2 to 7 inches, with an occasional specimen being longer.

Lesser Earless Lizard
(Holbrookia maculata)

DESCRIPTION

These lizards have no external ear openings, unlike most other lizards (Figure 20-17). They are one of the most widespread lizards in the western United States. The short, rather plump body ends in a tail about the length of the body and somewhat flattened. They are brownish gray, with four rows of black or dark brown spots along the back. Lesser earless lizards usually have a pair of black bars between their limbs and dark bars along the under-side of the tail.

(Source: © Rusty Dodson, 2010. Used under license from Shutterstock.com)

FIGURE 20-17 Lesser earless lizard.

Lesser Earless Lizard
(Holbrookia maculata)

DESCRIPTION ■

They prefer sandy and rocky areas with little vegetation. Lesser earless lizards are active and forage for insects as their main source of nourishment. They are not particularly wary and are quite **inquisitive.**

Racerunners
(Cnemidophorus)

DESCRIPTION ■

This is a fairly large and widespread group of lizards. Racerunners earned their name because of their often blinding speed (Figure 20-18). It is seldom possible for the human eye to follow the smaller individuals of this group if they are in a real hurry. They have a slender, graceful build, with a long tail that gradually tapers to a point. The longish head is much less blunt than that of most other lizards. The base color is an olive brown, and most species have four to six yellow, white, or gray stripes running from the back of the head down the body. Hind legs are sturdy and strong, appearing to skip and scurry along the ground.

(Source: Photo courtesy of John D. Willson.)

FIGURE 20-18 Racerunner.

These are primarily ground-dwelling species and are often observed in open areas. Because of their need to use their great speed they prefer dry, sandy areas without obstructions. Racerunners may also be seen on floodplains, short grass areas, and open woodlands. These lizards forage heavily on insects, probably capturing many with their great speed. Racerunners range from 6 to 12 inches on average and can be found throughout the southeastern and plains states.

Skinks
(Scincidae)

DESCRIPTION

This is a very large and widely distributed group of lizards (Figure 20-19). However, they are secretive and are not often seen unless deliberately searched for. The skinks are small to moderate in size, ranging from 5 to 14 inches long. They have a smoother, less-scaled appearance than most other lizards. Skinks generally have a base color of olive to dark brown, with various stripes and spots. Color is highly variable among species and among individuals within a species. Mature adults may exhibit no stripes or notable markings.

(Source: Photo courtesy of Texas Parks & Wildlife Department © 2002.)

FIGURE 20-19 Short-lined skink.

These lizards are primarily ground-dwellers but are capable of climbing trees or the sides of buildings. Some skinks burrow. All are difficult to observe. Most skinks require a damp environment, consequently, the western varieties are usually found only near a water source. The eastern varieties are fond of burrowing and foraging in leaf litter. Skinks eat insects, spiders, and other smaller lizards. The females are known to lay fairly large clutches of eggs, and many species incubate or tend to their eggs until they hatch.

Horned Lizards
(Phrynosoma)

DESCRIPTION

These lizards are often incorrectly referred to as "horned toads" (Figure 20-20). They have a wide, fat, toadlike body and generally have short, slender tails. Horned lizards have a series of spines or "horns" that are usually located on the head and down the back. Some species have additional spines on the throat. The most common of the horned lizards is the Texas horned lizard. Although it is found in several western and Plains states, most notably New Mexico, eastern Arizona, Colorado, Kansas, and Oklahoma, it is still known as the Texas horned lizard. Another fairly well-distributed species is the round-tailed horned lizard. Horned lizards are an overall drab sandy brown, with their spines or horns being quite distinctive. They have four well-developed, strong limbs and are quite fast.

(Source: Photo courtesy of Texas Parks & Wildlife Department © 2002.)

FIGURE 20-20 Texas horned lizard.

Horned lizards prefer dry, sandy areas that are usually arid or semi-arid. They are not commonly found in areas with a lot of vegetation. Horned lizards feed heavily on ants and other insects. There have been notable population declines in recent years, for which increased use of pesticides is suspected. Their decline has also been linked to a decline in the prey ant species on which horned lizards rely due to the proliferation of the red imported fire ants. Horned lizards range in size from 2 to 6 or more inches, including the tail.

Gila Monster
(Heloderma suspectum)

DESCRIPTION

This is the only poisonous lizard in the United States (Figure 20-21). Gila monsters have a very restricted range, being found only in the desert regions of extreme southern Utah and Nevada, Arizona, and far western New Mexico. Their venom is mainly neurotoxic, paralyzing the respiratory and circulatory systems. Gila monster bites are painful but seldom fatal to humans. They do not have as effective a delivery system for their venom as do most poisonous snakes. The Gila monster's venom is delivered through grooves in its teeth as it bites its victim, rather than being injected with fangs. They have a very powerful bite and hang on to victims **tenaciously.** Gila monsters are slow, deliberate creatures when undisturbed, but they can be surprisingly quick if bothered.

(Source: © Rusty Dodson, 2010. Used under license from Shutterstock.com)

FIGURE 20-21 Gila monster.

Gila monsters average 12 to 20 inches in length, with the occasional specimens somewhat longer. They are heavy bodied, with relatively short, powerful legs. Their tails are usually one-half to one-third of their body length. Gila monsters are a mixture of black, tan, and lighter colors, usually pinkish white or pale yellow. They are a curious mix of spots and blotches with the dark colors being dominant on some specimens and the light on others. They are egg-layers, like most lizards, and seem to feed on the eggs of lizards and ground-nesting birds, rodents, and insects.

SNAKES

Snakes, for the most part, are beneficial to humans. Many people fear snakes to the point of **hysteria,** but in most cases their fear is unfounded. Few species are poisonous. In fact, the majority kill their prey by swallowing it while it is still alive or by constricting it within the folds or coils of their bodies. Fewer people are killed in the United States each year by poisonous snakes than are killed by adverse reactions to wasp, hornet, and bee stings. Yet an unreasonable fear of snakes persists among many people.

Snakes differ from lizards and other reptiles in that they have no limbs, no movable eyelids, and no external ear openings. Snakes are also generally slender in build and have greatly expandable jaws. They usually have only a single row of scales on the belly and curved teeth. Snakes cannot hear in the same sense that humans and other mammals can. However, they can pick up vibrations through the ground. This allows them to know of the approach of something that they might not be able to see. Most snakes have excellent eyesight and a good sense of smell. Snakes smell with their forked tongues, which also allow them to taste. In order to grow, snakes shed their skins, often several times each year. This is accomplished by crawling out of the old skin, head first, once the new one is ready.

One feature of snakes that separates them from other reptiles is their expandable jaws (Figure 20-22). Snakes are unable to cut up or chew prey, so their prey must be swallowed whole. Snakes are able to accomplish this because the skin between the scales is very elastic and can stretch well beyond its normal size. Snakes do not use their teeth to chew. Teeth are used instead to grasp and hold the prey firmly. A few snakes are equipped with a set of fangs that inject venom into prey. In North America, the snakes with movable, hypodermic-type fangs are known as **pit vipers.** Some snakes, the coral snake being the only representative in North America, have short, permanently erect fangs. In order for a coral snake to deliver its highly toxic venom, it must bite and chew on its victim. The pit vipers have no such handicap and can strike and deliver a **potent** dose of venom in an instant.

Of the dozens of snake species in North America, the vast majority are nonpoisonous. Only four poisonous species are found in the United States, although some snakes have numerous subspecies. There are more than a dozen subspecies of rattlesnakes, for example, and several subspecies of copperheads. These snakes may be found in vastly different habitats, but they are all very similar in most other aspects of their existence. Most nonpoisonous snakes lead very similar lives as well and can be found in much the same habitat.

Habitat: Like most wildlife, snakes have definite preferred habitats. Along watercourses, in and around stone walls and rock outcroppings, rocky ledges, around and under fallen logs, and atop stumps and logs are all good places to find snakes. They also prefer the edges of marshes, lakes, and swamps. Piles of downed

FIGURE 20-22 This rattlesnake shows how snakes swallow their prey, in this case, a white rat.

(Source: Photo courtesy of Texas Parks & Wildlife Department © 2002.)

timber, brush piles, and large piles of leaves are also good snake habitat. Flower beds, particularly those with shrubs, can also be favorite spots. Some snakes are aquatic, living in and around ponds, lakes, rivers, streams, and other bodies of water. Most snakes, however, are **terrestrial,** preferring to live on land.

Most snakes are largely nocturnal, but many are active early and late in the day. Snakes are reptiles, so they like warmth, but they cannot take too much heat. Especially during the summer months, the midday hours find most snakes in the shade of a rock or in an old animal burrow. Snakes spend the winter months in hibernation and are often most visible just before the onset of hibernation. Snakes tend to have common denning sites during winter, that is to say, some sites are better suited for hibernation than others. Therefore large numbers of snakes can be found at a good den site. Such sites are used year after year, so long as they are undisturbed. When spring arrives, snakes are once more active and are often visible as they forage for food after the long winter fast.

Feeding Habits: Mice and rats make up a large part of the snake diet. Other small mammals, such as rabbits, are also taken. Ground-dwelling birds and especially their eggs are often taken. Insects are an important diet item for smaller snakes and young snakes. Some snakes eat other snakes, and in fact some are quite adept at it. Frogs, toads, and salamanders are also important food sources for snakes. There are no vegetarian snakes—all survive strictly on animal matter.

Snakes have three main eating methods. Nonpoisonous snakes either **constrict** victims in the coils of their bodies or catch prey and swallow them while they are still alive. Poisonous snakes inject venom into their victims and swallow them once their struggles have ceased. In all cases the prey is swallowed whole, almost always head first. Because the prey is often larger in diameter than the snake, the snake must have expandable jaws. Snakes are slow to digest their prey, and if a meal is particularly large, the snake often becomes lethargic until digestion is complete. It is not uncommon for a snake to consume prey that weighs more than half its own body weight.

Life Cycle: Snakes generally mate shortly after emerging from their winter dens. Most snakes in North America lay eggs, although all the poisonous varieties, except the coral snake, give birth to live young. Snake eggs are soft and leathery and are generally laid in a site that at least keeps them moist. Organic matter, decaying leaves, and areas under stones are often chosen for that reason. Clutches of eggs probably average 12 to 24 eggs, but more may be laid or, in rare instances, fewer. Young snakes are active and capable of fending for themselves immediately after hatching. Live-born young and those hatched from eggs receive no parental care to speak of. A few nonpoisonous snakes, such as the garter and water snakes, also give birth to live young. Litters of live-born young of poisonous snakes tend to be rather small, perhaps a dozen or so, while those of the nonpoisonous varieties tend to be larger, occasionally as many as 50.

Most snakes are thought to live fairly long lives, particularly the larger varieties. It is estimated that a young snake in the wild requires 3 years or so to attain adulthood. However, snakes are preyed on quite heavily. The eggs are sought after by the usual egg-eaters—skunks, raccoons, opossums, and, of course, other snakes. The young are killed and eaten by a vast array of predators, including crows, bobcats, hawks, and other snakes. Large bullfrogs probably take a fair number of the young of aquatic species of snakes. No snake, with the possible exception of a large member of a poisonous species, is immune to predation. As with most wildlife, humans are their greatest enemy.

COMMON NONPOISONOUS SNAKES

The category of nonpoisonous snakes includes well over 100 species, not counting a multitude of subspecies. We discuss only a few of the most widespread and common. All snakes in this family have broad scales on the belly and lack the pits behind the nostrils that poisonous pit vipers have. Nor do they have the permanently erect fangs of the coral snake. Many of these snakes have numerous subspecies in various parts of the United States. Identification of snakes is often difficult because of individual color variations. It is often difficult, if not impossible, to find two snakes of the same species that are marked in an identical manner. To identify many snake species, a snake identification text with color pictures is usually necessary, particularly when trying to identify subspecies of rattlesnakes, garter snakes, or hognose snakes, which have multiple subspecies.

Common Garter Snake
(Thamnophis sirtalis)

DESCRIPTION

This is a common and widespread member of the garter snake clan, being found from southern Canada south and east throughout the eastern two-thirds of the United States (Figure 20-23). There are numerous species of garter snakes, and one or more are found over virtually the entire United States. Most garter snakes range in size from 24 to 45 inches, with 26 to 30 being the average. The common garter snake usually has three stripes down its back that are typically greenish yellow. The base body color varies from black to brownish to green, or yellow and the belly is usually yellow or off-white. Garter snakes like habitats near water, and various amphibians, fish, mice, and the young of ground-nesting birds are their primary food sources.

(Source: Photo courtesy of Texas Parks & Wildlife Department.)

FIGURE 20-23 Garter snake.

Hognose Snakes
(Heterodon)

DESCRIPTION

All the species of hognose snakes have distinctive, upturned snouts, which is probably where they get their name (Figure 20-24). Color is highly variable, from solid brown, black, gray, or green to blotched with olive or brown. The belly of the western hognose is often black and that of the eastern is often yellow or whitish. The southern hognose is usually lighter in color and more likely to have a uniform gray belly. Hognose snakes puff up, hiss, and strike at potential threats. Should this fail to **intimidate** opponents, they often roll up on their backs and fake death, often with the tongue hung out. The distinct upturned snout and behavior when threatened are the best methods of identification in the field.

(Source: © Rusty Dodson, 2010. Used under license from Shutterstock.com)

FIGURE 20-24 Hognose snake.

The three major species of hognose snakes are the western, eastern, and southern. The eastern and western ranges probably overlap in portions of the central plains states, just as the southern and eastern probably overlap in the Gulf Coast states. One or more of these snakes is found from southern Canada throughout the eastern two-thirds of the United States to Mexico. The southern hognose is primarily a Gulf Coast variety, found from southern Mississippi to central North Carolina but generally absent from southern Florida.

Common King Snake
(Lampropeltis getulus)

DESCRIPTION ◾

This widespread species of snake is unusual in its immunity to the poison of our native poisonous snakes (Figure 20-25). This could explain why king snakes commonly prey on other snakes, particularly poisonous ones. Color is highly variable, ranging from yellow to dark blue to black, and they may have a banded, speckled, or striped body pattern. Length is even more variable, depending on the subspecies. These snakes are generally large and stout-bodied, obtaining lengths of 6 feet or more. They are all powerful snakes that kill their prey by constriction.

(Source: Courtesy of John D. Willson.)

FIGURE 20-25 Eastern king snake.

Four of the more widespread subspecies are the eastern king snake, the black king snake, the speckled king snake, and the desert king snake. Together with the common king snake, one or more of these species is found from southern New Hampshire to southeastern Arizona and as far north as Illinois.

Milk Snakes
(Lampropeltis)

DESCRIPTION

Some half dozen widely distributed subspecies of milk snakes are found in the United States (Figure 20-26). Color is extremely varied and the ranges of many subspecies overlap, making identification all the more difficult. They have slender bodies and are often marked with blotches of gray, red, or brown, which sometimes form rings around their bodies. Base body color is often gray, white, or yellowish.

These snakes range from Mexico northward to Montana and throughout the south and east to southern Maine. There are a variety of important subspecies, such as the eastern milk snake, the red milk snake, the central plains milk snake, the Louisiana milk snake, and the pale milk snake. Subspecies are found in the western United States and Mexico. This family

(Source: Photo courtesy of Texas Parks & Wildlife Department © 2002.)

FIGURE 20-26 Milk snake.

of snakes is very widely distributed. They are often secretive and difficult to locate. Milk snakes have a preference for eggs and young birds but also consume large quantities of rodents, lizards, other snakes, and frogs.

Coachwhip
(Masticophis flagellum)

DESCRIPTION ■

The coachwhip is a long, slender snake with a narrow head and rather large eyes (Figure 20-27). These snakes are large, reaching lengths of 5 feet. The base color ranges from black to gray to reddish. Their very long bodies have marks resembling a braided leather whip. Coachwhips are very fast, capable of outrunning most people. They are terrestrial and prefer warm, dry uplands; semiarid areas; pastures; fields; and open pine forests. Coachwhips range throughout most of Mexico northward into the southwestern United States and eastward to North Carolina. They are

(Source: Photo courtesy of Texas Parks & Wildlife Department © 2002.)

FIGURE 20-27 Western coachwhip.

generally absent along the Mississippi River. These snakes feed on other snakes, young birds, rodents, turtles, and lizards. They can inflict a nasty bite if handled.

Bull snake
(Pituophis sayi)

DESCRIPTION ■

This extremely large snake achieves lengths of 6 feet, and occasionally more (Figure 20-28). They are widespread, ranging from Texas north to Minnesota and as far east as Wisconsin and Illinois. They are primarily residents of the Great Plains states. The common gopher snake found in western states is a subspecies. Their base color is white, yellow, or gray and usually has numerous dark blotches. They have stout bodies, relatively small heads, and pointed snouts. If disturbed, they hiss and rattle their tails against the ground.

(Source: Photo courtesy of Texas Parks & Wildlife Department © 2002.)

FIGURE 20-28 Bull snake.

Bull snake
(Pituophis sayi)

DESCRIPTION ▶

Bull snakes and related subspecies are very beneficial to humans. They consume large quantities of rodents, especially gophers and ground squirrels. They also feed on birds and the young and eggs of ground-nesting birds. They are partial to roadsides, farm buildings (where they make excellent mousers), pastures and prairies, preferring dry, sandy habitats.

Rat snake
(Elaphe)

DESCRIPTION ▶

Another very large snake, the rat snake can be more than 6 feet in length (Figure 20-29). Color is again highly variable, with upper parts striped, blotched, or solid black. All the rat snakes are strong-bodied constricting snakes that are prone to climb. They have a tendency to bite if caught, and they hiss and vibrate their tails if threatened. They are widespread, from southern New England to Texas and north to southern Ontario, Wisconsin, and Nebraska. Rat snakes are often called "chicken snakes," possibly because of the adults' strong preference for chicken eggs. There are four major sub-species found in the Southeast: gray rat snake, yellow rat snake, Texas rat snake, and black rat

(Source: Photo courtesy of Texas Parks & Wildlife Department © 2002.)

FIGURE 20-29 Texas rat snake.

snake. These snakes prefer woodlands, hollow logs, and fence-rows. They are good climbers and often raid bird nests. Young rat snakes prefer lizards and frogs, while the adults prefer rodents, eggs, and birds.

Water Snakes
(Natrix)

DESCRIPTION

These snakes are never found far from water (Figure 20-30). They are largely aquatic and seldom leave the water. All are good swimmers and subsist primarily on aquatic prey, such as fish, frogs, and toads. The common water snake is widely distributed, ranging from Oklahoma to southern Canada and eastward throughout the eastern and southern United States. There are several subspecies, with the yellow-bellied water snake probably having the widest distribution, being found in Louisiana, Texas, Oklahoma, Arkansas, Kansas, and New Mexico. Water snakes are usually dark, dull gray, or brown, with

(Source: Courtesy of U.S. Fish and Wildlife Service.)

FIGURE 20-30 Water snake.

darker blotches down their bodies and cream or lighter-colored bellies. They probably average between 2 and 3 feet in length, with strong, muscular bodies. The head is distinct from the remainder of the body, not blending with it, as in the slimmer snakes.

POISONOUS SNAKES

There are two types of poisonous snakes in North America. The coral snake has small, permanently erect fangs. Pit vipers have larger, hollow fangs that fold back against the top of the mouth when not in use. Pit vipers deliver their venom much like a hypodermic needle. The hollow fangs are connected to venom sacs, and when the snake strikes the fangs descend. The skin and muscle tissue of the victim are penetrated, and venom is injected. The venom of the pit vipers causes tissue destruction and is not quite as dangerous as that of the coral snake, which is a **neurotoxin.** Because it works on the central nervous system, coral snake venom has a fairly rapid effect on the heart and lungs of its victim. This makes coral snake venom very deadly. However, the coral snake is relatively timid and small and does not have as effective a delivery system for its venom. Because its fangs are short, it must literally chew on its victim to deliver a fatal dose. The larger pit vipers can deliver a large amount of venom in an instant. All the poisonous snakes in North America give birth to live young, except the coral snake, which lays eggs.

Eastern Coral Snake
(Micrurus fulvius)

DESCRIPTION ◗

The eastern coral snake is our most common and widespread coral snake, though several subspecies are found in the Southwest (Figure 20-31). Coral snakes have conspicuous alternating rings of black, yellow, and red around their bodies. While several species of snakes resemble the coral snake, their rings are seldom complete as are the coral snakes'. No other snake has the red-touching-yellow color combination and the black face mask of the coral snake. The old saying, "Red on yellow kill a fellow" still holds true. Coral snakes are slender, rarely exceed 30 inches

(Source: Photo courtesy of Texas Parks & Wildlife Department © 2002.)

FIGURE 20-31 Coral snake.

in length, and have a smooth appearance. The flat, blunt head is not distinct from the body. The eastern coral snake ranges from eastern Mexico through the eastern half of Texas and through the coastal states to North Carolina. Several subspecies are found in the southwestern United States and western Mexico.

Coral snakes feed mainly on toads, frogs, and small snakes. They are secretive and spend a great deal of their time underground in old animal burrows, crevices, and small caves. They are not particularly aggressive and can seem quite placid, which is how problems can arise with coral snakes. They are beautifully colored and easy to catch. This can result in tragedy. Most fatal coral snake bites occur to young children who pick up the snakes to play with them. However, fatal snake bites, and coral snake bites in particular, are rare.

Copperhead
(Agkistrodon contortrix)

DESCRIPTION

Copperheads are quite wide-spread, with one or more subspecies being found from Massachusetts through the southeastern states to Texas and as far north as Kansas, Nebraska, and southern Iowa (Figure 20-32). They are members of the pit viper family and seem to prefer dry, sandy, upland habitats. They are not normally large snakes, probably averaging 20 to 30 inches, with the occasional specimen up to 36 inches. Because they are pit vipers, they have broad, triangular heads and distinct pits between the nostrils and eyes. They are an overall copper-brown, with dark bands or blotches on the body and pinkish, red-brown, or copper-

(Source: Photo courtesy of Texas Parks & Wildlife Department © 2002.)

FIGURE 20-32 Copperhead.

colored heads. The major subspecies are the northern copperhead, Osage copperhead, southern copperhead, and the broad-banded copperhead. Copperheads are not generally aggressive and usually only bite when startled. However, they do become more aggressive in hot weather. They are extremely well camouflaged in most of their habitat and are very difficult to see when lying still, which is what they spend most of their time doing. Copperheads feed heavily on rodents, birds, lizards, toads, and frogs. They give birth to several litters of live young, usually in August, September, or October.

Water Moccasin or Cottonmouth
(Agkistrodon piscivorus)

DESCRIPTION ▪

The water moccasin is a largely aquatic snake, living in and around water (Figure 20-33). They live in small ponds, swamps, and backwaters and seem to prefer still, stagnant water. They are large snakes, 3 to 5 feet in length and larger in diameter than many other snakes. They are generally dark brown or olive in color, with the young being lighter and possibly exhibiting some dark blotches. Water moccasins are widespread, ranging from central Texas throughout most of the southeastern states. The common nickname, "cottonmouth," arises from the white lining in its mouth.

(Source: Photo courtesy of Texas Parks & Wildlife Department © 2002.)

FIGURE 20-33 Western cottonmouth.

Water moccasins feed heavily on aquatic prey such as frogs, as well as small mammals that venture to the water's edge to drink. They are often seen lying on half-submerged logs, apparently sunbathing. Some authorities contend that the water moccasin is a nonaggressive species; others say the opposite. My personal observations lead me to believe that they can be aggressive, particularly during the spring and early summer. However, most snakes avoid humans if given the opportunity.

RATTLESNAKES

Description: More than a dozen species of rattlesnakes and a number of subspecies fall under two genera, *Sistrurus* and *Crotalus*, found in North America. This makes the rattlesnake our most widespread poisonous snake. One or more of the rattlesnakes are found from the New England area across the southern two-thirds of the states to California. They are especially well represented in the southeastern and southwestern United States. One subspecies, the western rattlesnake, is found from southwestern Canada south across many of the western states and into Mexico. Rattlesnakes come in great variety of colors and

FIGURE 20-34 Western diamondback rattlesnake.

sizes. Most have a gray to tan to red-brown body color, with some form of spots, splashes, or blotches down the back. In some species, such as the diamondback, these markings have a definite shape (Figure 20-34). All rattlesnakes, unless injured, have a "rattle," which consists of a series of bony segments at the end of the tail that are linked together. When the tail vibrates, it produces the characteristic "buzz" that no one wants to hear! All rattlesnakes have the characteristic broad head and pits between the nostrils and eyes of the pit viper family. A few of the more widespread species of rattlesnakes are the diamondback, massasauga, sidewinder, western, and pygmy. Adult rattlesnakes range in size from 18 to more than 80 inches and have thick, powerful bodies. Rattlesnakes consume large quantities of rodents, as well as birds and small mammals such as cottontail rabbits. Rattlesnakes can be found nearly anywhere, but most avoid wetlands and areas with heavy human populations. They seem to prefer fencerows; timbered and scrublands in the Southeast; and desert, semiarid, and scrub brush areas in the Southwest. Rocky areas, crevices, and overhangs are also favorite spots. Because of the quantity of venom that most rattlesnakes can deliver, especially large specimens, they can be very dangerous. Hikers, campers, and sightseers traveling in rattlesnake country should be extremely cautious. Rattlesnakes, like the vast majority of all snakes, are normally shy and retreat if given an opportunity. However, if they are pressed or feel threatened, they will defend themselves.

SUMMARY

Reptiles are often misunderstood. Snakes in particular are **reviled** by many people and are often killed **indiscriminately.** This is a dismal state of affairs because snakes are by and large very beneficial. They consume a tremendous number of rodents, which more than offsets the loss of a few hen eggs. Turtles are often derided as fish stealers by people who fish, but they are a necessary part of a complete aquatic ecosystem. Lizards are extremely beneficial, especially to the home gardener, because of the great quantity of insects they consume. Reptiles have been around for thousands of years and some, such as the alligator, have changed little. For the most part we have healthy populations of reptiles that will probably stay healthy for the foreseeable future. However, some species of reptiles are sensitive to human intrusion. These sensitive species need to be watched carefully.

REVIEW QUESTIONS

Fill in the Blank

Fill in the blanks to complete the following statements.

1. Reptile blood is not actually cold, therefore, _____ is a more appropriate term than coldblooded.

2. During cold weather most reptiles are forced to _____ and seek shelter from cold weather.

3. There are fewer than _____ species of reptiles found north of Mexico.

4. Turtles, _____ snakes, and are all reptiles.

5. Most reptiles reproduce by laying _____ .

6. Perhaps the most readily identifiable reptile in North America is the _____ .

7. Alligators are _____ , surviving on a diet consisting largely of meat.

8. A turtle's shell is typically divided into two parts, the carapace and the _____ .

9. Each turtle has four well-developed _____ , unlike some other reptiles.

10. Identification of individual species of _____ can be very difficult, particularly among the aquatic species.

11. The common _____ is a large turtle, common throughout much of the South and Southeast.

12. The _____ is the largest freshwater turtle in North America.

13. Painted turtles, as one might guess from the name, are well marked and _____ .

14. Many species of _____ turtles are found in the southeastern United States.

15. The _____ may be the most familiar turtle in America.

16. There are several species of spiny and smooth _____ turtles.

17. Soft-shell turtles are highly _____ , leaving their watery homes only to nest and to bask in the sun.

18. The _____ is a true terrestrial, land-dwelling tortoise.

19. _____ are the largest single group of reptiles.

20. _____ lizards prefer semiarid, rocky habitat and are most often found there.

21. The _____ is commonly known as a chameleon, although it is not a true chameleon.

22. There are several species of _____ lizards, one or more of which can be found across much of America.

23. _____ earned their name because of their often blinding speed.

24. The _____ is the only poisonous lizard in the United States.

25. In North America, snakes with movable, hypodermic-type fangs are known as _____ .

26. There are no _____ snakes. They all survive strictly on animal matter.

Short Answer

1. Name five characteristics common to most reptiles.

2. List the four species of poisonous snakes native to North America. How are they different from each other?

3. List the types of habitat where you might encounter one or more species of lizards. Where in your area might you find some?

4. In what types of habitat might you expect to find one or more of the aquatic species of turtles? Where in your area might you find some?

Discussion

1. Discuss the ways snakes are beneficial to humans. In your opinion, what can be done to change the average person's opinion of snakes?

2. In what ways are lizards beneficial to humans? Discuss things you might do to attract lizards to a backyard or other area.

Learning Activities

1. Discuss the preferred habitat, feeding habits, reproductive traits, and predators of five species of lizards found in your area.

2. Discuss the preferred habitat, feeding habits, reproductive traits, and predators of five species of snakes found in your area.

CHAPTER 21

Amphibians

OBJECTIVES

After completing this chapter, you should be able to

- Describe some common characteristics of amphibians.
- Identify the type of habitat where amphibians might be found.
- Describe some common habits of amphibians.
- List some of the common amphibians found in the United States.
- Describe some of the differences between various types of amphibians.

INTRODUCTION

Like reptiles, amphibians are cold-blooded or ectothermic and are therefore not found at the North and South poles. Amphibians occupy all other landmasses but are generally most abundant around the tropics. They form one of the most abundant classes of vertebrates worldwide, with about 2,500 known species. They have visible limbs and moist skin, unlike reptiles, which tend to have dry, scaly skin. The reproductive habits of reptiles and amphibians are where the major differences between the species exist. Amphibians must have moist places to lay their eggs; either wet soil or water, and generally lay hundreds or even thousands of small eggs enclosed in a jellylike capsule. Reptiles lay fewer, larger eggs enclosed in a leathery or hard shell, and the eggs need not be laid in water or especially moist areas. Most amphibians return to fresh water to complete their life cycle because they have aquatic larvae. In North America the requirement for water is probably one of the limiting factors on their range. There are approximately 170 species of toads, frogs, and salamanders in North America. Amphibians breathe with **gills**, lungs, through their skin, or by a combination of these methods. All the amphibians have gills at some stage of their lives. The young of many species of amphibians take the form of tadpoles when first hatched. Most of these tadpoles later transform into adults that spend most of their lives on land. However, some species spend their entire lives on land, and others spend their entire lives in the water. Both frog and toad tadpoles have tails that are later absorbed. Of the adult amphibians, only salamanders usually have tails. Rarely do amphibians drink; they normally absorb water through their skin from the moist soil or water surrounding them. The orders of amphibians discussed here are salamanders, frogs, and toads.

FROGS AND TOADS

The terms *frog* and *toad* can often be confusing, mainly because there are no real scientific rules to distinguish between the two. However, as a general rule toads have rougher, warty skin, and most frogs have relatively smooth skin. Although both groups have well-developed legs, the hind legs of a frog are generally longer than those of a toad. For this reason toads tend to hop, and frogs leap. Frogs also tend to be more aquatic than toads, which are more terrestrial than frogs.

Description: Identification of toads and frogs is very difficult. There are great variations in color among individuals of the same species. Such factors as temperature, **humidity**, and environment all affect the color of toads and frogs. As a general rule the higher the humidity and cooler the temperature, the darker the color. The female as a general rule is larger than the male, and during the breeding season may have an enlarged belly due to the eggs she is carrying. Tadpoles exhibit tails, but these are absorbed before maturity, hence adult toads and frogs generally lack this feature. Toads are usually more squat or compact than frogs, but neither has a **discernible** neck. Toads tend to be rounder than frogs and may flatten themselves against the ground when closely approached. Toads are also noted for burrowing, especially in loose soils. Both frogs and toads are noted for blunt snouts and large, prominent eyes. All toads and frogs have very good hearing, and most have an obvious **tympanum**, or ear covering, on both sides of the head. Toads may have a pair of obvious glands behind the eyes. Male toads and frogs often have pads on their fingers, thumbs, chest, or forelegs, used for clasping the female during mating. The males also tend to have darker throats than the females. Most frogs and toads are nocturnal, but some, particularly some of the toads, are **diurnal**. Almost everyone has heard frogs and toads make various croaks and calls, but most probably do not realize that their **vocalizations** may be the surest way of identifying the different species. Although there are some silent species of toads and frogs, most utilize several different vocalizations. These are made with a vocal sac located under the throat, which is inflated with air to produce sounds. Mating calls are often given in chorus by the males and may be heard for miles. Frogs and toads use the calls to identify others of their species and to identify potential mates. Toads and frogs use various calls in addition to their mating calls, such as territorial calls, distress calls, and release calls. Males and unreceptive females, on being grabbed by a breeding male, utter a release call. Territorial calls are used at any time of the year to declare territory.

Habitat: Frogs are found almost exclusively in wet areas, either in the water or under cover. Marshes, ponds, lakes, slow-moving streams, and prairie potholes are all good spots to find frogs. The shallow areas and edges of larger ponds and lakes are usually excellent frog habitat. Toads are generally more terrestrial than frogs and may be found quite some distance from standing water. They prefer pasture, mountain, or prairie situations. An excellent spot to find toads is around a well-watered lawn or flower bed.

Feeding Habits: In sharp contrast to their adult form, most tadpoles are vegetarian, feeding on tiny aquatic plant life. Adults are carnivorous and feed mostly on insects and other **invertebrates**. Bullfrogs are known to eat anything they can swallow, including small birds and young bullfrogs. Most toads and frogs have teeth only in the upper jaw, and these are used more for grasping the prey than for chewing. Prey is caught with the aid of the long and sticky tongue and is swallowed largely intact.

Life Cycle: The vast majority of frogs and toads begin life as eggs, which hatch into tadpoles and then develop into adults. However, live-bearing species are known. Most frogs and toads lay their eggs in water, although a few species lay theirs on land. Eggs are generally laid in shallow water, where they attach to aquatic vegetation. The mating season may occur from March until late summer but is usually in April, May, or June. Females may lay anywhere from a few hundred eggs to many thousands. Female bullfrogs are known to lay as many as 20,000 eggs at one time and are thought to be capable of laying a million eggs during one breeding season. Males generally attach themselves to a receptive female, holding their position for over a day if necessary, but often only for a few minutes. When the female **extrudes** her eggs, the males fertilize them and go on their way. Once the eggs are laid, parental involvement in the development of the young usually ends. Many species of toads, especially those living in desert regions, may mate and lay their eggs in shallow pools of rainwater. The eggs of these species must develop quickly.

The eggs usually hatch into tadpoles in a fairly short time. The rate depends on such factors as the species involved and water temperature. The tadpoles of some species, such as the bullfrog, may take 2 years to develop into adults. However, most species are faster, developing in a year or less. As a general rule, the warmer the water temperature, the faster the eggs and tadpoles develop. Tadpoles have tails and breathe through gills, which are covered by a flap of skin. When the tadpoles develop into adults they absorb their tails and develop lungs, and then four limbs appear. As young adults, frogs and toads spend more time along the edges of ponds and streams and less time in deep water. Adult toads and frogs spend the cold winter months in hibernation, usually burrowing into soft soil or mud. It is not uncommon to turn up a toad or two during early spring in the process of tilling a garden or flower bed.

While some toads and frogs are known to live long lives, the vast majority perish as eggs, tadpoles, or young adults. The eggs and tadpoles are eaten by a variety of predators, from minnows to black bass to adult frogs. Young frogs and toads are preyed on by raccoons, a variety of fish, and snakes. Snakes, especially the aquatic species, take a large number of frogs, while the more terrestrial snakes prey heavily on the toads. All but perhaps the largest bullfrogs are subject to heavy predation. Of course, humans have been known to enjoy a batch of frogs' legs from time to time as well.

Many frog and toad populations are spotty, and few species enjoy blanket distribution over a large area. Some families of toads and frogs are more broadly distributed, and these are the ones we have covered here in some detail. However, it must again be noted that frogs and toads are very difficult for anyone less than an expert to identify. The lengths given below are from the snout to the vent, or rear, of the average specimen. Many of the frogs and toads discussed below have more than one common name. Some have several. Because common names vary greatly, you may recognize some of these species by another name.

TRUE TOADS

The true toads belong to the family Bufonidae, and one or more of these toads can be found virtually throughout the United States. Members of this family have thick skins and the typical toad appearance, squat and warty. They are largely terrestrial, and hopping is their method of locomotion. Because of their thick skins, members of this family are more likely to be found at higher elevations and in desert areas.

Great Plains Toad
(Bufo cognatus)

DESCRIPTION

This toad ranges from the southern portions of western Canada south and eastward to New Mexico and west Texas (Figure 21-1). Males are smaller than females, a common trait among toads, and the adults have obvious blotches on the back and sides. These toads are greenish, grayish, or a yellow brownish color on the upper body, and the belly is light in color. The belly is usually unmarked and the legs are green-spotted. These toads can be found in a great variety of habitats, such as prairies, deserts, irrigation ditches, and pools of rainwater. They average 2 to 4 inches in length.

(Source: Photo courtesy of Texas Parks & Wildlife Department © 2002.)

FIGURE 21-1 Great Plains toad.

Texas Toad
(Bufo speciosus)

DESCRIPTION ◼

The Texas toad is found from west Texas northward to central Montana and westward to southeastern California (Figure 21-2). These toads are usually olive to grayish brown and are covered with small warts. Average length is 2 to 3.5 inches, and the males are smaller than the females. They are well adapted to dry conditions and prefer sandy soils. They are abundant over their range and may be found in arid plains areas, short-grass prairies, and cultivated areas. They use stock ponds, irrigation ditches, and temporary rainwater pools as water sources.

(Source: Photo courtesy of Andy Jacobs)

FIGURE 21-2 Texas toad.

Western Toad
(Bufo boreas)

DESCRIPTION ◼

These toads are very widely distributed, ranging from the Rocky Mountains west to the Pacific and from southern California to southern Alaska (Figure 21-3). They are usually dull greenish, blackish, grayish, or brownish above and whitish below, with black spots. Females are significantly larger than males. These toads can be found from the mountains to sea level in a variety of habitats but are less common in forested areas. Western toads are quite large, ranging from 2.5 to 5 inches in length.

(Source: © Elemental Imaging, 2010. Used under license from Shutterstock.com)

FIGURE 21-3 Western toad.

Woodhouse's Toad
(Bufo woodhouseii)

DESCRIPTION

This toad is widespread from southern New Mexico to southern Montana, westward to southeastern California and as far east as western North and South Dakota, Nebraska, and Kansas (Figure 21-4). They have smoother skin than many toads and are usually blackish, greenish, grayish, or brownish yellow above, with yellowish bellies. They range in length from 2 to 4 inches and are commonly found in gardens, marshes, irrigated areas, and fields.

(Source: Courtesy of U.S. Fish and Wildlife Service. Photo by Gary Stolz.)

FIGURE 21-4   Woodhouse's toad.

Spadefoot Toads
(Scaphiopus)

DESCRIPTION

Members of the spadefoot family have vertical eye pupils and usually have smoother skin than other toads (Figure 21-5). The name comes from the spade-shaped appendage on its hind feet. These spades are usually black and are of a bony or hornlike material. Spadefoots have a reputation as burrowers. Most spadefoot toads breed in shallow, temporary puddles and pools of water that develop after heavy rains. As an adaptation to their breeding habitat,

(Source: Photo courtesy of John D. Willson.)

FIGURE 21-5   Spadefoot toad.

spadefoot offspring may develop from the egg to larval stage to young adult in just 12 days. Spadefoots are primarily nocturnal and terrestrial and are widespread, particularly in deserts. They tend to prefer loose, sandy, or gravelly soils.

5

Western Spadefoot
(Scaphiopus hammondi)

DESCRIPTION ■—

This toad is widespread, ranging from west Texas
to southeastern British Columbia. The upper body
is usually greenish gray or brown with scattered
spots of darker color. Warts are often tipped in
reddish orange, especially in the juveniles. The tip
of the snout is upturned slightly, and it has large,
prominent eyes. Western spadefoots can be found
in short grass prairie areas, hills, and alkaline flats
in semiarid regions. They are small toads, averaging
1.5 to 2.5 inches in length. Males are slightly smaller
than females and have a dark throat.

(Source: U.S. Fish & Wildlife Service.
Photo Credit: James Bettaso)

FIGURE 21-5a Western Spadefood
(Spea hammondii).

Plains Spadefoot
(Scaphiopus bombifrons)

DESCRIPTION ■—

The plains spadefoot has a conspicuous bony lump
between its eyes, and its forehead protrudes. The
belly is whitish and the upper body usually greenish
or grayish, spotted with darker color. Males tend
to have darker throats than females and have dark
pads on their fingers. The plains spadefoot ranges
from the Great Plains northwest to Montana and
southwest to west Texas. They are smallish toads,
averaging 1.5 to 2.5 inches in length.

Eastern or Holbrook's Spadefoot
(Scaphiopus holbrooki)

DESCRIPTION ■—

This spadefoot is widespread in the eastern half
of the United States. Its range extends from the
extreme edge of southeastern Texas north and
east to Ohio and Massachusetts. It is usually brown,
with smoother skin than some toads, but with some
scattered warts. The throat and breast is whitish,
which turns to a grayish color on the belly. It has the
broad feet and burrowing habits of a typical spadefoot. Adults average 2 to 3 inches in length.

American Toad
(Bufo americanus)

DESCRIPTION

This toad is also known as the hop toad or northern toad and ranges over most of the northeastern United States, except some of the southern coastal states (Figure 21-6). Adults are an olive-brown to reddish color, with short, round bodies. There are usually several pairs of dark spots down the back, numerous spots on the sides, and a few on the belly. It is generally lighter in color on the undersides. They average 2 to 4 inches in length and are common in gardens and other cultivated areas. These are burrowing toads and during the winter or the heat of summer they often retreat underground.

FIGURE 21-6   American toad.

(Source: © Francis Bossé, 2010. Used under license from Shutterstock.com)

Southern Toad
(Bufo terrestris)

DESCRIPTION

These are strictly southeastern toads, ranging from the Mississippi River east to Florida and North Carolina (Figure 21-7). They are abundant throughout their range and seem to be especially fond of cultivated fields. However, they may be found in virtually any land habitat within their range. They vary from gray to black or even reddish. The adult has a prominent knob on the back of the head. They are smallish toads, averaging 1.5 to 3.5 inches.

FIGURE 21-7   Southern toad.

(Source: Photo courtesy of John Wilson.)

TRUE FROGS (Rana)

These members of the Ranidae family are generally long-legged and slimmer than most toads. They have large, webbed hind feet and usually two prominent ridges on the back, one down each side. The skin is usually fairly smooth and the fingers lack the discs or pads commonly found on toads. They also lack the head ridges or knobs that are common among toads. True frogs have large, prominent eardrums or tympanum, and the females are known to lay large numbers of eggs.

Bullfrog
(Rana catesbeiana)

DESCRIPTION

This is perhaps our most wide-spread and common frog, ranging from Texas to Minnesota east across most of the eastern two-thirds of the United States (Figure 21-8). Bullfrogs have also been widely introduced in many western states. They are the largest frogs in North America, averaging 3.5 to 8 inches in length. Long and powerful hind legs make them capable of amazing leaps. They are usually a dark olive green, with darker spots and whitish bellies. They possess a deep, hoarse voice and emit a series of loud croaks. Adults are very aggressive and feed on anything they can catch and swallow.

(Source: Photo courtesy of Texas Parks & Wildlife Department.)

FIGURE 21-8 Bullfrog.

Leopard Frogs
(Rana)

DESCRIPTION

There are several species of leopard frogs, all closely related and difficult to tell apart. The two most common are probably the northern leopard frog *(Rana pipiens)* and the southern leopard frog *(Rana sphenocephala)* (Figure 21-9). Leopard frogs are very widely distributed, with one or more species being found over most of the United States. Their overall appearance is also very similar. However, they are in fact several closely related species, not simply variations of a single species. The various species of leopard frogs can often be found with overlapping ranges and using the same breeding grounds. However, little or no hybridization appears to occur.

(Source: © Gilles DeCruyenaere, 2010. Used under license from Shutterstock. com)

FIGURE 21-9 Leopard frog.

Leopard frogs are normally 2.5 to 4.5 inches long, and an overall dark green to brown color. The name arises from the irregular oval and round dark spots on the upper body, which gives them a leopard-like appearance. The under parts are generally white, and the legs often have dark bands around them. They have smooth skin and a narrow or slender body. Leopard frogs prefer marshes but can also be found in grasslands, ponds, and slow-moving streams.

Green Frog
(Rana elamitans)

DESCRIPTION

This is a fairly common frog in the eastern half of the United States (Figure 21-10). Green frogs range from southeastern Canada through the eastern United States (except south Florida) and west to Minnesota and east Texas. It has pale under parts that may be spotted and a green to brown upper body, sometimes spotted with black. The female has a white throat with spots, and the male has a yellow throat. They are generally solitary frogs except during the mating season. They prefer swamps, ponds, streams, and the edges of lakes.

(SOURCE: © LEN RUE JR.)

FIGURE 21-10 Green frog.

Tree Frogs and Allies
(Hyla)

DESCRIPTION ▶

These are abundant, widespread frogs that are generally small (Figure 21-11). Members of this group are usually thinner and narrower at the waist than other frogs. They normally have rounded pads at the tips of their toes, and most species are good climbers. However, some species are aquatic or terrestrial and are poor climbers. A few species are burrowers.

(Source: Photo courtesy of Texas Parks & Wildlife Department © 2002.)

FIGURE 21-11 Green tree frog.

Chorus Frogs
(Pseudacris)

DESCRIPTION ▶

The United States has a number of species of chorus frogs (Figure 21-12). They are widespread, with one or more species being found over most of the United States. They are generally gray, green, or brown on the upper body, but color is highly variable. Most chorus frogs have three dark stripes down the back, although the center stripe may be broken. Under parts are usually whitish, occasionally with a few dark spots. Both males and females have fairly smooth skin. Chorus frogs are small, ranging from 0.5 to 2 inches in length. They prefer swampy ground, marshes, damp woodlands, and stream or river bottoms. Chorus frogs are largely terrestrial and are often spotted on the ground or in low bushes. The vocalizations of these frogs sound more like a series of chirps than croaks. This can be an important factor when attempting to identify species.

(Source: © Sarah Theophilus, 2010. Used under license from Shutterstock.com)

FIGURE 21-12 Chorus frog.

Northern Cricket Frog
(Acris crepitans)

DESCRIPTION

These frogs get their name from their calls, which sound like a series of cricket chirps (Figure 21-13). They are widespread and common within their range, which is from east Texas to Nebraska and east throughout the eastern half of the United States. They are small, averaging only 0.6 to 1.5 inches in length, but are known for their tremendous leaping ability. Upper body color is highly variable, but generally green, gray, and various shades of brown to almost black. Lower body color is usually whitish. There is a distinct dark triangle between the eyes and a light stripe down the back. They prefer the grassy edges of streams, ponds, swamps, and marshes. They are quite shy and difficult to approach, usually making a dash for water if frightened. They are terrestrial frogs and are not good climbers.

FIGURE 21-13 Northern cricket frog.

(Souce: Photo courtesy of John D. Willson.)

Salamanders
(Ambystoma)

DESCRIPTION

Salamanders can be extremely difficult to identify for a variety of reasons. For one, they lack distinguishing features, such as a definite color or scale pattern (Figure 21-14). They are also small, secretive, and difficult to find, unless one catches them for close examination, which is difficult at best. There may be differences in color or pattern, but these are often variations within a species. Just as there is a wide variety of human hair and eye colors, so there are wide variations in color patterns within species of salamanders. To further complicate matters,

FIGURE 21-14 Salamander.

(Source: Courtesy of U.S. Fish and Wildlife Service. Photo by C.K. Dodd, Jr.)

salamanders appear to have difficulty identifying members of their own species. It is not uncommon for salamanders to choose a mate from a different, but usually closely related species, and thus produce hybrid offspring. These hybrids are usually very difficult to identify positively. However, there are three main items to look for when attempting to identify a salamander. The color and color pattern, although highly variable, are still important. Some salamanders have a **gular fold**, or a flap of skin across the throat, which can help distinguish species. The last major feature to look for is the number of **costal grooves** on the sides. These are vertical grooves, or lines, on the sides of salamanders.

The salamanders are the only other amphibians besides frogs and toads. They are easily distinguished from other amphibians because they keep their tails as adults, unlike the vast majority of toads and frogs, which absorb their tails before adulthood. If a salamander loses its tail through an accident or an encounter with a predator, it can regenerate or re-grow it. It can also regenerate other lost limbs. Adult salamanders have four short legs, with four toes per foot. Unlike toads and frogs, salamanders are essentially voiceless. Most salamanders have smooth, moist skin and no external ear openings. Salamanders also do not have claws, which most toads and frogs possess, and they never have more than four toes on the front feet.

HABITAT: Salamanders live in a variety of habitats, but almost all are moist. Some species are mostly terrestrial; others are arboreal. Some species spend their entire lives in water. Salamanders can take quite a lot of variation in altitude and may be found from sea level to 13,000 feet or more. Although salamanders are amphibians, and therefore ectothermic, they prefer cool weather to hot. During hot weather they seek out cool places under logs or rocks or in holes. If the winters are cold they hibernate underground but may remain active in their underground homes. Salamanders are found under rocks, logs, leaf litter, the loose bark on decaying trees, and other cool, dark places. During the breeding season salamanders can be observed near water, where the female deposits the eggs. Most species of salamanders are nocturnal and thus are seldom seen during the day.

FEEDING HABITS: There are no vegetarian salamanders—they are strictly carnivorous. They feed on a variety of terrestrial and aquatic insects and other small invertebrates. They also feed on the eggs and larvae of various invertebrates, as well as the larvae of small vertebrates, including salamander larvae.

LIFE CYCLE: Unlike frogs and toads, fertilization of most salamander eggs is an internal process. An interesting aspect of salamander reproduction is that mating and egg-laying need not occur in the same season. The sperm appear to remain **viable**, and the females are able to carry it for a number of months. Eggs may be laid on land or in water, with most species laying them in water. The eggs may be laid in masses, in strings, or singly. Should they be laid on land, they are suspended from the roof of a moist hideaway. After an appropriate period of time, eggs deposited in water will hatch into tadpoles with gills. Those laid on land pass the larval period while still in the egg and hatch possessing only **vestiges** of gills, which are soon shed. The amount of time spent in the larval stage by the aquatic species varies greatly, from a few days to over a year. While young salamanders, eggs, and larvae develop at different rates, lower **ambient** temperatures slow these growth rates.

Salamanders are thought to lead fairly short lives. They are eaten by nearly anything that can catch them. The eggs, young, and adults of the aquatic species are eaten by a wide variety of fishes, as well as snakes, turtles, and larger frogs. The terrestrial species are prey for everything from house cats, to birds, to snakes, bobcats, and raccoons. Because of their high mortality rate, females normally lay large quantities of eggs.

SUMMARY

We have discussed a few of the most common species of amphibians. Amphibians are difficult to identify, and in the case of salamanders rarely seen, amphibians are however, an important group of vertebrates. Amphibians, particularly frogs and toads, consume large quantities of insects and are generally beneficial to humankind. They are also prey for a large and diverse group of predators. Frogs and toads are an especially important food source for snakes, particularly the aquatic species.

Amphibians can be quite sensitive to human interference or disturbance. Many species, especially the salamanders, are secretive by nature, and entire populations can be endangered without our realizing it. Many species and subspecies also have limited distributions. As with most wildlife, human encroachment and destruction of habitat are the amphibians' greatest enemies.

REVIEW QUESTIONS

Fill in the Blank

Fill in the blanks to complete the following statements.

1. Like reptiles, amphibians are "cold-blooded," or _____.

2. Amphibians are one of our most abundant classes of _____.

3. Unlike reptiles, amphibians have _____ skin.

4. There are about _____ known species of amphibians worldwide.

5. Amphibians breathe through the use of _____, lungs, through their skin, or by a combination of these methods.

6. Amphibians rarely drink. They normally _____ water through their _____.

7. Of the adult amphibians, only _____ usually have tails.

8. Toads tend to hop, whereas frogs _____.

9. Frogs are usually more _____ than toads.

10. Name two factors that affect the color of toads and frogs.

    (a) _____

    (b) _____

11. Female frogs and toads are usually _____ than the males.

12. Toads are noted for _____, especially in loose soils.

13. All toads and frogs have very good _____.

14. Most frogs and toads are _____, but some are diurnal.

15. Toads and frogs make vocalizations with the use of a _____.

16. _____ calls are often given in chorus by the males and may be heard for miles.

17. Name three uses, other than mating, for which frogs and toads use calls.

    (a) _____

    (b) _____

    (c) _____

18. List five areas that are good spots to find frogs.

    (a) _____

    (b) _____

    (c) _____

    (d) _____

    (e) _____

19. Toads are generally more _____ than frogs and can often be found some distance from _____.

20. Most tadpoles are _____.

21. Adult frogs and toads are _____ and feed mostly on _____.

Short Answer

1. Describe the life cycle of a typical toad or frog, from beginning to adult forms. Include such information as habitat, predators, and feeding habits.

2. Why are salamanders difficult to identify? Be specific.

3. List three toads and three frogs. Give color and size for each.

4. Describe the life cycle of a typical salamander, from beginning to adult. Include such information as habitat, predators, and feeding habits.

Discussion

1. Are amphibians beneficial to humans? Discuss why and how they are beneficial.

2. Most amphibians require fresh water to complete their life cycle. Discuss how development, pollution, and other human activities may affect their populations. What can be done to curb these effects?

Learning Activities

1. Visit the **herpetorium** at the nearest zoo. List the common species of amphibians found in your area and provide a short description of their habits and preferred habitat.

CHAPTER 22

Freshwater Habitats and Their Management

OBJECTIVES

After completing this chapter, you should be able to

- Identify the different types of freshwater habitats.
- Describe management techniques for freshwater habitats.
- List some of the common life forms that live in freshwater habitats.
- Describe some of the characteristics of healthy aquatic habitats.
- Identify potential problems in freshwater habitats.

INTRODUCTION

Water is vital to all life forms on Earth. Without water, life as we know it cannot exist. Fresh water is available in limited supply (Figure 22-1). Although 70 percent or so of the Earth is covered with water, 97 percent of this amount is found in the oceans. Therefore, 97 percent of our water is salt water, which is not suitable for use as drinking water, in agriculture, or in most industries. Of the remaining 3 percent, more than one-third is frozen in polar ice caps and glaciers. Thus, less than 2 percent of the water on Earth is usable fresh water. Industry, agriculture, domestic users, and hydroelectric plants all use great volumes of water. It is estimated that the population of the United States drinks more than 100 million gallons of water per day. This amount does not include the water used for showers, dishwashing, washing clothes, watering

(Source: Courtesy of U.S. Fish and Wildlife.)

FIGURE 22-1 Plants and animals depend on fresh water for survival.

lawns, or washing cars. Every time you flush a toilet you use 1 to 2.5 gallons of water. If we add up all our domestic uses of fresh water, each person in the United States uses about 150 gallons of water per day! Multiply that amount by our population of more than 300 million, and the volume of water we use each day is staggering. Add to this amount the estimated 2.5 billion gallons of fresh water that industry uses every day and the billions of gallons used to irrigate farmland and one begins to realize the sheer volume of water we Americans consume. It is estimated that Americans use twice as much water as do people in other industrialized nations. With limited amounts of fresh water available, it is apparent that we must protect and conserve water. As our population continues to grow, can we continue to waste enormous volumes of fresh water?

FISH AND WILDLIFE

Water is vital to hundreds of species of fish and wildlife. The number and variety of wild creatures found in and around freshwater habitats is unequaled by any other single type of habitat. Forests, **deserts**, and prairies all provide wildlife with habitat, but none of these support the number and variety of fish, amphibians, reptiles, **insects**, and mammals that can be found in and around a healthy wetland. It has been estimated that the average wetland supports hundreds of birds, amphibians, and reptiles; as well as some 20 species of small mammals; 7 species of large animals; and countless other species of wildlife. In addition, there are usually countless numbers of invertebrates and a variety of fish species. It is impossible to overestimate the value of our wetlands to wildlife.

Some species of fish and wildlife require specific habitats to survive. Many species have evolved over thousands of years to fill a particular **niche** in an ecosystem, and the environmental conditions necessary for their survival are quite specific. When we alter their habitat, they are unable to adapt to the changes in time. This can cause a reduction in their numbers or even extinction of the species. We alter freshwater habitats in a variety of ways. We dam rivers and streams, drain marshes and swamps, and pollute fresh water with **sewage**, industrial waste, pesticides, and petroleum products. Our usage of fresh water also can lower the water levels in some habitats enough to cause fish and wildlife serious problems.

POLLUTION

Controlling pollution in freshwater habitats is probably the toughest problem facing scientists today. For decades cities and industries dumped chemicals, polluted waste water, and raw sewage into many of the nation's waterways. The pesticides and herbicides that farmers use to produce our abundant and cheap

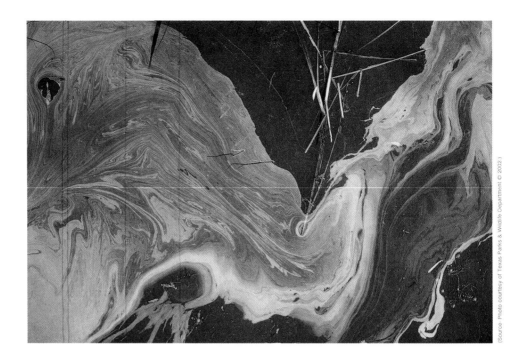

FIGURE 22-2 Oil, gasoline, diesel, and other petroleum products are very toxic to most fish and wildlife. Slicks like this one are all too common on lakes and streams.

food supply also wash into streams, rivers, lakes, and reservoirs. As our population has grown, so have the number of houses and subdivisions. Herbicides, pesticides, fertilizers, and other chemicals used on millions of lawns now pose a serious problem. By the late 1960s and early 1970s many of our historic and scenic waterways were a national disgrace (Figure 22-2). Lake Erie, for example, had been called biologically dead, and it was feared that it would remain so forever. The historic Potomac River, flowing through our nation's capital and beside which George Washington's estate lay, was a cesspool. It was so polluted that the stench drove tourists away and swimming was prohibited. In the past 30 years or so great strides have been made in protecting our fresh water from pollution and cleaning up already polluted waterways. However, pollution is still a very serious problem. The Water Pollution Control Act was passed in 1972. The act provided funding to identify sources of water pollution and develop techniques to control its flow into our rivers, streams, lakes, and reservoirs. Congress amended the act in 1977 and established some lofty goals for the program. The stated goals of the act were to make all the waters in the nation fishable and swimmable by 1985. Although these goals were not met, great progress was made. For instance, Lake Erie, thought by many to be lost forever as a productive body of water, was again supporting game fish. More than 45,000 miles of streams that had been monitored as part of the program were in noticeably better condition in 1985. In 1985 the Clean Water Act authorized the expenditure of $20 billion to continue cleaning up our waterways and to safeguard those not already polluted. It is essential that our fresh water be protected for our use and for the hundreds of species of fish and wildlife who share our habitat. It should be noted that most members of the industrial community have been very supportive of efforts to clean our waterways. Untreated chemicals are no longer dumped straight into the nearest river. Wastewater left over from the production of such essential commodities as paper and steel is now treated before being returned to its source. Cities no longer dump untreated, raw sewage into streams and rivers. The agriculture industry has developed new, safer pesticides that do not persist in nature as some have in the past. For example, traces of the pesticide DDT, which was banned in the United States in 1972, were still being found in some fish in the late 1980s. Great care is also taken to ensure that pesticides are applied properly and containers either reused or disposed of correctly. The 1985 farm bill also established the Conservation Reserve Program (CRP). This program paid farmers to set aside some of their more highly erodible soils and plant them in permanent vegetative cover. It is estimated that the CRP has reduced the use of pesticides by approximately 60 million pounds per year. The grasses and trees planted as part of the program have reduced soil erosion by significant amounts, perhaps by as much as 700 million tons per year. Although the complete benefits of the CRP have not yet been estimated, it has most certainly saved millions of tons of topsoil. In doing so, the CRP has had tremendous positive effect on our rivers, streams, and lakes. Each ton of soil left in place is a ton that does not end up as silt and sediment in our waterways. In addition, many pounds of pesticides are caught up and reduced to harmless compounds before they ever reach our fresh water. However, with today's high prices for grain, more and more land is being removed from the CRP when the contracts expire. Pollution from increased use of herbicides and pesticides on grain crops may well lower our water quality once again.

Another important source of water pollution and pollution in general, is litter. Plastic, aluminum cans, fishing line, glass, and plastic bottles take decades, perhaps hundreds of years to decay. In addition to being an eyesore, they can be serious hazards for wildlife and fish. There are numerous accounts of small mammals becoming entangled in thoughtlessly discarded plastic rings that hold six-packs of canned drinks. These animals are usually doomed to a terrible death by **strangulation** or starvation. Thankfully, these wildlife death

FIGURE 22-3
Chemicals that control weeds and pests can be dangerous to the environment. It is important that they be used properly.

(Source: Courtesy of Utah Agricultural Experiment Station.)

traps are largely a thing of the past. One quart of used motor oil, improperly disposed of, can pollute hundreds of gallons of fresh water. Many homeowners, in their attempt to create the "perfect" lawn, are overly zealous in their use of lawn fertilizers and chemicals. The resulting runoff pollutes streams, rivers, ponds, lakes, and reservoirs (Figure 22-3). Fertilizers and lawn chemicals are formulated in such a manner as to be safe if applied properly and in accordance with their instructions. Needless to say, this is the only manner in which they should ever be applied.

We as individuals can do much to reduce everyday litter and pollution. If we police ourselves and take great care in how we use and dispose of household items such as fertilizers, chemicals, and old motor oil, we can greatly reduce the amount of waste and pollution we create. Today, most Americans practice some sort of recycling, whether it be plastics, aluminum, or paper. It is each individual's responsibility to see to it that we do not trash and pollute the Earth.

DESTRUCTION OF FRESHWATER HABITATS

We have destroyed, and continue to destroy, thousands of miles of stream and river habitat. When we dam a stream or a river, the land around it is altered forever (Figure 22-4). We also degrade and destroy stream and river habitats by developing their banks. When we bulldoze the vegetation along a waterway to make way for a road or subdivision, the vegetation is often destroyed. This usually results in erosion. Erosion usually results in cloudy, muddy waterways, which are very detrimental to many aquatic life forms. Fish and other aquatic life forms that have evolved in clear, cool streams often cannot adjust to the new conditions and

FIGURE 22-4 Large hydroelectric dams like this one provide cheap electricity and drinking water. However, they bury thousands of acres of wildlife habitat under water and damage or destroy river ecosystems for miles upstream from the dam.

(Source: Photo courtesy of Texas Parks & Wildlife Department © 2002.)

perish. The silt fences and hay bales we see around sensitive areas of new construction sites today are most certainly an improvement over previous decades. They undoubtedly help improve our water quality by keeping silt and sediment from flowing into our rivers and streams. Overgrazing of streamside vegetation can also result in erosion; this is particularly a problem along many western riparian waterways.

Overuse is also a serious problem, particularly in the water-starved West. Many rivers have been dammed, channeled, or diverted and their water levels reduced to the point that they no longer exist. The mighty Colorado is a good example. It once fed millions of gallons of fresh water into the Gulf of California each year. Today it barely reaches the gulf. Millions and millions of gallons are diverted to cities and to irrigation projects throughout the Southwest. As a result of our overuse and development of the Colorado River, many native aquatic species have suffered and some may even be on the road to extinction. In the more arid portions of our country, the decisions on how to use fresh water will likely be settled in court.

FRESHWATER HABITATS

Freshwater habitats for wildlife and fish take various forms. The most obvious freshwater habitats are lakes, rivers, and reservoirs. However, from the standpoint of productivity, small ponds, swamps, streams, and marshes provide habitat for a greater number and variety of wild creatures than do large impoundments.

Reservoirs are large **impoundments** we have created to provide water and hydroelectric power for many large metropolitan areas. They also provide recreational opportunities in the form of such activities as boating, swimming, fishing, and water skiing. Lakes are also large bodies of water, either naturally occurring or artificial, that offer similar recreational activities and provide water and power for some towns and cities. Although we have created millions of acres of lakes and reservoirs by damming streams and rivers, these habitats are generally not as productive as natural wetlands. In addition, when a large lake or reservoir is constructed, thousands of acres of prime wildlife habitat are usually buried under water. Fish and invertebrates that thrived in the river or stream that was dammed may find that the lake created is unsuitable or, at best, marginal habitat. In addition, large areas of lakes and reservoirs may have water too deep to be useful to many species.

Rivers and streams drain excess surface water and provide excellent wildlife and fish habitat. The low-lying areas, or bottoms, associated with streams and rivers generally teem with wildlife. Rivers and streams periodically overflow their banks and deposit layers of fine, rich soil eroded from areas upstream on the surrounding bottoms. For this reason river bottoms support a large and healthy plant community. Wildlife can find food, cover, water, and space, all suitably arranged, in the average river or stream bottom. For this reason river bottoms are the preferred habitat for many species.

Small ponds may occur naturally or be created by humans. They are generally much shallower than lakes or reservoirs, which allows more sunlight to penetrate to the bottom and thus promotes more aquatic plant growth. This growth provides habitat for many species of fish and amphibians that cannot find suitable habitat in a lake or reservoir. In addition, these ponds, often known as "stock tanks," are more appealing to many species of waterfowl because of the more readily available food supply and generally more sheltered water. In arid areas where naturally occurring surface water was once rare, these stock tanks provide habitat for thousands of migrating waterfowl.

Marshes and swamps are low-lying areas. Swamps may have few, if any, large areas of open water, but water is always present. Such areas support a tremendous variety of plants that usually cannot survive in a more arid setting. These are some of the best wildlife homes, but they are also extremely sensitive to pollution, low water levels due to excess irrigation or drainage, and increased **sedimentation** resulting from construction.

Healthy freshwater habitats support remarkable amounts of life. Everything from tiny plankton to fish, amphibians, reptiles, and thousands of insects and other invertebrates make their homes in aquatic habitats. A healthy aquatic ecosystem contains large numbers of each of these life forms. However, the balances of environmental conditions necessary to support all this life are quite delicate (Figure 22-5). For aquatic plant life to flourish and grow, water must contain adequate amounts of the proper nutrients. The plants must also receive sunlight in order to carry on **photosynthesis**, an essential process for survival. Aquatic plants are vital to the health of freshwater habitats. They produce oxygen in the water as a by-product of photosynthesis, just as land-based trees and plants provide oxygen in the air. Aquatic plant life also provides food and shelter to many lower life forms. Tadpoles, the fry of many fishes, a variety of aquatic insects, and other invertebrates all seek shelter from their many predators among aquatic plant growth. Smaller fish and crustaceans feed on aquatic plants and the lower life forms that seek shelter among

(Source: © Delmar Cengage Learning.)

FIGURE 22-5
Environmental conditions must be balanced in order to support life.

them. They are in turn preyed on by larger fish, amphibians, and reptiles.

Aquatic habitats can be more complex than terrestrial ecosystems. Anything that affects one facet of the ecosystem generally sets off a chain of events that affects each of the life forms in the system. For example, if sediment from erosion clouds the water, it can reduce the amount of sunlight that reaches aquatic plants. Plants then produce less oxygen and are less healthy. In severe cases aquatic plant life dies out. When this happens, all the lower life forms that depend on the plant life decrease in number. The numbers of higher life forms also decrease because they have less to eat and less oxygen to breathe. Pollutants and sewage also use up available oxygen; this can cause large-scale die-offs of many fish species.

Healthy aquatic habitats contain a balance of many different life forms, which coexist in a complex ecosystem. It is the task of scientists to maintain the balance within a variety of freshwater habitats.

AQUATIC HABITAT MANAGEMENT

Aquatic habitat management is a very complex issue. Scientists must monitor rivers, streams, lakes, and reservoirs continually. The oxygen content, toxic substances, acidity, and dissolved salts are just a few of the items that scientists and health officials must monitor. Often the first sign that something is amiss in an aquatic ecosystem is a lack of or a die-off of the larger fishes. When this occurs, there is usually something out of balance in the system.

Sewage and wastewater must now be treated before being returned to its source. Excessive runoff and erosion are carefully managed to help prevent sediment buildup and pesticide contamination. As in most endeavors, an ounce of prevention is usually worth a pound of cure. Stopping pollution, be it chemicals or eroded soil, is far better than treating a river, lake, or stream after it has been contaminated. Every American can dispose of trash properly and recycle such items as plastic, paper, and aluminum. In most artificial ponds, lakes, and reservoirs that are properly constructed, there are few management problems (Figure 22-6). Landowners and wildlife officials decide which fish species to stock in an impoundment and how many to stock. If chosen carefully, the fish species should complement each other and not be overly competitive. Care must be taken not to stock a species that might quickly overpopulate an impoundment. Bluegills, for example, are prolific breeders and so will often overpopulate an impoundment, especially smaller ponds and lakes. Commonly stocked species, particularly in southern waters, include largemouth or black bass, channel catfish, and one of the species of sunfish, often redear.

FIGURE 22-6 Large impoundments can give some fish and amphibian species new homes and provide recreational opportunities for people.

(Source: Photo courtesy of Texas Parks & Wildlife Department © 2002.)

SUMMARY

We actually have adequate amounts of fresh water, although some of it may not be where human populations need it. We still have millions of acres of wetlands and thousands of miles of unspoiled rivers and streams.

But we must stop draining our wetlands at the rate of thousands of acres each year. We must also consider all the benefits and dangers of each new dam we construct—before it is built.

REVIEW QUESTIONS

Fill in the Blank

Fill in the blanks to complete the following statements.

1. _____ is vital to all life forms on Earth.

2. Ninety-seven percent of the water found on Earth is found in the _____.

3. Fresh water makes up about _____ percent of the water on Earth, and of this amount more than _____ is frozen in the polar icecaps.

4. List three major users of freshwater.

    (a) _____

    (b) _____

    (c) _____

5. It is estimated that each American uses about _____ gallons of water each day.

6. It is estimated that _____ in the United States uses about 2.5 billion gallons of water each day.

7. Water is also vital to hundreds of species of _____ and _____.

8. It has been estimated that the average wetland supports more than _____ species of small game mammals.

9. List three ways in which we alter wetland habitat.

    (a) _____

    (b) _____

    (c) _____

10. List three common pollutants of freshwater habitats.

    (a) _____

    (b) _____

    (c) _____

11. Pollution of our water had become so bad by the early 1970s that the historically important _____ River was a cesspool.

12. The first legislation to protect our water, _____, was passed in 1972.

13. In 1985 a new bill, known as the _____, was passed to protect fresh water.

14. List three things that are now done to help reduce pollution in fresh water.

(a) _____

(b) _____

(c) _____

15. The pesticide _____ could still be found in some fish in the 1980s, even though it had been banned in _____.

16. It is estimated that the CRP reduced pesticide use by _____ million pounds per year.

17. List five different types of freshwater habitats.

(a) _____

(b) _____

(c) _____

(d) _____

(e) _____

18. List three ways we destroy or damage freshwater habitats.

(a) _____

(b) _____

(c) _____

19. Aquatic plants provide essential _____ to freshwater habitats as a by-product of the photosynthesis process.

20. _____ and _____ use up the available oxygen in the water.

21. Healthy aquatic habitats contain a balance of many different _____.

Short Answer

1. Why is it necessary to conserve and protect our freshwater supplies?

2. How can we protect and conserve freshwater supplies?

3. Discuss one method or program that helps prevent the runoff of herbicides and pesticides into our streams, lakes, and rivers.

Discussion

1. Discuss the factors that make wetlands such productive wildlife habitat. How do these same wetlands, swamps, marshes, and so forth help protect our freshwater supplies?

2. What are some ways we might better manage our aquatic habitats? How would you implement management changes, if any?

Learning Activities

1. Take a walk along the shore of a river or lake in your area. Note all the different types of litter or trash. Report your findings to your class.

2. With the assistance of your instructor or other adult sponsor(s), return to the site you used for activity 1 and remove all the litter.

3. Spend an hour observing the wildlife that use a local freshwater habitat. Note the number and variety of species observed. Report your findings to your class.

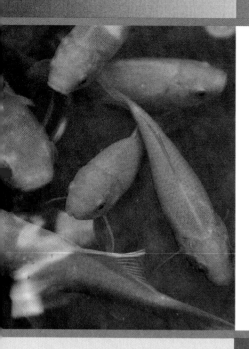

CHAPTER 23

Freshwater Fishes

OBJECTIVES

After completing this chapter, you should be able to:

- Describe some of the common characteristics of freshwater fishes.
- Identify the types of habitats that various species might use.
- Describe some of the habits of common freshwater fishes.
- List some of the most common species of fishes found in the United States.
- Describe some differences between the species of freshwater fish.

INTRODUCTION

Including the saltwater fishes, there are more than 20,000 species of fish in the world. This is by far the largest group of vertebrates, outnumbering all other vertebrates combined. More than 350 species of freshwater fishes are found in the United States. We will discuss the most common, economically important species.

Almost all fish are covered with scales and get their oxygen through their gills. Fish use the oxygen that is dissolved in the water. Should the amount of dissolved oxygen fall below necessary levels, fish begin to die. Many species of fish are gregarious, at least early in their lives, and often group together in **schools**. Some, such as the black bass, become solitary as adults. Like most other animals, fish have definite hours in which they prefer to work. Some fish are largely nocturnal, and others are diurnal. Nocturnal fish tend to forage with the use of sharp senses of smell and taste, whereas diurnal species tend to use taste and sight. Some species, such as the pike, may use sight almost entirely to locate food.

Although all fish breathe through the use of gills and almost all have some type of scales covering their bodies, they come in a surprisingly wide variety of shapes. Some fish are plant eaters, but many are carnivorous. Most of the game fishes fall into the carnivorous group. The habitat utilized, feeding habits, reproductive habits, and life cycles of most fish species are very similar.

Habitat: Some fish prefer streams and rivers; others prefer lakes and ponds. Many do well in most aquatic habitats. However, some are specialized and can only be found in specific types of aquatic habitat. The type of habitat preferred is included in the information on each species.

Feeding Habits: Fish have a variety of feeding habits, often dictated by the stage of life they are in. The young of most fish feed on tiny organisms known as **plankton**. Plankton may be either animal-like (zooplankton) or plantlike (phytoplankton). As fish grow they begin to eat larger prey, such as insects and the **fry** of most species of fish. Larger fish feed on large insects, smaller fish, amphibians such as salamanders, and, the occasional mouse or duckling.

Life Cycle: Fish are egg-layers, usually producing tremendous numbers of offspring. These numbers help to offset the large variety of hazards the eggs and young are exposed to. There are a few species of live-bearing fishes, and they generally produce fewer offspring. Some species, such as the carp, may produce up to a million eggs. In the species that produce live young, the eggs are retained in the **oviduct** of the female. **Sperm** is introduced by the male, and the eggs develop and hatch in the female. Although these fish produce far fewer offspring, their chances of survival are much better. The vast majority of freshwater fishes are egg-layers and utilize one of two methods for rearing their young. Perhaps the most common method is for the female to lay her eggs randomly on suitable spawning beds. Eggs are then fertilized by one or more attending males. The fertilized eggs are left to develop and receive no further parental care. Most of these random egg-layers produce huge quantities of eggs, often many thousands.

The second common method of reproduction for freshwater fishes is nest building. Fish that use this method usually clear an area and make a small depression with their tails. These nests are usually in loose gravel or sandy areas. The eggs are laid in this depression and are then fertilized by the male. The males often remain behind to guard the nest and the young fry when they first hatch. Some species of fish lay and fertilize their eggs, cover them with gravel, and leave them to fend for themselves. Most young fish receive little parental assistance.

Most fish lead short lives. The young and eggs are eaten by a variety of other fish, amphibians, reptiles, mammals, and birds. Adults are taken by larger fish, large turtles, mammals (such as raccoons), and birds (such as great blue herons). Of course, humans have always used fish as a major source of food.

Minnows
(Cyprinidae)

DESCRIPTION

Minnows make up the largest group of freshwater fishes found in North America. Most minnows are difficult to identify, but they are usually small and lack scales on their heads. Their bodies are usually covered with small scales and their tails are forked. A single dorsal fin is normally located in the middle of the back. Most species do not have teeth in their jaws. Instead, teeth are located in their **pharynx** or throat. Many species can be positively identified only by counting the number of pharyngeal teeth.

Golden Shiner
(Notemigonus crysoleucas)

DESCRIPTION

These minnows have a distinct silver or golden sheen and may grow to 11 inches (Figure 23-1). They prefer clear water, with abundant aquatic vegetation. They feed on plant materials, insects, and plankton. Golden shiners are generally found from southern Canada across the eastern two-thirds of the United States, excluding the western half of Texas. These fish are often raised to be fed to hatchery fish. There are numerous subspecies.

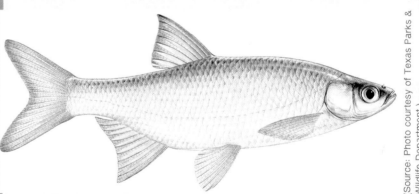

FIGURE 23-1 Golden shiner.

(Source: Photo courtesy of Texas Parks & Wildlife Department.)

Goldfish
(Carassius auratus)

DESCRIPTION

These fish were originally natives of Asia but have been widely introduced in the United States (Figure 23-2). Goldfish are generally present from Maine to Montana and from there southward. They prefer habitat that is thick with aquatic vegetation. Goldfish feed on plant material, insects, crustaceans, and snails and may grow to 16 inches in length.

FIGURE 23-2 Orlando variety of goldfish.

(Source: © Orla, 2010. Used under license from Shutterstock.com)

Fathead Minnow
(Pimephales promelas)

DESCRIPTION

These minnows are relatively small, reaching only 3 inches (Figure 23-3). They have short, stout bodies and seem to prefer ponds, smaller lakes, creeks, and small muddy streams. Fathead minnows are common and widespread from the Rocky Mountains east to the Appalachians. They are very adaptable and are a favorite species for many larger fish. They are also a favorite bait of anglers and are widely used for fish **propagation**. There are numerous subspecies.

FIGURE 23-3 Fathead minnows.

(Source: Courtesy of U.S. Fish and Wildlife Service. Photo by Duane Raver.)

Carp
(Cyprinus carpio)

DESCRIPTION ▪

Carp is another Asian native that has been widely introduced in North America (Figure 23-4). Originally brought from Europe, carp can now be found from Nova Scotia to southern Manitoba and southward throughout much of the United States. Unlike most minnows, carp get big, sometimes weighing 50 pounds or more and exceeding 3 feet in length. They have large scales, usually with a dark spot on the base of each scale. They have a ruddy back with a lighter-colored belly and usually have 16 or more rays in the dorsal fin. Carp prefer warm, often shallow water, and are particularly fond of lakes.

FIGURE 23-4 Carp.

(Source: Photo courtesy of Texas Parks & Wildlife Department © 2002.)

SHAD

Members of this genus are generally shiny and silvery. They usually have jaws of equal length or the lower jaw projects slightly. Shad lack teeth and their jaws are quite weak. They usually have short dorsal fins.

Gizzard Shad
(Porosoma cepedianum)

DESCRIPTION ▪

This is our most common and widespread freshwater shad (Figure 23-5). They range from western New York to southern Minnesota, south to west Texas, and throughout the eastern half of the United States. Gizzard shad may grow to 15 inches and prefer shallow water in rivers, reservoirs, and lakes. They feed primarily on plankton and are a favorite food fish of many larger carnivorous fishes. They have a certain tolerance for **turbid** water and perhaps because of this have been on the increase in our reservoirs in recent years.

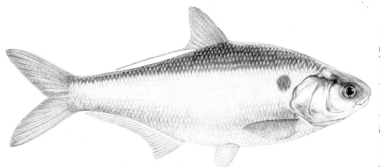

FIGURE 23-5 Gizzard shad.

(Source: Photo courtesy of Texas Parks & Wildlife Department.)

GAR

(Lepisostedae)

Gar are very old fishes and are often referred to as "living fossils" because most of the related species are extinct. They are long, torpedo-shaped fish, with large, diamond-shaped scales. The jaws are long, thin, and beaklike and are armed with rows of strong, needle-sharp teeth. Gar prefer shallow rivers and lakes with abundant aquatic vegetation. They can use oxygen from the atmosphere and are often observed basking on or near the surface.

Longnose Gar

(Lepisosteus osseus)

DESCRIPTION ■

This is the only gar found in North America with a snout twice the length of its head. These gar may grow to 5 feet in length and weigh in excess of 45 pounds. Their distribution is spotty, but they can generally be found from Vermont to Montana and southward to Florida and Mexico.

Alligator Gar

(Lepisosteus spatula)

DESCRIPTION ■

The alligator gar is our largest gar, with the rod-and-reel record weighing 279 pounds (Figure 23-6). They may reach a length of 7 feet or more. They have a shorter, broader snout than the L0ngnose gar, and they have a double row of teeth in their upper jaw. They are found in larger rivers in the South, particularly the Mississippi River drainage.

(Source: Photo courtesy of Texas Parks & Wildlife Department © 2002.)

FIGURE 23-6 Alligator gar.

SALMONS (Salmonidae)

Economically, the salmons are very important fishes. Entire fishing industries in Alaska and the Pacific Northwest, worth millions of dollars annually, are based on the salmon (Figure 23-7). The members of this group of fish are anadromous, that is, they spend the majority of their time at sea, but return to inland rivers and streams to reproduce. Shortly after they spawn, salmon die. They are carnivorous fish, with long bodies and heads that lack scales. Salmon have large mouths and well-developed teeth. Their dorsal fins usually have fewer than 15 rays, and they prefer waters of 70°F or cooler. Three species of salmon are especially abundant on the Pacific coast: coho, chinook, and sockeye. They are widespread, ranging from coastal Alaska to Baja California. The chinook is the largest of the group, weighing up to 125 pounds and reaching 58 inches or so in length. The coho is next, weighing up to 30 pounds and reaching 3 feet or more. The sockeye is usually the smallest, reaching 32 inches and weighing up to 16 pounds. Each of these salmon have silver bellies but somewhat different upper-body color patterns. The coho is a greenish blue to metallic blue, with abundant small spots or speckles. Breeding male coho have a bright red stripe. The sockeye is also greenish blue, with many fine speckles. Males are often reddish and females may have yellow or dark red blotches. The chinook is greenish to very dark blue or blackish, with many prominent spots.

The Atlantic salmon is the species most commonly found in the Northeast, ranging from Maine to Labrador. They are also found in some deep, cold landlocked lakes in the region. The non-landlocked Atlantic salmon may weigh up to 75 pounds, but the rod-and-reel record for the landlocked variety is 28.5 pounds. The Atlantic salmon is silver to steel blue, with large dark spots that often have blue rings or halos around them. They also have a deeply forked tail.

FIGURE 23-7 Salmon.

(source: Courtesy of U.S. Fish and Wildlife Service.)

TROUT

Trout are widespread and widely introduced across much of North America. They are perhaps our most heavily propagated fish, with more time and money spent raising them than nearly any other fish. Trout are carnivorous fish and are widely sought after by anglers. There are five species of major importance in North America: rainbow, brown, cutthroat, lake, and brook trout.

Lake Trout
(Salvelinus namaycush)

DESCRIPTION

This is our largest freshwater trout, with a rod-and-reel record of over 60 pounds. They have dark bodies with pale or red spots (Figure 23-8). Lake trout inhabit large, deep lakes in the northern Rocky Mountains, Great Lakes area, and virtually throughout Canada. The tail is deeply forked and the dorsal fin has pale spots.

(Source: Courtesy of U.S. Fish and Wildlife Service. Photo by Eric Engbretson.)

FIGURE 23-8 Lake trout.

Rainbow Trout
(Salmo gairdneri)

DESCRIPTION

Rainbow trout may reach a length of 3 feet and a weight of 35 pounds (Figure 23-9). It has a straight tail, with abundant spots, a pink or light red stripe on the sides, and dark spots on the body. They prefer swift, clear water, with adequate vegetative cover. Rainbow trout feed heavily on insects and aquatic insect larvae. They were widely introduced in much of North America and are widespread in the western third of the United States and along western coastal Canada. They are also found from Minnesota across southern Canada to Maine and south into New York.

(Source: Photo courtesy of Texas Parks & Wildlife Department © 2002.)

FIGURE 23-9 Rainbow trout.

Brown Trout
(Salmo trutta fario)

DESCRIPTION

Brown trout prefer slower waters than the rainbow, but otherwise their habitats are very similar (Figure 23-10). Brown trout grow to 2 feet and longer and may weigh 35 pounds or more. They have reddish orange spots on their sides and backs, ringed with lighter color to form halos, similar to the Atlantic salmon. Their range is similar to the rainbow's, except that there is a narrow band across southern Canada that connects the east and west areas. Brown trout do not range as far north as the rainbow. Brown trout also feed heavily on insects and their larvae.

(Source: Photo courtesy of Texas Parks & Wildlife Department © 2002.)

FIGURE 23-10 Brown trout.

Cutthroat Trout
(Salmo clarki)

DESCRIPTION

The cutthroat is usually a small trout, commonly reaching 15 inches (Figure 23-11). However, the rod-and-reel record is 41 pounds. They prefer cold lakes, rivers, and streams and can be found in the Colorado, Arkansas, Missouri, Platte, and Rio Grande river drainages. Cutthroats are excellent table fare and are highly prized by anglers. They have a distinct pink or light red streak on the underside of the lower jaw.

(Source: Courtesy of U.S. Fish and Wildlife Service.)

FIGURE 23-11 Cutthroat trout.

Brook Trout
(Salvelinus fontinalis)

DESCRIPTION

Brook trout have an unforked tail and dark bodies with abundant blue-bordered red spots (Figure 23-12). They prefer cold water in small ponds and streams. They are widespread in the Rocky Mountain states and from northeastern Canada south into Minnesota, the Great Lakes area, New York, and down the Appalachians. They are very cautious and are easily scared away. Brook trout are much sought after by anglers and may weigh up to 14 pounds.

(Source: Courtesy of U.S. Fish and Wildlife Service. Photo by Eric Engbretson.)

FIGURE 23-12 Brook trout.

PIKE

Pikes are carnivorous, and the larger species of this group are important game fish. They are characterized by long, torpedo-shaped bodies and forked tails. The pike has a snout that resembles a duck's bill and is full of sharp teeth. They feed largely on frogs and other fish and prefer ponds, lakes, and streams with abundant aquatic vegetation. They generally move into shallow water during the summer and spend the winter in deep water.

Redfin Pickerel
(Esox americanus)

DESCRIPTION

Redfins are widespread, ranging from Nebraska east to New Hampshire and south to the Gulf of Mexico. They are smallish members of the pike clan, averaging 12 inches. They are usually a medium dark green, with dark bands on the sides.

Northern Pike
(Esox lucius)

DESCRIPTION

This is a larger member of the group, weighing 40 pounds on occasion and reaching lengths greater than 4 feet (Figure 23-13). Body is olive colored, with light, oblong spots on a darker background. They also have dark spots on their fins. They are found from Nebraska north to Alaska and across much of north and northwest Canada.

(Source: Photo courtesy of Texas Parks & Wildlife Department © 2002.)

FIGURE 23-13 Northern Pike.

Muskellunge
(Esox masquinongy)

DESCRIPTION

These members of the pike family are dark greenish above and paler below (Figure 23-14). They are also quite large, reaching a length of 5 feet and a weight of almost 70 pounds. Muskellunge also have dark spots or bars on sides and fins, with a lighter background. They typically spend their summers in deep water and move into shallow water in the fall.

(Source: © Len Rue Jr.)

FIGURE 23-14 Muskellunge.

CATFISH

A fairly large family of fish, catfish are of great importance to anglers and to the **aquaculture** industry. Some members of this family have become the principal species raised for the restaurant and processed fish trades. Catfish have scale-less skin and broad, flat heads. Both the dorsal and pectoral fins have a strong spine that can inflict a painful wound if the fish is handled carelessly. Catfish lack the sharp teeth of many species. Their teeth are more like short, stiff bristles. Catfish are largely bottom feeders and appear to be omnivorous in their feeding habits. They are also very **prolific**, with females laying up to 20,000 eggs, usually in late spring and summer. Although there are several dozen species of catfish, we will discuss three of the more widely distributed species: channel, blue, and flathead catfishes.

Channel Catfish
(Ictalurus punctatus)

DESCRIPTION

These catfish have deeply forked tails and a distinct bluish silver body color, often with dark spots (Figure 23-15). They may grow to 50 pounds or more and are a favorite with anglers. Channel catfish prefer larger bodies of water, such as reservoirs and large rivers. They are widespread east of the Rocky Mountains and have been widely introduced elsewhere around the United States.

(Source: Courtesy of U.S. Fish and Wildlife Service. Photo by Eric Engbretson.)

FIGURE 23-15 Channel catfish.

Blue Catfish
(Ictalurus fureatus)

DESCRIPTION

The blue catfish also has a deeply forked tail and is off-white to bluish in color, with no body spots (Figure 23-16). The eyes are rather small and the lower jaw is slightly shorter than the upper jaw. These are large fish, with an occasional one weighing in excess of 90 pounds. Blue catfish are very tasty and are in much demand with anglers. They are found in the large rivers of the Mississippi Valley and have been widely introduced in reservoirs and other impoundments.

(Source: Photo courtesy of Texas Parks & Wildlife Department.)

FIGURE 23-16 Blue catfish.

Flathead Catfish
(Pylodictus olivaris)

DESCRIPTION

Flatheads can be very large, weighing 100 pounds or more, and reaching lengths of 5 feet (Figure 23-17). It has an elongated, flat head, hence the name "flathead," very small eyes, and a large mouth. It does not have the forked tail of the channel or blue catfish; it is rather rounded. Flatheads prefer the deepest pools of slow-moving rivers and streams. They range from western Pennsylvania to South Dakota and south to the Gulf Coast.

(Source: Photo courtesy of Texas Parks & Wildlife Department.)

FIGURE 23-17 Flathead catfish.

SUNFISH FAMILY

Sunfish are a large family of fish that includes black basses and crappies. Most members of this family are either pan or game fishes. These fish were once largely restricted to east of the Rocky Mountains, but they have been widely introduced all over North America. They prefer warm water lakes, streams, and smaller rivers, and one or more species can typically be found from the Gulf of Mexico to southern Canada. All of these species are nest builders, with the male usually doing the job and guarding the eggs and young. These carnivorous fishes usually have completely joined dorsal fins.

Smallmouth Bass
(Micropterus dolomieui)

DESCRIPTION

These bass have a small notch between the dorsal fins and are olive green (Figure 23-18). They usually have 13 to 15 dorsal rays and lack the stripe of dark spots on their sides that the largemouth bass exhibits. Smallmouths prefer rocky, clear lakes and rivers and may reach 10 pounds or more. They feed primarily on fish and are one of our favorite freshwater game fish. They range from Quebec and Minnesota south to Alabama and Arkansas and have been widely introduced elsewhere.

(Source: Photo courtesy of Texas Parks & Wildlife Department © 2002.)

FIGURE 23-18 Smallmouth bass.

Largemouth Bass
(Micropterus salmoides)

DESCRIPTION ■

Largemouth bass have a distinctive deep notch that nearly separates their dorsal fins (Figure 23-19). The back is usually olive, with silver sides and belly. A dark band on each side becomes broken as the fish grows older. Largemouths normally have 12 to 15 dorsal-fin rays and an especially large mouth. Exceptional specimens may grow to 20 pounds or more, and trophy largemouth bass are one of our most sought-after fish. Largemouth's can be found in small ponds, lakes, and reservoirs and feed on fish, frogs, crayfish, insects, ducklings, and anything else they can swallow. They are widely distributed across much of the eastern United States, ranging from Maine to west Texas and north to Minnesota.

(Souce: Photo courtesy of Texas Parks & Wildlife Department © 2002.)

FIGURE 23-19   Largemouth bass.

Redear Sunfish
(Lepomis microlophus)

DESCRIPTION ■

A true sunfish, the redear has a deeper, thinner body than either of the basses (Figure 23-20). They are quite widespread, ranging from the Gulf up the Mississippi Valley to Iowa and Indiana, and they have been widely introduced in other areas. Redear sunfish grow to about 10 inches and are olive green above and brass brown below. The rear edge of the gill cover is red, and they generally have no distinct spots on the dorsal fins. They prefer clear, quiet waters with some vegetation and do well in most lakes, ponds, or reservoirs.

(Source: Photo courtesy of Texas Parks & Wildlife Department © 2002.)

FIGURE 23-20   Redear sunfish.

Bluegill Sunfish
(Lepomis macrochirus)

DESCRIPTION

The bluegill is a very widespread sunfish, ranging over most of the eastern half of the United States (Figure 23-21). Bluegills are popular fish among beginning anglers, especially children, because they are prolific and aggressive and therefore relatively easy to catch. They may grow as large as 15 inches and weigh as much as 4 pounds, although most are smaller. The back is olive green, with orange and blue on the sides. Bluegills have rather small mouths and eat large numbers of insects, crayfish, and small fish. They prefer warm waters with dense vegetation and are most commonly found in ponds and lakes.

(Source: Courtesy of U.S. Fish and Wildlife Service. Photo by Eric Engbretson.)

FIGURE 23-21 Bluegill.

Longear Sunfish
(Lepomis megalotis)

DESCRIPTION

This sunfish inhabits a broad expanse of middle America, ranging from southern Wisconsin east to western New York and south to the Florida panhandle and most of Texas (Figure 23-22). Longears may grow to be 9 inches or so, and their color is highly variable but often quite bright. The sides are often spotted, and the belly may be bright red or yellow. It has a small mouth and seems to prefer insects as its main source of food.

(Source: Photo courtesy of Texas Parks & Wildlife Department.)

FIGURE 23-22 Longear sunfish.

Black Crappie
(Pomoxis nigro maculatus)

DESCRIPTION

The black crappie is widespread, ranging from eastern North Dakota south across the eastern portions of the plains states to the eastern third of Texas and east to New England and southern Quebec (Figure 23-23). It may reach a length of 15 inches, with a maximum weight of about 5 pounds. The body is greenish with dark or black blotches or spots. Black crappie prefer still, quiet, clear waters, with sand or mud bottoms and abundant submerged vegetation. They feed heavily on smaller fish and insects. Black crappie have been widely distributed and stocked by state game departments but do not do well in direct competition with white crappie.

(Source: Photo courtesy of Texas Parks & Wildlife Department.)

FIGURE 23-23 Black crappie.

White Crappie
(Pomoxis annularis)

DESCRIPTION

Twelve inches and 5 pounds is about as large as these prolific crappie get (Figure 23-24). They have a greenish silver body with faint, darker vertical bars or stripes. White crappie can be found in a great variety of habitats, from ponds to lakes and **bayous**. They prefer habitats with lots of aquatic vegetation and are quite tolerant of turbid water. White crappie feed on insects, crustaceans, and small fish, often feeding in schools. They are abundant, popular, and much stocked by state conservation departments. They can generally be found from the Gulf Coast to southwestern New York and westward to eastern South Dakota.

(Source: Photo courtesy of Texas Parks & Wildlife Department © 2002.)

FIGURE 23-24 White crappie.

Walleye
(Stizostedion vitreum)

DESCRIPTION ■

The walleye is our only large perch, reaching 3 feet and 20 pounds or so (Figure 23-25). It is a dark olive, and fish less than 12 inches have irregular dark bars across the sides. Larger specimens usually have a dark spot at the base of the last dorsal spine and lack obvious vertical stripes or bars on their sides. Walleyes are distributed from northwest Canada south to the lower Mississippi River Valley and eastward to the Atlantic. They prefer the cold water of deep lakes and large rivers. Walleyes are a very popular game fish and are widely sought by commercial and sports anglers.

(Source: Photo courtesy of Texas Parks & Wildlife Department © 2002.)

FIGURE 23-25 Walleye.

White Bass
(Roccus chrysops)

DESCRIPTION ■

The white bass has two separate dorsal fins and a lower jaw that is longer than its upper jaw (Figure 23-26). It is an overall silvery color, with seven prominent stripes on each side. They may grow to a length of 12 inches and a weight of 3 pounds. They normally travel in schools and feed on insects, crayfish, smaller fish, and plankton. Although they have been widely introduced elsewhere, their general distribution is from Mexico north to Minnesota and east to western New York.

(Source: Photo courtesy of Texas Parks & Wildlife Department © 2002.)

FIGURE 23-26 White bass.

Striped Bass
(Roccus saxatilis)

DESCRIPTION

Striped bass is actually a saltwater species that has been widely introduced in larger impoundments and reservoirs (Figure 23-27). Striped bass may reach 4 or 5 feet in length and weigh more than 100 pounds in their natural saltwater homes, but they do not attain this large size in fresh water. They are still quite large and are popular game fish in areas where they have been introduced. Striped bass have the streamlined bodies of a typical saltwater fish and are dark olive green to bluish black above, with silver on the belly and lower sides. They have six to nine, often seven, prominent horizontal black stripes down each side. The lower jaw projects beyond the upper and the mouth is quite large. They are carnivorous and feed on virtually any smaller fish.

(Source: Photo courtesy of Texas Parks & Wildlife Department © 2002.)

FIGURE 23-27 Striped bass.

SUMMARY

We have been guilty of polluting habitats and overfishing certain species in the past, but with wise management and conservation practices many previously endangered species are on the mend. However, several species are still in danger, most notably the salmons. Pollution is still a problem in some areas and has the potential to be an enormous problem throughout the country. We have created millions of acres of habitat for lake-dwelling species while destroying thousands of miles of stream and river habitat for others.

REVIEW QUESTIONS

Fill in the Blank

Fill in the blanks to complete the following statements.

1. Including the saltwater fishes, there are more than _____ species of fish in the world.
2. More than 300 species of _____ are found in the United States.

3. The _____ are the largest group of freshwater fishes found in North America.

4. _____ were originally native to Asia but have been widely introduced in the United States.

5. The _____ minnow is a favorite bait of anglers and is widely used for fish propagation.

6. Originally introduced from Europe, _____ can now be found from Nova Scotia to southern Manitoba and southward throughout much of the United States.

7. The _____ is our most common and widespread shad.

8. The _____ is the only gar with a snout twice the length of its head found in the United States.

9. The _____ is our largest gar, with a rod-and-reel record of 279 pounds.

10. Economically, the _____ are very important fishes.

11. Our largest freshwater trout, the _____, has a rod-and-reel record of over 60 pounds.

12. The rainbow trout prefers swift, _____ water, with adequate vegetative cover.

13. The _____ trout prefers slower waters than the rainbow, but otherwise its habitat preference is very similar.

14. Catfish are largely bottom feeders and appear to be _____ in their feeding habits.

15. _____ catfish prefer larger bodies of water, such as reservoirs and large rivers.

16. Smallmouth bass prefer rocky, _____ and rivers and may reach 10 pounds or more.

17. Trophy _____ are one of our most sought-after fish.

18. The _____ is a widespread sunfish, ranging over most of the eastern half of the United States.

19. _____ prefer still, quiet, clear waters, with sand or mud bottoms and abundant submerged vegetation.

20. The _____ is our only large perch, reaching 3 feet and 20 pounds.

21. The _____ is actually a saltwater species that has been widely introduced in larger impoundments and reservoirs.

Short Answer

1. List the habitat needs of most freshwater fish. In what ways do humans affect freshwater habitats?
2. How can we minimize or eliminate our impact on freshwater fish habitats?

Discussion

1. In your opinion, what can be done to preserve existing freshwater habitats and to clean up damaged habitats?
2. Freshwater fish are very important resources, from both economic and recreational standpoints. What can be done to protect freshwater fish populations from such things as pollution, overharvesting, and other threats? How can this be accomplished?

Learning Activities

1. Select two species of freshwater fish commonly found in your area and not covered in this chapter. Using your local library and classroom or home resources, give a complete description of each species and its preferred habitat, feeding habits, and life cycle.

Section
THREE

Careers

CHAPTER 24

Careers in Wildlife and Fisheries Management

OBJECTIVES

After completing this chapter, you should be able to

- List some careers available in the fields of wildlife and fisheries management.
- Describe the educational requirements for careers in wildlife and fisheries management.
- Identify the characteristics of some wildlife and fisheries careers.
- Explain the importance of careers in wildlife and fisheries management.

INTRODUCTION

The protection of our natural resources is one of our most important tasks. With an increase in government regulations that protect our environment, career opportunities in conservation and environmental services should expand in the future. The estimated revenue generated by the environmental services industry in the United States is in excess of $100 billion per year and is expected to continue growing at a rate of 10 percent or more annually.

There are many occupations related to conservation and the environment. Dozens of career options await those who are interested in the conservation and environmental field. Some of the more common occupations are discussed in this chapter. We have refrained from discussing salaries in most cases because they are highly variable and often change significantly from year to year and from one area to the next.

SELECTING A CAREER

Choosing a career is a very important process, and a choice should be made only after much thought and research. In the conservation and environmental fields, one factor to consider is whether you wish to work indoors or outdoors (Figure 24-1). Almost everyone enjoys being outdoors in the spring and early summer, but many conservation positions require being outdoors virtually all year. Pleasant spring temperatures often give way to summer temperatures of 100°F or more. The mild weather of fall often gives way to bitter winter cold. People who work outdoors day in and day out must learn to cope with rain, snow, cold, and heat. One must also consider other common outdoor annoyances, such as spiders, snakes, wasps, bees, ants, and allergies. If you would rather be dry and comfortable all or most of the time, you might be better off choosing a conservation or environmental career that allows you to spend the bulk of your time indoors.

Another important factor to consider when selecting a career in the conservation and environmental field is whether you are a "people" person or prefer to work with wildlife. Many of us are more skilled at working with animals than we are at dealing with people. Other people enjoy working with people and possess good communication skills. In fact, in today's world, most employment opportunities require good communication skills and the ability to get along with a variety of other people. Employers expect employees to work hard and work smart. Employees must be well mannered and well groomed, regardless of the place of employment or job description. A professional work ethic means

FIGURE 24-1
Conservation and environmental research are fields in which work is performed both indoors and outdoors.

(Source: Courtesy of USDA #036.)

arriving at work on time, dressed and groomed in an appropriate manner, and being prepared to work long hours if necessary.

Safety can also be an issue, particularly when working with wildlife. In many cases the animals being worked with are sedated or restrained, or both. Although most wild animals would not normally injure a person, they might do just that when they feel threatened or cornered. This is the type of situation a wildlife biologist or technician may encounter. Whenever you handle wildlife, whether to attach a radio or GPS transmitter, weigh the animal, or take blood samples, the risk of injury is present for the person and the animal. Paying special attention to safety training and precautions is very important. Most wild animals are not accustomed to being near people, much less being touched and handled by them. They become frightened and are more likely to defend themselves in such situations. Working with animals of any kind, wild or domestic, requires great patience. These are important factors to consider when making a career choice.

Another factor to consider when making a career choice is the level of education required for a particular career. Many of the careers in the natural resource management field require education beyond high school. Some careers require a 4-year college degree, and others require only 2 years of college or technical school. Others require trade skills, such as welding, plumbing, or carpentry. The ability and desire to adapt to changes is very important in today's work environment. Although a college degree is not necessarily required for some occupations, it will almost certainly help in your advancement in most jobs. Ultimately, what you are seeking is a career, not merely a job. Life is too short to be in a profession you do not enjoy. With more than 100 different occupations within the conservation field, selecting one that appeals to you should not be difficult.

CAREER OPTIONS

Careers in the conservation and environmental field are extremely varied, including research scientist, field biologist, engineer, and animal caretaker. Many clerical and trade positions are available with the U.S. Fish and Wildlife Service and most state wildlife agencies. Professional positions such as wildlife biologists, fishery biologists, engineers, and research scientists require at least a 4-year college degree. Most enforcement positions, such as game wardens and customs inspectors, require at least some college education. While most clerical positions do not require a college education, many administrative ones will. There are also many jobs available in the computer field. With the increased use of **technology** in the wildlife field, more and more jobs are becoming available for computer programmers, analysts, and specialists. Jobs are available for personnel specialists and public affairs officers in many state, federal, and private companies. Most state wildlife departments, as well as the U.S. Fish and Wildlife Service, use people in a variety of technical positions, most often as aides or assistants. The U.S. Department of Agriculture, particularly the Natural Resource Conservation Service, also has many conservation careers available. Educational requirements for these positions vary widely. Most federal and state wildlife management areas employ a variety of craftspeople, such as carpenters, electricians, and plumbers, to keep the facilities in proper working order. A great

(Source: Courtesy of U.S. Fish and Wildlife. Photo by John and Karen Hollingsworth.)

FIGURE 24-2 Wildlife biologists enjoy working outdoors and are directly involved with diverse habitats and wildlife.

variety of jobs are available, with various degrees of education required. We will now take a more in-depth look at a few of the positions available in wildlife science.

WILDLIFE BIOLOGISTS

Wildlife biologists are directly involved with the study of wildlife and their habitat (Figure 24-2). They use aerial and ground surveys to examine animal populations, monitor the status and trends of waterfowl migrating across America, and reconstruct wildlife habitats, such as tall-grass prairies and wetlands. Wildlife biologists may also work with conservation officials from around the world to monitor animals that are of mutual management concern, such as seals, polar bears, walruses, and various species of waterfowl.

Wildlife biologists tend to spend a good portion of their time outdoors—hiking, climbing, walking, observing, and handling wildlife. They plan and implement management and conservation programs for many different species. They also evaluate and, if necessary, develop plans to renew wildlife habitat. However, because of the diversity and **complexity** of wildlife there are usually one or two individuals in a department who specialize in white-tailed deer, upland game birds, waterfowl, songbirds, endangered species, and so forth. This helps ensure that there is an expert on most of the many different species of wildlife at each state wildlife department. Anyone interested in becoming a wildlife biologist should enjoy working with animals because most wildlife biologists spend a great deal of time working directly with wildlife. Wildlife biologist positions require at least a 4-year college degree.

CONSERVATION OFFICERS

Those responsible for enforcing wildlife and conservation laws are usually called game wardens or conservation officers. They spend many hours outdoors in the fields and woods, enforcing a wide variety of laws. They may also speak to many groups of people during the course of a year. Conservation officers are often used as resource personnel for school programs. Many states require conservation officers to have some college courses, but many do not require a 4-year degree. However, it is often very difficult to get accepted into a states training academy without a degree, typically in law enforcement or biology. Competition for the available jobs can be fierce. Conservation officers must be in top physical shape and be comfortable with a wide variety of duties. In addition to enforcing wildlife and conservation regulations, conservation officers are often called on to enforce boating and water safety regulations on lakes and rivers. Conservation officers should enjoy working with people and wildlife. They are often the first called when someone discovers wildlife, such as a skunk in the basement, in a suburban neighborhood.

FIGURE 24-3 Outdoor recreation specialists help coordinate the use of land and waterways for educational and recreational activities.

(Source: Courtesy of U.S. Fish and Wildlife Service. Photo by Jim Williams.)

OUTDOOR RECREATION SPECIALISTS

Outdoor recreation specialists are the people who often greet us as we enter a state or national wildlife refuge or monument (Figure 24-3). Outdoor recreation specialists plan programs and activities at many of the nation's wildlife refuges and monuments. They are also employed in a similar capacity by many state wildlife departments. They plan nature walks, educational programs, and tours. They assist refuge managers and biologists in implementing management programs and coordinating the use of lands and waterways. Most of these positions do not require a college degree but do require a desire and ability to work with people, as well as good communication skills. The majority of these are outdoor positions, at least during the spring and summer months, so a desire to work outdoors is a real plus. Some of these positions may also be seasonal. A seasonal summer position may in fact be a good way to see if this might be a good career path for you.

COMPUTER SPECIALISTS AND TECHNICIANS

With the continued increase in the use of technology in wildlife sciences, the need for computer specialists and technicians can only grow (Figure 24-4). Computer specialists are vital support personnel for many wildlife agencies. Computerized records are now the norm, and the storage and maintenance of these records is of utmost importance. Computer simulations and projections of such things as habitat loss and population trends within a species are becoming vital to the successful management of species. Radio tracking devices now send their signals to satellites, which in turn may beam them

(Source: Photo courtesy of Texas Parks & Wildlife Department © 2002.)

FIGURE 24-4 Data collected by wildlife management personnel is stored and analyzed using computers.

directly to a computer being used to track the species. The maintenance and programming of all this high-tech equipment requires technicians and computer specialists. These jobs are usually entirely indoors and may not require a college education.

SUMMARY

While we have only discussed four careers in any detail, there are dozens of career possibilities in the wildlife conservation and environmental field. The type of position you choose depends on your desire to work indoors or outdoors, whether you wish to work with people or animals or both, and whether you wish to pursue a college education. Good employees will continue to be in demand in this field for the foreseeable future. Most jobs in the past have been with state and federal agencies, but as government regulations on industry increase, more and more opportunities will become available with private industry. Many timber, mining, and oil and gas companies now employ environmental scientists and wildlife biologists. Their responsibilities may include everything from the preparation of environmental impact statements, to the management of wildlife on company properties, to reclamation work after a mining operation. Many companies are also hiring campus managers for their corporate sites. These people are responsible for supervising the maintenance of existing animal and plant communities and enhancing them. Most private corporations are very aware of their public image and are sensitive to the public perception of their environmental policies.

SUMMARY (*continued*)

Students interested in a career in the natural resource conservation field should consider wildlife-related supervised agricultural experience programs. Raising quail, pheasants, or chukars; assisting a local biologist with fieldwork; and other such tasks provide valuable experience in natural resource conservation. This sort of experience can also give you an idea of where your interest is within the conservation field. Should you choose a wildlife science or environmental career, the future of America's wildlife and natural resources could be in your hands.

REVIEW QUESTIONS

Fill in the Blank

Fill in the blanks to complete the following statements.

1. The estimated revenue generated by the environmental services industry in the United States is in excess of _____ billion.

2. In the conservation and environmental fields one factor that must be considered is whether you want to work _____ or _____.

3. Careers in the conservation and environmental field are extremely _____.

4. Wildlife biologists are directly involved with the study of wildlife and their _____.

5. _____ usually spend a good portion of their time outdoors—hiking, climbing, and walking and observing and handling wildlife.

6. Those responsible for enforcing wildlife and conservation laws are often called _____.

7. Conservation officers should enjoy working with _____ and wildlife.

8. Outdoor _____ are the people who often greet us when we enter a state or national wildlife refuge or monument.

9. Outdoor recreation specialists plan nature walks, _____, and tours for a wide variety of facilities.

10. _____ specialists are vital support personnel for many wildlife agencies.

11. Most jobs in the past have been with _____ and federal agencies.

12. More and more opportunities will become available with _____.

Short Answer

1. List some of the factors to consider before selecting a conservation career.

2. In what ways might a conservation professional, such as a wildlife biologist or game warden, be able to assist you in selecting a conservation career?

Discussion

1. Is the natural resource management field important? Why? Why not?

2. Select a career in the natural resource and conservation field. Discuss the qualifications one must have in order to pursue this career. What courses might you take at your high school to prepare you for a career as a wildlife or fisheries biologist?

Learning Activities

1. Using the appendices at the back of this text, write your individual state wildlife and conservation department and the U.S. Fish and Wildlife service to request career opportunity information.

2. Contact your state's nearest conservation agency office. Ask for a conservation professional, such as a public information officer, game warden, or biologist, to speak to your class about careers. Be sure that you have your instructor's permission.

Useful Web Site

<http://www.fws.gov>

Glossary

A

abnormal   not normal; not typical or average

abundant   plentiful; in ample supply

accelerate   to speed up or move faster; to cause to progress more quickly

accipiter   a group of hawks with short wings and long tails

accumulate   to collect or gather together over some time

acquisition   something gained or acquired; the act of acquiring

adapt   to adjust to new or changed circumstances

adaptable   able to change or adapt to a variety of situations or environments

adequate   sufficient; suitable

aesthetic   relating to or dealing with beauty

alfalfa   a perennial, leguminous herb used as forage and hay

alteration   being altered; changed

alternative   something that can be chosen or used instead

ambient   surrounding; on all sides; can be applied to the prevailing temperature

amphibian   any of a group of vertebrates, including frogs, toads, and salamanders, that begin their life cycle in water and are unable to regulate their own body temperature

anadromous   fish that swim or "run" upstream, from the sea to spawn such as salmon

appalling   causing shock, horror, or dismay

aquaculture   the cultivation of water plants or animals for human consumption

aquatic   something (plant or animal) that grows in water

arable land   land that is suitable for producing crops

arboreal   living in trees or adapted to living in trees

archery   the practice of shooting with a bow and arrow

artificial insemination   the placing of sperm into a female using artificial means

attribute   quality or characteristic

avian   having to do with birds

avian cholera   a contagious bacterial disease, virtually unknown in wild birds before 1944

B

banded birds   birds, typically waterfowl, that have had a band or tag placed around one or both legs so that they may be identified later

barren   empty; unproductive; lacking essentials for life

bayou   small stream; sluggish inlet or outlet of a lake or river in the southern United States

beaver   a member of the rodent family, with soft brown fur and a flat, paddle-shaped tail

beneficial   helpful

biological   connected with biology; the nature of living matter; pertaining to plants and animals

bird banding   the practice of marking birds with a band, usually just above the foot, for identification purposes; done primarily to migratory birds

bottoms   low-lying land along a watercourse, usually a river or stream

brackish   water that has more salt than fresh water but not as much salt as ocean water; typical of marshes near the sea

browse   tender shoots of shrubs and trees eaten by certain herbivores

C

camouflaged   well hidden or difficult to see; blending with the environment in such a manner as to appear inconspicuous

canid   member of the dog family, such as wolves and coyotes

Cape Horn   southernmost point of South America

captive breeding   breeding wild species in a controlled environment; often done now in the hope of eventually returning the offspring to natural habitat

carnivore   an animal (or plant) that eats animals

carnivorous   flesh-eating

carrion   dead animals; dead and decaying flesh

catapulted   launched; leaped

catkins   a drooping spike or raceme on flowers that lack petals, eaten by some species of birds

census   a population count of people, or wildlife specie

chameleon   any of a group of lizards having eyes that move independently and having the ability to change color rapidly

channelization   dredging of bays and rivers to provide a channel deep enough for ship travel

chemical pollution   pollution resulting from chemicals such as pesticides, fertilizers, and petroleum products

climate   the prevailing or average weather conditions of an area

cloaca   region of the gut in birds, reptiles, and amphibians into which the intestinal, urinary, and genital tracts empty

closed season   a period during which it is unlawful to take or kill a certain species of wildlife; protects wildlife from overharvesting

clutch   a nest of eggs or a brood of chicks

colonies   groups of individuals nesting in close proximity to one another

comb   fleshy crest or area on the heads of some birds

commercial   having to do with trade; applied to fish, it indicates species that are caught and sold on a large scale for profit

commercialize   to use mainly for profit

complex   not simple; complicated; involved

complexity   complicated nature; the condition of being difficult to understand or being composed of many interrelated things

coniferous   cone-bearing (tree or shrub, usually an evergreen)

conquer   to gain control or possession of by winning a war

conservation   the practice of protecting natural resources from waste

conservationist   a person who advocates the conservation of natural resources

conserve   to save; to keep from being wasted

conspicuous   easy to see; obvious; attracting attention

constrict   to make smaller or narrower by squeezing

contagious   spread by direct or indirect contact

cooperation   joint effort; working together; the act of cooperating

costal groove   vertical furrow on the flanks of salamanders

coulee   a deep ravine or gully, usually dry, that has been formed by running water

coverts   small feathers that cover the base of larger feathers on a bird's wing and tail

crayfish   any of a number of freshwater crustaceans somewhat resembling small lobsters

crustacean   any member of a class of arthropods, including lobsters, shrimps, crabs, and crayfish, having a hard outer shell, usually living in water, and breathing through gills

curare   a medicine derived from several South American plants used mainly as a muscle relaxant

D

DDT   the first of the long-lasting chlorinated hydrocarbon insecticides; accumulates in certain fatty tissues of some organisms

deciduous   shedding leaves annually; generally hardwood trees

decimate   to kill or destroy a large part of

degradation   making something less valuable

demise   ceasing to exist; death

demobilization   the process of discharging soldiers from the armed forces and sending them home

dependent   relying on something else for support

derived   to obtain from a specific source

desalination   to remove salt from

desert   an area that receives little water; dry, often sandy region

detection   being found out

detrimental   causing damage or harm to

devastate   to lay waste; to bring to ruin

dew   water condensed on the surfaces of cool objects, especially at night

dewlap   a loose fold of skin hanging from the throat of cattle and certain other animals

dike   an embankment used to hold water in an area

diminutive   small; tiny

disastrous   having extremely damaging results

discernible   recognizable as separate or different; clearly visible

distributed   scattered; widely spread out

diurnal   active during daylight

diversity   variety; difference

dominant   ruling; prevailing

down   fine, soft feathers under the outer feathers of adult birds

draws   low-lying areas of land, usually formed from water runoff, through which water may drain during wet periods

drought   prolonged period of inadequate or below-normal rainfall

duck plague   a viral disease first associated with domestic ducks and not common in wild birds

Dust Bowl   a large region in the southern United States that suffered badly from wind erosion in the 1930s

dwindling   decreasing in number, size, or area

E

ecological resources   living resources such as forests, wetlands, and wildlife

economic value   monetary value of something

ecosystem   a system made up of a community of plants and animals that coexist and are dependent on one another

ectothermic   an ectothermic animal maintains its body temperature by absorbing heat from its environment

egret   a heronlike bird, usually having long white plumes (feathers)

elevation   a high place or position; height above sea level

eliminate   to get rid of

elusive   hard to see; seldom seen; secretive

encroach   to trespass or intrude on the rights or property of another

encroachment   trespassing or advancing beyond proper limits

endangered species   any species of plant or animal in danger of extinction throughout all or a major portion of its range

energy   the capacity for doing work; power

environment   all the conditions, such as air, water, and soil, that affect life; the physical factors that influence the existence or development of a living organism

eradicate   to get rid of; destroy completely; wipe out

erodible   something that is susceptible to erosion

erosion   the wearing away and removal of topsoil by the actions of wind and water

essential   absolutely necessary; indispensable

estuary   an inlet or arm of the sea, such as the area where a river flows into the sea

ethics   a code of morals of a person or group of people

evaluate   to measure the worth or value of something

evergreen   a plant that does not lose its leaves with the changing of the seasons

evolution   the development of a species or organism

evolve   to develop or work out

excise tax   a tax placed on goods for a domestic market

exclusive   belonging to no other; not shared

exploit   to take advantage of something or someone; to use or develop something in order to gain a benefit

exportation   the act of sending an item to another country

extensive   covering a large area; vast; having a wide scope or effect

exterminated   to destroy, kill, or get rid of something completely

extinction   the dying out of a species

extraction   removal or taking out

extrude   to push or force out

F

fauna   the animals of a region or area

feign   to pretend; imitate; make a false showing

ferocious   fierce; savage

fertilizer   a material put on or in the soil to improve the quality or quantity of plant growth

finite   having definite limits; a limited supply of something

fledged   having acquired the feathers necessary for flight

flourish   to grow vigorously; to succeed, thrive, prosper

fluctuate   to rise and fall; to change continually

food fishes   fish species used for food by people

food plots   areas of crops, usually fairly small, planted or left unharvested only to benefit wildlife

forage   food for animals, especially that taken by browsing or grazing

forbs   legumes, herbs, and other nongrasses eaten by herbivores

forest   a thick growth of trees and underbrush covering an extensive tract of land

forest resources   timber used for the production of building materials, paper, and so forth

fossil fuels   fuels that occur naturally in the earth, such as coal, natural gas, and petroleum (oil)

fragile   easily broken, destroyed, or damaged

fragmented   broken up; a detached, isolated, or incomplete part

fry   the young of various fish or other animals

fungi   plural of fungus; fungi include mildews, molds, mushrooms, and yeasts

fur-bearer   an animal with a coat of fur, such as a raccoon, mink, and beaver

G

game animals   wild birds or animals that are hunted for sport and food

gander   a mature male goose

gangling   tall; awkward; having a lanky build

gestation   period during which the female carries her young

gills   organs used for breathing by most animals that live in water

girdle   to cut away the bark and cambium in a ring around a tree, thereby killing it by interrupting the circulation of water and nutrients

gladiator   in ancient Rome, a professional fighter who fought other men or wild animals for the entertainment of spectators

gosling   a very young or newly hatched goose

Great Depression   period from 1929 to 1939 marked by decreased business activity, wide-spread unemployment, and poverty

Great Plains   large region stretching from Texas to Southern Alberta, Canada and east from the Rocky Mountains for approximately 400 miles

gregarious   sociable; tending to associate with others of one's kind

gular fold   fold of skin across the posterior section of the throat in salamanders

guzzler   artificial watering structure often con-structed in arid areas to provide water for wildlife

H

habitat   in this text, refers to an animal's home; good wildlife habitat has food, water, cover, and space, all suitably arranged

hardiest   toughest; most able to withstand hardship

hardwood   any tree yielding a tough, compact, tight-grained wood; usually broadleaf trees

harrier   a hawk that hunts by flying low over land

hatchery   a place where fish are hatched and raised to a releasable size

hazardous   risky; perilous; dangerous

heath hen   an extinct bird of the grouse family; a relative of the prairie chicken

herbivore   an animal that eats only plants or plant material

heritage   tradition; something acquired from a predecessor

herons   any of a large family (Ardeidae) of wading birds that have long necks and legs and a long bill; usually found along rivers, in or around marshes or ponds, or in other wetland habitat

herpetorium   building or structure normally housing reptiles and amphibians

hibernate   to spend the winter in a dormant state

hierarchy   a form of organization in animals in which different members of a group have different levels of status, which affects feeding and mating

horrendous   horrible; frightful; terrible

hospitable   a climate or region that is not difficult to live in

humidity   the amount or degree of moisture in the air; dampness

hunt   to pursue game for the purpose of capturing or killing it for food and sport

hybrid   the offspring produced by the mating of two individuals of different varieties or species

hybridization   producing hybrids; cross-breeding

hysteria   any outbreak of wild, uncontrolled excitement or fear

immigrants   people who move into a new country or region

imperative   absolutely necessary; urgent; compelling

import   to bring in an item from another country

impound   to gather or enclose water for human uses; impoundments are human-made structures used to gather water

inaccessible   impossible to reach

inadequate   not sufficient; not equal to what is required

inclement   rough; severe; stormy

incubation   period when birds incubate or sit on their eggs

incubator   a mechanical device that provides the environmental conditions necessary for eggs to hatch

indiscriminate   not making careful choices; random

inexhaustible   cannot be used up; cannot be exhausted

inhospitable   not offering protection; barren; forbidding

inland fisheries   areas, mainly rivers and lakes, in the interior of a country where fish may be caught

inlet   a narrow strip of water extending into land from an ocean, lake, river, and so forth

innovative   new and original

inordinate   too great; too many; excessive

inquisitive   curious; eager to learn

insatiable   incapable of being satisfied

insects   large class of small arthropod animals, which as adults have bodies divided into head, thorax, and abdomen, with three pairs of legs on the thorax

instinct   an inborn tendency

institution   a thing or organization long established in one place

instrumental   important or helpful in making something happen

intentional   something done on purpose

international   involving several countries or nations

intimidate   to make afraid

invasive   to intrude or infringe on another

invertebrates   lacking a spinal column

iridescent   having or showing shifting changes of color when seen from different angles

irreplaceable   not replaceable; one of a kind

irrigation   supplying a crop with water by artificial means, such as by the use of sprinklers, canals, or ditches

ivory the hard, white substance that makes up the tusks of walruses, elephants, and other tusked animals

J

juvenile young or immature

L

languish to live under distressing conditions

latitude distance north and south of the equator, measured in degrees

lease a legal contract allowing someone to use property in return for rent

legumes a family of plants with many valuable food and forage species, such as peas, soybeans, and peanuts

lethargic abnormally drowsy or sluggish

levee an embankment, usually earthen, built along the course of a river, designed to prevent the surrounding land from flooding during times of high water

license a document or permit that indicates that permission has been given

lichens a rather large group of plants composed of fungus and algae growing in a symbiotic relationship; typically growing in crustlike patches on rocks, trees, and soil, particularly in northern latitudes

livelihood a person's means of making a living

M

malnourished improperly nourished; inadequate nutrition, generally from a poor diet

manufacturer the maker of a good or product

marine mammals mammals that live in the sea or coastal area, such as porpoises and manatees

market hunter a hunter who shot wildlife for money in America from 1850 to 1920

marsupial a mammal species, the female of which has an external pouch that contains teats and in which the young complete development

mast nuts, such as acorns and beechnuts, used as food by animals

migration a group of animals moving together, generally for the purpose of finding food and suitable habitat

migratory moving from one region or country to another

migratory birds birds that move from one region or country to another, usually seasonally; many songbirds, shorebirds, and waterfowl are migratory

mineral resource an inorganic substance, such as iron, copper, or gold, that occurs naturally in the earth

moderating avoiding extremes; staying within reasonable limits

modicum a small portion; a limited quantity

mollusks any of a large number of invertebrate animals, such as snails, slugs, and oysters

monitoring watching; checking on

monogamous having only one mate

mortality the number of deaths in a given period

morphine a medicine derived from opium and used to treat pain

mottes small groves of trees

N

natural resource items such as coal, water, arable land, oil, and minerals that are found in nature and of use to humankind

naturalistic imitating nature; looking natural

navigation the act of plotting a course and traveling from one point to another

neotropical [in this case] referring to bird species which spend a portion of the year in temperate zones and a portion in tropical zones; for example, songbirds which spend the spring and summer in the United States and the fall and winter in Mexico, Central America, and/or South America

neurotoxin a poison that destroys nerves or nervous tissue

niche a place particularly suited to the animal, thing, or person in it

nightjar any of a group of slender, graceful birds that perch in a horizontal position and feed at dusk or after dark on insects caught in flight

nocturnal   active during the night; animals that are nocturnal do most of their hunting, traveling, and prowling at night

nongame species   species of wild animals that are not regarded as game animals, such as songbirds, raptors, and rodents

nonindigenous   not native to or naturally occurring in a region or country

nonnative species   animals that were not originally found in North America, such as pheasants, aoudad sheep, and starlings

nonrenewable   not capable of being renewed, reestablished, or rejuvenated

nutrition   the level of nourishment that an animal is receiving

O

omnivorous   eating both plant and animal matter

opportunistic   adaptable; able to take advantage of a situation

optimum   the best condition (for the growth and reproduction of a species)

overpopulate   to populate an area too heavily for the available resources

oviduct   the tube or duct through which eggs pass from the ovary to the uterus or to the outside

P

palate   the roof of the mouth

palmate   shaped like a hand with the fingers spread wide

peat   partly decayed, moisture-absorbing plant material used in the horticulture industry

pelage   hair or fur covering an animal

penicillin   a common antibiotic used to treat humans

perilous   dangerous

periodically   occurring from time to time

persecution   constant harassing for the purpose of causing injury

pesticide   a substance used to control plant, animal, or insect pests

pharaoh   a title given to rulers of ancient Egypt

pharynx   the part of the alimentary canal leading from the mouth to the esophagus

philosopher   a person who studies philosophy; a person who philosophizes

photosynthesis   the production of organic substances, chiefly sugars, from carbon dioxide and water, occurring in green plant cells supplied with enough light to allow chlorophyll to aid in the transformation of the radiant energy into chemical form

pit viper   any one of a family of poisonous snakes with an obvious pair of dimples, or "pits," between the nostrils and eyes

plankton   microscopic animal and plant life found floating in oceans and fresh water; used as food by fish

playa lake   a shallow lake that may be dry except for periods after rainfall

plume   a feather, particularly a large, showy one

poacher   person who takes wildlife illegally for personal gain

poaching   hunting or fishing illegally

polluted   unclean, impure, contaminated

polygamous   the practice of a male mating with more than one female

potent   strong; effective or powerful in action

potential   what can, but has not yet, come into being; what might be possible

poult   a young fowl, particularly a young turkey

predator   an animal that survives by killing and eating other animals

preference   preferring; being preferred; liking something better

preserve   to maintain and protect

prioritize   to list a series of items in order of importance

processing   turning something in a raw state, such as wheat, into a finished product, such as bread

prodigious   extremely large; huge; enormous

productivity   abundant production; ability to produce in quantity

prolific   producing many young; abundant reproduction

promiscuous   mating indiscriminately with as many mates as possible

propagation   reproduction of a plant or animal

proponent   one who supports a cause

prosperity   economic well-being; the condition of being successful

proximity   closeness; nearness

pugnacious   eager and ready to fight; combative

pulpwood   wood used in making pulp for paper

Q

quality   the measure of how good something is

quantity   an amount or number of something; the measurable property of something

quinine   a medicine extracted from bark used in the treatment of malaria

R

range   area over which an animal may be found

receptive   receiving or tending to receive; open to a potential mate's sexual advances

reclamation   the recovery of a wetland or marsh; reclaiming or being reclaimed

recycling   using something again and again

reforestation   the planting of trees on previously harvested land

refuge   an area set aside for protection (of wildlife)

regulate   to control or direct according to a rule, principle, or system

regurgitate   to return partially digested food to the mouth; typical method used by many birds for feeding their young

reintroduce   to release captive-bred animals (or wild stock trapped in another area or state with an adequate population of the species) into suitable habitat to reestablish a population

remnant   what's left over; remaining

renewable   capable of being reestablished

renowned   famous; great fame or reputation

replenish   to fill or build up again

restocking   the act of returning wildlife to areas where they formerly lived

restoration   being restored; brought back to its former state

reviled   treated as vile; hated

roosts   places where birds rest, particularly at night

rural   related to country rather than urban or suburban areas

S

sagebrush   herblike, bushy plants native to the western and southwestern United States

salmonid   any of several species of salmon

sanctuary   an area of safety; area where animals are protected

school   a large number of fish of the same kind swimming or feeding together

secretive   concealing one's actions

sedges   a group of grasslike plants often found in wet areas

sediment   soil or organic matter carried and deposited by water

sedimentation   the depositing or formation of sediment by water

segregate   to separate from others; to set apart

semblance   outward form or appearance; likeness, image, or copy

sewage   waste matter, typically human waste, carried off by sewers and drains

shellfish   any aquatic animal with a shell, such as lobsters, crayfish, and clams

significant   important; having special meaning

sloughs   narrow backwaters or swamps, typically the abandoned channel of a river

snout   projecting nose and jaws of an animal

sod   a layer of earth containing grass and its roots

soil conservation   the protection of fertile topsoil from erosion and the replacement of nutrients in the soil

Soil Conservation Service   government agency charged with promoting the protection of the nation's soils

songbirds   a group of birds known to call, or "sing," with a series of musical tones

spawning   the act of producing eggs or young; method of reproduction used by fish, mollusks, crustaceans, and other species

species a general class of animals that can reproduce with one another

speculum a distinctive patch of color on the wings of some birds, especially ducks

sperm male germ cell secreted by the male reproductive organs

sport fish fish species valued by sport fishermen and women for their sporting characteristics

sport hunter a person who hunts for food and enjoyment

sport hunting hunting for recreation

spur a bony protuberance on the lower leg of some birds; used as a weapon by males during territory or breeding disputes

starvation to die from a lack of food

status current state or condition

strangulation process of cutting off the flow of blood or oxygen

subsidence to sink or settle to a lower level

subsistence means of support or livelihood

subspecies a natural subdivision of a species that shows small but consistent variations from other subspecies

succumb to give way; to yield; to die

supplemental something added, usually to make up for a deficiency

surplus excess; in terms of wildlife population, the offspring produced in excess of what is needed to maintain a population or that the habitat will support

swallow any of a group of birds noted for their forked tails, graceful flight, and habit of feeding on insects caught while in flight

synchronize to happen together; to cause to agree; to make synchronous

T

tag a permit often required to legally harvest certain game animals

taxidermist person who preserves and stuffs the hides of animals so as to make them appear lifelike

technique any method or manner of accomplishing something

technology the science or study of applied sciences and practical arts used to improve or provide humankind with needed or wanted things

tenacious holding firmly; persistent

terrace a raised mound of earth with sloping sides, used to divert or hold water

terrestrial living on land, rather than in air, trees, or water

territory a particular area occupied by an animal, usually for breeding or nesting purposes, and defended against any intruders

thrush any of several small- to medium-size birds of the family Turdidae

topsoil the upper layer of soil; the richest, most productive soil layer

tradition a long-established custom or practice

tuber a short, thickened underground stem; the Irish potato is a tuber

tundra treeless plain of the arctic regions of Europe, Asia, and North America; the ground below the immediate surface remains frozen all year

turbid muddy or cloudy from having sediment stirred up

tympanum a thin, round or oval external ear covering; prominent in many frogs

U

unenviable not desirable

ungulate a mammal having hoofs

unintentionally not done on purpose; not intentionally

uplands normally dry areas surrounding a wetland

USDA United States Department of Agriculture

V

variation changing or varying in form or substance

vegetarian an animal that consumes only vegetable products and no animal products

vegetated wetlands generally inland, freshwater marshes, swamps, or other low-lying wet areas covered with lush vegetation

vegetative   pertaining to plants; of vegetation or plants

vertebrates   a large group of animals that have a segmented spinal column, including all mammals, birds, fish, reptiles, and amphibians

vestige   a trace; bit; an atrophied or rudimentary organ

viable   able to live; likely to survive

vigorous   strong; powerful or forceful

vocalize   to speak or sing

vocation   the work in which a person is employed

voracious   having a huge appetite

vulnerable   open to attack or damage

W

waterfowl   a bird that frequents water, especially one that swims; ducks and geese are waterfowl

watershed   the area drained by a river or river system

wattle   a fleshy, wrinkled, often brightly colored piece of skin that hangs from the throat of some birds, such as turkeys

wetland   areas containing much soil moisture, such as swamps, bogs, and prairie potholes

wilderness   an uncultivated, uninhabited area

wildlife   all nondomesticated animals

wildlife management   the care of wildlife and its environment in such a manner as to ensure the continuation of the species

windmill   a device that utilizes the wind to turn large vanes to provide power for pumping water

woodpecker   any of a group of birds known for their habit of drilling in trees for insects or to excavate nesting cavities

Z

zoology   the science of animals; a branch of biology dealing with the structure, life, classification, and growth of animals

Appendix A
U.S. Fish and Wildlife Service Offices

Director, U.S. Fish and Wildlife Service Department of the Interior
Washington, D.C. 20240

Regional Director, U.S. Fish and Wildlife Service
1002 NE Holladay St.
Portland, Oregon 97232-4181

Regional Director, U.S. Fish and Wildlife Service
P.O. Box 1306
Albuquerque, New Mexico 87103

Regional Director, U.S. Fish and Wildlife Service
Federal Building, Ft. Snelling
Twin Cities, Minnesota 55111

Regional Director, U.S. Fish and Wildlife Service
Federal Building, Room 1200
75 Spring Street SW
Atlanta, Georgia 30303

Regional Director, U.S. Fish and Wildlife Service
One Gateway Center, Suite 700
Newton Corner, Massachusetts 02158

Regional Director, U.S. Fish and Wildlife Service
P.O. Box 25486, Denver Federal Center
Denver, Colorado 80225

Regional Director, U.S. Fish and Wildlife Service
1011 East Tudor Road
Anchorage, Alaska 99503

U.S. Fish and Wildlife Service Web Site
http://www.fws.gov

U.S.D.A. Forest Service
http://www.fs.fed.us

Bureau of Land Management
http://www.blm.gov

National Park Service
http://www.nps.gov/parks.html

Appendix B

State Conservation and Natural Resource Agency Offices

By using the United States Fish and Wildlife Service Web site, http://www.fws.gov, you can also find all 50 state Web sites. Another Web site with links to each of the state wildlife conservation agency Web sites is http://www.smarthunter.com.

Alabama
Dept. of Conservation & Natural Resources
Division of Fish and Game
64 N. Union St.
Montgomery, Alabama 36130
(205) 242-3465

Alaska
Dept. of Fish & Game
Division of Wildlife Conservation
P.O. 25526
Juneau, Alaska 99802-5526
(907) 465-4190

Arizona
Game & Fish Department
2221 West Greenway Rd.
Phoenix, Arizona 85023
(602) 942-3000

Arkansas
Game & Fish Commission
No. 2 Natural Resources Dr.
Little Rock, Arkansas 72205
(501) 223-6300

California
Dept. of Fish & Game
Wildlife Management Division
1416 Ninth St.
Sacramento, California 95814
(916) 653-7664

Colorado
Division of Wildlife
6060 Broadway
Denver, Colorado 80216
(303) 297-1192

Connecticut
Dept. of Environmental Protection, Wildlife Div.
79 Elm Street
Hartford, Connecticut 06106-5127
(203) 566-4683

Delaware
Dept. of Natural Resources & Environmental Control
Div. of Fish and Wildlife
89 Kings Highway
P.O. Box 1401
Dover, Delaware 19903
(302) 739-5297

Florida
Game & Fresh Water Commission
620 South Meridian St.
Tallahassee, Florida 32399-1600
(904) 488-4676

Georgia
Dept. of Natural Resources
2109 U.S. Highway 78 S.E.
Social Circle, Georgia 30279
(404) 918-6415

Hawaii
Dept. of Land & Natural Resources
Division of Forestry & Wildlife
1151 Punchbowl St., Room 325
Honolulu, Hawaii 96813
(808) 587-0166

Idaho
Dept. of Fish & Game
600 South Walnut St.
P.O. Box 25
Boise, Idaho 83707
(208) 334-3700

Illinois
Dept. of Conservation
Lincoln Tower Plaza
524 S. Second St.
Springfield, Illinois 62701-1787
(217) 782-6384

Indiana
Dept. of Natural Resources
Division of Fish & Wildlife
402 W. Washington, Room 273
Indianapolis, Indiana 46204
(317) 232-4080

Iowa
Dept. of Natural Resources
Wallace State Office Bldg.
Des Moines, Iowa 50319
(515) 281-5918

Kansas
Dept. of Wildlife & Parks
512 S.E. 25th Avenue
Pratt, Kansas 67124
(316) 672-5911

Kentucky
Dept. of Fish & Wildlife Resources
1 Game Farm Road
Frankfort, Kentucky 40601
(502) 564-4336

Louisiana
Dept. of Wildlife & Fisheries, I & E Div.
P.O. Box 98000

Baton Rouge, Louisiana
70898-9000
(504) 765-2925

Maine
Dept. of Inland Fisheries & Wildlife
284 State St., State House Station 41
Augusta, Maine 04333
(207) 287-2871

Maryland
Fish Heritage & Wildlife
Administration
Tawes State Office Bldg. E1
580 Taylor Ave.
Annapolis, Maryland 21401
(410) 974-3195

Massachusetts
Division of Fisheries & Wildlife
100 Cambridge St., 19th Floor
Boston, Massachusetts 02202
(617) 727-3151

Michigan
Wildlife Div., Michigan Dept. of
Natural Resources
P.O. Box 30444
Lansing, Michigan 48909-7944
(517) 373-1263

Minnesota
Dept. of Natural Resources
Division of Fish and Wildlife
500 Lafayette
St. Paul, Minnesota 55155-4007
(612) 296-6157

Mississippi
Dept. of Wildlife Conservation
P.O. Box 451
Jackson, Mississippi 39205
(601) 362-9212

Missouri
Dept. of Conservation
P.O. Box 180
Jefferson City, Missouri 65102
(314) 751-4115

Montana
Dept. of Fish, Wildlife & Parks
1420 East Sixth Ave.
P.O. Box 200701
Helena, Montana 59620-0701
(406) 444-2535

Nebraska
Game & Parks Commission
2200 N. 33rd St.
P.O. Box 30370
Lincoln, Nebraska 68503
(402) 471-0641

Nevada
Division of Wildlife
1100 Valley Rd.
P.O. Box 10678
Reno, Nevada 89502
(702) 688-1500

New Hampshire
Fish & Game Department
2 Hazen Dr.
Concord, New Hampshire 03301
(603) 271-3421

New Jersey
Dept. of Environmental Protection
Division of Fish, Game & Wildlife
CN-400, 501 E. State Street
Trenton, New Jersey 08625
(609) 292-2965

New Mexico
Dept. of Game & Fish
P.O. Box 25112
Santa Fe, New Mexico 87504
(505) 827-7911

New York
Dept. of Environmental Conservation
Fish & Wildlife Division
50 Wolf Road
Albany, New York 12233-4750
(518) 457-5400

North Carolina
Wildlife Resources Commission
512 N. Salisbury St.
Raleigh, North Carolina 27604-1188
(919) 733-3391

North Dakota
Game & Fish Department
100 N. Bismarck Expressway
Bismarck, North Dakota 58501
(701) 221-6300

Ohio
Dept. of Natural Resources, Div. of Wildlife
1840 Belcher Drive
Columbus, Ohio 43224-1329
(614) 265-6300

Oklahoma
Dept. of Wildlife Conservation
1801 N. Lincoln
P.O. Box 73512
Oklahoma City, Oklahoma 73105
(405) 521-3851

Oregon
Dept. of Fish & Wildlife
P.O. Box 59
Portland, Oregon 97207
(503) 229-5400

Pennsylvania
Game Commission
2001 Elmerton Ave.
Harrisburg, Pennsylvania 17110-9797
(717) 787-6286

Rhode Island
Dept. of Environmental Management
Division of Fish & Wildlife
Gov't Center, 4808 Tower Hill Rd.
Wakefield, Rhode Island 02879
(401) 789-3094

South Carolina
South Carolina Dept. of Natural Resources
P.O. Box 167

Columbia, South Carolina 29202
(803) 734-3888

South Dakota
Dept. of Game, Fish & Parks
523 East Capital
Pierre, South Dakota 57501
(605) 773-3485

Tennessee
Tennessee Wildlife Resources Agency
P.O. Box 40747
Nashville, Tennessee 37204
(615) 781-6500

Texas
Texas Parks & Wildlife Dept.
4200 Smith School Rd.
Austin, Texas 78744
(512) 389-4800

Utah
Division of Wildlife Resources
1596 West North Temple
Salt Lake City, Utah 84116
(801) 538-4700

Vermont
Fish & Wildlife Dept.
103 S. Main Street, Bldg. 10 South
Waterbury, Vermont 05671-0501
(802) 241-3700

Virginia
Dept. of Game & Inland
Fisheries
4010 W. Broad St.
P.O. Box 11104
Richmond, Virginia 23230-1104
(804) 367-1000

Washington
Dept. of Fish & Wildlife
600 Capitol Way N.
Olympia, Washington 98501-1091
(206) 753-5700

West Virginia
Wildlife Resources Division
State Capitol Complex, Bldg. 3
Charleston, West Virginia 25305
(304) 558-2771

Wisconsin
Bureau of Wildlife Management
P.O. Box 7921
Madison, Wisconsin 53707
(608) 266-1877

Wyoming
Game & Fish Department
5400 Bishop Blvd.
Cheyenne, Wyoming 82006
(307) 777-4601

Appendix C
Private Conservation Organizations

There are hundreds, perhaps thousands, of private organizations involved in some form of wildlife or habitat conservation. This appendix lists organizations that actively promote the conservation of wildlife and their habitat through education, scientific management, and the purchase and protection of habitat. Undoubtedly, many worthy organizations are not included in this appendix. If you want a more detailed list, the National Wildlife Federation publishes a conservation directory annually. This directory can be obtained by contacting the Federation at the address listed in this appendix.

American Fisheries Society
5410 Grosvenor Lane, Suite 110
Bethesda, Maryland 20814
http://www.fisheries.org/

Boone and Crockett Club
Old Milwaukee, 250 Station Dr.
Missoula, Montana 59801
http://www.boone-crockett.org/

Delta Waterfowl Foundation
102 Wilmot Road, Suite 410
Deerfield, Illinois 60015
http://www.deltawaterfowl.org/

Ducks Unlimited, Inc.
One Waterfowl Way
Long Grove, Illinois 60047
http://www.ducks.org/

Foundation for North American Wild Sheep
720 Allen Ave.
Cody, Wyoming 82414
http://www.fnaws.org/

National Wild Turkey Federation, Inc.
The Wild Turkey Building
P.O. Box 530

Edgefield, South Carolina 29824-0530
http://www.nwtf.org/

National Wildlife Federation
1400 Sixteenth Street, NW
Washington, D.C. 20036
http://www.nwf.org/

The Nature Conservancy
1815 North Lynn Street
Arlington, Virginia 22209
http://www.tnc.org/

Pheasants Forever, Inc.
P.O. Box 75473
St. Paul, Minnesota 55175
http://www.pheasantsforever.org/

Quail Unlimited, Inc.
P.O. Box 10041
Augusta, Georgia 30903
http://www.qu.org/

Rocky Mountain Elk Foundation
P.O. Box 8249
Missoula, Montana 59807
http://www.rmef.org/

The Ruffed Grouse Society
451 McCormick Road
Coraopolis, Pennsylvania 15108
http://www.ruffedgrousesociety.org/

Sierra Club
730 Polk Street
San Francisco, California 94109
http://www.sierraclub.org/

Trout Unlimited
National Headquarters
SE, Suite 250
Vienna, Virginia 22180-4959
http://www.tu.org/

Whitetails Unlimited, Inc.
P.O. Box 422
Sturgeon Bay, Wisconsin 54235
http://www.whitetailsunlimited.org/

The Wilderness Society
900 17th Street NW
Washington, D.C. 20006-2596
http://www.wilderness.org/

Wildlife Forever
12301 Whitewater Drive, Suite 210
Minnetonka, Minnesota 55343
http://www.wildlifeforever.org/

The Wildlife Legislative Fund of America and the Wildlife Conservation Fund of America
801 Kingsmill Parkway
Columbus, Ohio 43229
http://www.wlfa.org/

World Wildlife Fund
1250 24th Street NW
Washington, D.C. 20037
http://www.wwf.org/

Index

Note: The notation 'f' following the locators refer to figures cited in the text